Communications and Control Engineering

Communications and Control Engineering is a high-level academic monograph series publishing research in control and systems theory, control engineering and communications. It has worldwide distribution to engineers, researchers, educators (several of the titles in this series find use as advanced textbooks although that is not their primary purpose), and libraries.

The series reflects the major technological and mathematical advances that have a great impact in the fields of communication and control. The range of areas to which control and systems theory is applied is broadening rapidly with particular growth being noticeable in the fields of finance and biologically inspired control. Books in this series generally pull together many related research threads in more mature areas of the subject than the highly specialised volumes of *Lecture Notes in Control and Information Sciences*. This series's mathematical and control-theoretic emphasis is complemented by *Advances in Industrial Control* which provides a much more applied, engineering-oriented outlook.

Indexed by SCOPUS and Engineering Index.

Publishing Ethics: Researchers should conduct their research from research proposal to publication in line with best practices and codes of conduct of relevant professional bodies and/or national and international regulatory bodies. For more details on individual ethics matters please see:
https://www.springer.com/gp/authors-editors/journal-author/journal-author-helpdesk/publishing-ethics/14214

Daoyi Dong · Ian R. Petersen

Learning and Robust Control in Quantum Technology

Springer

Daoyi Dong
School of Engineering and Information Technology
University of New South Wales
Canberra, ACT, Australia

Ian R. Petersen
School of Engineering
Australian National University
Canberra, ACT, Australia

ISSN 0178-5354 ISSN 2197-7119 (electronic)
Communications and Control Engineering
ISBN 978-3-031-20244-5 ISBN 978-3-031-20245-2 (eBook)
https://doi.org/10.1007/978-3-031-20245-2

This Springer imprint is published by the registered company Springer Nature Switzerland AG
The registered company address is: Gewerbestrasse 11, 6330 Cham, Switzerland

To my wife Dandan and children Henry and Charlotte

Daoyi Dong

In memory of my parents Frank and Gwenyth

Ian R. Petersen

Preface

Controlling quantum phenomena has become a central goal in quantum physics and physical chemistry since the establishment of quantum mechanics. In the last three decades, the rapid development of quantum technology provides a strong drive to develop systematic quantum control theory. A main goal of quantum control theory is to establish a firm theoretical footing and develop systematic methods for the active manipulation and control of quantum systems, which will enable quantum technology such as quantum computers, quantum sensors and quantum networks practical. The two main themes of this monograph are quantum learning control and quantum robust control. In quantum learning control, we explore different machine learning algorithms for control design in various applications such as quantum state transfer, quantum gate generation and quantum information compression. Two common objectives we consider are to achieve optimal control and enhance robustness performance. In quantum robust control, we aim at analyzing robust stability of quantum systems and designing controllers to achieve improved robustness for quantum systems with uncertainties.

There are several excellent books which provided comprehensive introductions to various topics of quantum control. For example, the book "Introduction to Quantum Control and Dynamics" by D'Alessandro [1] provides an introduction to the analysis and control of quantum dynamics using Lie algebra and Lie group theory. Controllability and open-loop control design of quantum systems have been extensively covered. The book "Quantum Measurement and Control" by Wiseman and Milburn [3] provides a comprehensive treatment of modern quantum measurement and measurement-based quantum control. In particular, this standard reference presented an elegant cover to quantum feedback control with theoretical and experimental developments which are suitable for both physicists and control theorists who are interested in relevant topics. The monograph *Linear Dynamical Quantum Systems* by Nurdin and Yamamoto [2] provides an in-depth treatment of modeling and control of linear dynamical quantum systems from a system-theoretic point of view where linear algebra and quantum stochastic calculus were used as main tools. Our research monograph complements these existing quantum control books and focuses on quantum learning control and quantum robust control. We mainly present some relevant results on these two topics from our groups and collaborators in the last fifteen years. There exist a number of excellent results in the emerging areas of quantum learning control and quantum robust control from many other research groups. Some are briefly mentioned while the others may have been missed.

This monograph is organized as follows: Chapter 1 provides a general introduction to the background and organization of this monograph. Chapter 2 includes a brief introduction to quantum mechanics, quantum systems and quantum control. Chapter 3 discusses control, discrimination and classification of inhomogeneous quantum ensembles. Chapter 4 presents a series of results on sampling-based learning control for quantum state transfer and quantum gate control. Chapter 5 introduces several results on applying machine learning for quantum control design. Chapter 6 presents the results on sliding mode control of quantum systems. Chapter 7 discusses robust stability and performance analysis of several classes of stochastic quantum systems. Chapter 8 explores H^{∞} control and fault-tolerant control of quantum systems. Chapter 9 provides concluding remarks. In the end of each chapter, further reading is suggested. The monograph is aimed at graduate students and researchers who have a background in control theory, quantum physics or applied mathematics and are interested in the topics of quantum control and machine learning. It is also a useful reference for researchers and engineers who need to consider the robustness and reliability in engineering quantum systems and developing quantum technology.

This monograph has benefited from the collaboration with many collaborators in the last fifteen years. We would like to acknowledge scientific interaction of these collaborators including (in alphabetical order) Chunlin Chen, Jiangchao Chen, Qing Gao, Parth Girdhar, Zhibo Hou, Changjiang Huang, Matthew James, Sen Kuang, Yanan Liu, Hailan Ma, Mohamed Mabrok, Franco Nori, Hendra Nurdin, Yu Pan, Bo Qi, Herschel Rabitz, Chuan-Cun Shu, Valery Ugrinovskii, Igor Vladimirov, Shi Wang, Yuanlong Wang, Chengzhi Wu, Kangda Wu, Rebing Wu, Chengdi Xiang, Guoyong Xiang, Xi Xing, Shota Yokoyama, Hidehiro Yonezawa, Qi Yu, Guofeng Zhang and Wei Zhang. This monograph has benefited from many of other collaborators who may be not included in the list. Hailan Ma and Chunlin Chen deserve particular thanks for their great help in assisting the authors for preparing most figures. We would like to thank Oliver Jackson from Springer for inviting us to write this monograph and providing support during the writing process. We would also like to acknowledge financial support for our research in this monograph from the Australian Research Council, the U.S. Office of Naval Research Global, the Air Force Office of Scientific Research and the Alexander von Humboldt Foundation of Germany.

Canberra, Australia
May 2022

Daoyi Dong
Ian R. Petersen

References

1. D'Alessandro D (2007) Introduction to quantum control and dynamics, Chapman & Hall/CRC
2. Nurdin HI, Yamamoto N (2017) Linear dynamical quantum systems, Springer International Publishing
3. Wiseman HM, Milburn GJ (2010) Quantum measurement and control. Cambridge University Press, Cambridge

Contents

Abbreviations and Notation

Abbreviations

ABB	Approximate Bang-Bang
DC	Direct Current
DE	Differential Evolution
DRL	Deep Reinforcement Learning
ES	Evolutionary Strategy
FGH	Fourier–grid-Hamiltonian
FPQL	Fidelity–based Probabilistic Q-Learning
fs	Femtosecond
FWHM	Full Width at Half Maximum
GA	Genetic Algorithm
GD	Gradient method
GHZ	Greenberger–Horne–Zeilinger
GRAPE	GRadient Ascent Pulse Engineering
HD	Homodyne detector
HWP	Half-Wave Plate
LMI	Linear Matrix Inequality
MME	Markovian Master Equations
*MS*_DE	DE algorithm with mixed strategies
*msMS*_DE	Multiple-samples and mixed-strategy DE
NBS	Non-polarizer Beam Splitter
NMR	Nuclear Magnetic Resonance
OPO	Optical Parametric Oscillator
PBS	Polarizer Beam Splitter
POVM	Positive Operator-Valued Measure
PS	Phase Shifter
QEC	Quantum Ensemble Classification
QL	Q-learning
QSDE	Quantum Stochastic Differential Equations
QWP	Quarter-Wave Plate

RL	Reinforcement Learning
RSC	Reduced State Consensus
SLC	Sampling-based Learning Control
SLM	Spatial Light Modulator
SMC	Sliding Mode Control
SME	Stochastic Master Equations
SPCM	Single Photon Counting Module
SPDC	Spontaneous Parametric Down-Conversion
SQUID	Superconducting QUantum Interference Device
TBQCP	Two-point Boundary-value Quantum Control Paradigm
TD	Temporal Difference
TL	Transform Limited
TOF-MS	Time-of-Flight Mass Spectrometry
TPA	Two Photon Absorption

Notation

$\mathbb{H}$	A Hilbert space
X^T	Transpose of a matrix or vector X
x^*	Conjugate of a complex number or vector x
$X^\dagger$	Transpose and conjugate of X
$\|\psi\rangle$	A ket, a unit complex vector on a Hilbert space
$\langle\psi\|$	A bra, the conjugate transpose of the vector $\|\psi\rangle$
$\langle a\|b\rangle$	Inner product between two complex vectors $\|a\rangle$ and $\|b\rangle$
$\|b\rangle\langle a\|$	Outer product between two complex vectors $\|a\rangle$ and $\|b\rangle$
$\hbar$	The reduced Planck constant
i	The imaginary unit $\mathrm{i} = \sqrt{-1}$
ρ	A density matrix or density operator, a quantum state
$\sigma_x, \sigma_y, \sigma_z$	Pauli operators $\sigma_x = \begin{bmatrix} 0 & 1 \\ 1 & 0 \end{bmatrix}, \sigma_y = \begin{bmatrix} 0 & -\mathrm{i} \\ \mathrm{i} & 0 \end{bmatrix}, \sigma_z = \begin{bmatrix} 1 & 0 \\ 0 & -1 \end{bmatrix}$
$A \otimes B$	Tensor product between A and B
$\mathrm{Tr}(X)$	Trace of the matrix X
$\mathrm{Tr}_A(X)$	Partial trace of X over A
$\mathbb{R}$	The set of real numbers
$\mathbb{C}$	The set of complex numbers
$\mathbf{Z}$	The set of integers
$\mathrm{diag}(J_1, J_2, \ldots, J_n)$	A block diagonal matrix with $J_1, J_2, \ldots, J_n$ on its diagonal blocks
$\nabla J(u)$	The gradient of J with respect to u
$\Im(x)$	The imaginary part of a complex number x
$\Re(x)$	The real part of a complex number x
$\mathbb{A}_v(J_i)$	The average value of J_i

$\mathscr{D}_{\mathrm{SMC}}$	Sliding mode domain
$X^{\#}$	The vector of adjoint operators for a vector of operators X or the complex conjugate matrix for a complex matrix X
$\mathbb{W}$	A set of perturbation Hamiltonians
$\langle X \rangle$	The quantum expectation of X
$\mathbf{1}_n$	The n-dimensional matrix with all of its elements being 1
$\mathrm{sgn}x$	The signum function of a real number x

Chapter 1
Introduction

Abstract This chapter presents a brief introduction to quantum technologies and quantum control. The scope and structure of this monograph are introduced and the topics of each chapter are outlined.

1.1 Quantum Technology

The establishment of quantum mechanics is one of the most exciting achievements of the twentieth century. Quantum mechanics provides a general framework to describe and understand the physical world involving quantum systems such as molecules, atoms, electrons and photons, and is also useful to help explain how the universe formed and why stars shine. Quantum systems have many unique characteristics such as quantum entanglement and quantum coherence, different from classical (non-quantum) systems. Such unique characteristics can provide advantages for developing powerful quantum technologies [15]. In recent decades, quantum technology has received considerable attention from both scientists and engineers because of its powerful potential as a future technology [1]. Several active areas for developing quantum technology include quantum information processing [19], quantum metrology and quantum simulation. Quantum information technology (e.g., quantum cryptography, quantum communication and quantum computation) has many important potential applications due to its advantages over traditional information technology [6]. For example, quantum communication can achieve communication which is more secure than classical communication technology. Quantum computers can take advantage of unique quantum effects of entanglement and coherence to speed up the solutions of some classical problems and can even solve some difficult problems that classical computers cannot [3, 19, 24]. In quantum metrology, the principles of quantum mechanics set the ultimate limit on the accuracy of sensing, and quantum resources can be effectively utilized to achieve high-precision sensing [4, 5, 11]. Quantum simulation aims at developing quantum simulators which

D. Dong and I. R. Petersen, *Learning and Robust Control in Quantum Technology*, Communications and Control Engineering,
https://doi.org/10.1007/978-3-031-20245-2_1

are devices designed for specific purposes to provide insight on complex many-body problems and have potential for wide application in, e.g., molecular chemistry, condensed-matter physics, atomic physics and drug discovery [8, 16].

1.2 Quantum Control

Controlling quantum systems has become a central task in the development of quantum technologies, and quantum control has witnessed rapid progress in the last three decades; for the overview, see, e.g., the survey papers [2, 7, 9, 13, 17, 21] or the monographs [10, 23]. The general goal of quantum control is to actively change and control the dynamics of quantum systems for achieving given objectives [1, 14] (e.g., rapid state transfer, high-fidelity gate operation, optimal quantum circuits, robust qubit control). Several fundamental issues in quantum control include investigating the controllability of quantum systems, analyzing stability, optimality and robustness of quantum control systems, designing controllers to achieve expected performance and implementing quantum control for various applications under practical conditions. Controllability is concerned with what control targets can be achieved and the controllability of finite-dimensional systems has been well addressed [10]. For controller design, optimal control theory [13], Lyapunov control approaches [22], quantum learning control [12] and quantum feedback control [23] have been developed in manipulating quantum systems for achieving various control objectives. Among various control design approaches, learning control is recognized as a powerful method for many complex quantum control tasks and has achieved great success in laser control of molecules and other applications since the approach was presented in the seminal paper [18]. Many quantum control tasks may be formulated as an optimization problem and a learning algorithm can be employed to search for an optimal control field satisfying a desired performance condition. With the rapid advancement of machine learning, various state-of-the-art machine learning algorithms can be integrated in control design for efficiently accomplishing quantum control tasks. Learning control of quantum systems is a major focus of this monograph. In developing practical quantum technology, reliability and robustness are basic requirements since we are unable to obtain accurate descriptions of system variables for most practical quantum systems and the existence of various noises is unavoidable. These inaccuracies can be modeled as different types of uncertainties (e.g., Hamiltonian uncertainties, imprecise controls), and one expects to develop systematic control theories and methods to enhance reliability and robustness. Robust control of quantum systems is the other major focus of this book.

1.3 Scope and Structure of This Monograph

This monograph presents a collection of results on learning control and robust control of quantum systems. We mainly focus on presenting some relevant results from our groups and collaborators. There exist a number of excellent results in the areas of quantum learning control and quantum robust control from many other research groups. Some are briefly mentioned while the others may not be covered in this monograph.

The structure of this monograph is outlined as Fig. 1.1. In particular, this chapter provides a general introduction to the background and organization of this monograph. Chapter 2 includes a brief introduction to quantum mechanics, several classes of quantum systems and quantum control. Chapter 3 discusses control, discrimination and classification of inhomogeneous quantum ensembles. Chapter 4 covers a series of results on sampling-based learning control for quantum state transfer and quantum gate generation with uncertainties. Chapter 5 presents several results on applying machine learning to achieve quantum control design including numerical and experimental results using differential evolution to various quantum control tasks and an introduction to reinforcement learning for quantum control. Chapter 6 presents results on sliding mode control of two-level and multi-level quantum systems and open quantum systems. Chapter 7 discusses robust stability and performance analysis of several classes of stochastic quantum systems using the small gain theorem and the Popov approach. Chapter 8 investigates H^∞ control and fault-tolerant control of quantum systems based on measurement-based quantum feedback and quantum coherent feedback. Chapter 9 provides concluding remarks. At the end of each chapter, further reading is suggested.

Chapter 3 through Chap. 6 deal with the Schrödinger picture of quantum mechanics while Chaps. 7 and 8 mainly consider the Heisenberg picture. Chapter 3 through Chap. 5 mainly investigate learning control of quantum systems where quantum robust control and quantum optimal control are two main objectives. Chapter 6 through Chap. 8 focus on robust control of quantum systems. Chapters 3 and 4 mainly consider open-loop quantum control while Chap. 5 involves open-loop quantum control and closed-loop learning control. Chapters 7 and 8 consider typical quantum feedback control while Chap. 6 can be looked as lying at the interface between quantum open-loop control and quantum feedback control. Chapters 3 and 6 involve spin and atomic systems while Chap. 4 involves quantum superconducting systems and molecular systems. Chapters 7 and 8 mainly consider quantum optical systems while Chap. 5 involves spin systems, molecular systems and quantum optical systems.

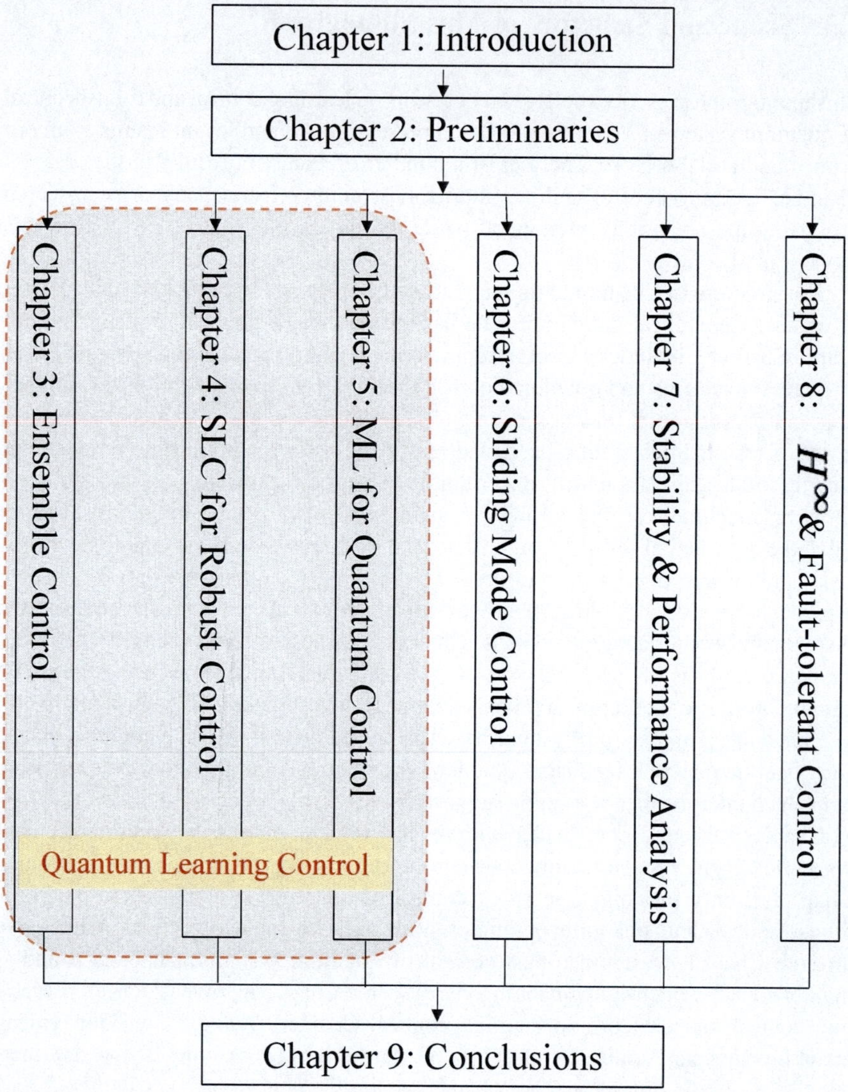

Fig. 1.1 Structure of this monograph (SLC: sampling-based learning control; ML: machine learning)

1.4 Summary and Further Reading

This chapter provides a brief introduction to the relevant background and explains the main focuses and the relationship between different chapters.

Further reading may include [2, 13] for a comprehensive introduction to quantum control, [23] for a comprehensive coverage of quantum measurement and quantum feedback, [10] for a detailed discussion on controllability and open-loop control of quantum systems and [20] for the control of linear quantum systems.

References

1. Acín A, Bloch I, Buhrman H, Calarco T, Eichler C, Eisert J et al (2018) The quantum technologies roadmap: a European community view. New J Phys 20:080201
2. Altafini C, Ticozzi F (2012) Modeling and control of quantum systems: an introduction. IEEE Trans Autom Control 57(8):1898–1917
3. Arute F, Arya K, Babbush R, Bacon D, Bardin JC, Barends R et al (2019) Quantum supremacy using a programmable superconducting processor. Nature 574:505–510
4. Bao L, Qi B, Dong D (2021) Fundamental limits for reciprocal and nonreciprocal non-Hermitian quantum sensing. Phys Rev A 103:042418
5. Bao L, Qi B, Dong D (2022) Exponentially-enhanced quantum non-Hermitian sensing via optimized coherent drive. Phys Rev Appl 17:014034
6. Bennett CH, Divincenzo DP (2000) Quantum information and computation. Nature 404:247–255
7. Brif C, Chakrabarti R, Rabitz H (2010) Control of quantum phenomena: past, present and future. New J Phys 12:075008
8. Buluta I, Nori F (2009) Quantum simulators. Science 326(5949):108–111
9. Chen Z, Dong D, Zhang C (2005) Quantum control: an introduction. University of Science and Technology of China Press, Hefei (in Chinese)
10. D'Alessandro D (2021) Introduction to quantum control and dynamics. Chapman & Hall/CRC
11. Degen CL, Reinhard F, Cappellaro P (2017) Quantum sensing. Rev Mod Phys 89:035002
12. Dong D (2021) Learning control of quantum systems. In: Baillieul J, Samad T (eds) Encyclopedia of systems and control. Springer-Verlag, London Ltd, pp 1090–1096
13. Dong D, Petersen IR (2010) Quantum control theory and applications: a survey. IET Control Theor Appl 4:2651–2671
14. Dong D, Petersen IR (2022) Quantum estimation, control and learning: opportunities and challenges. Ann Rev Control 54:243–251
15. Dowling JP, Milburn GJ (2003) Quantum technology: the second quantum revolution. Ohilosophical Trans R Soc A 361:1655–1674
16. Georgescu IM, Ashhab S, Nori F (2014) Quantum simulation. Rev Mod Phys 86:153–185
17. Glaser SJ, Boscain U, Calarco T, Koch CP, Köckenberger W, Kosloff R, Kuprov I, Luy B, Schirmer S, Schulte-Herbrüggen T, Sugny D, Wilhelm FK (2015) Training Schrödinger's cat: quantum optimal control. Eur Phys J D 69:279
18. Judson RS, Rabitz H (1992) Teaching lasers to control molecules. Phys Rev Lett 68:1500–1503
19. Nielsen MA, Chuang IL (2010) Quantum computation and quantum information. Cambridge University Press, Cambridge
20. Nurdin HI, Yamamoto N (2017) Linear dynamical quantum systems. Springer International Publishing, Berlin
21. Rabitz H, De Vivie-Riedle R, Motzkus M, Kompa K (2000) Whither the future of controlling quantum phenomena? Science 288(5467):824–828

22. Wang X, Schirmer SG (2010) Analysis of Lyapunov method for control of quantum states. IEEE Trans Autom Control 55:2259–2270
23. Wiseman HM, Milburn GJ (2010) Quantum measurement and control. Cambridge University Press, Cambridge
24. Zhong H-S, Wang H, Deng Y-H, Chen M-C, Peng L-C, Luo Y-H et al (2020) Quantum computational advantage using photons. Science 370(6523):1460–1463

Chapter 2
Introduction to Quantum Mechanics and Quantum Control

Abstract This chapter presents a brief introduction to the quantum mechanics postulates and several classes of quantum systems, and outlines quantum control models and various quantum control design methods including quantum optimal control, quantum Lyapunov control, measurement-based feedback control, coherent feedback control, quantum learning control and quantum robust control.

2.1 Introduction

Quantum mechanics provides a framework for describing our physical world and helps extend our capability for understanding various physical systems. Controlling quantum phenomena has become an implicit goal in much research on quantum physics and chemistry since the establishment of quantum mechanics [21, 158]. One of the main goals in quantum control is to establish a firm theoretical footing and develop systematic methods for the active manipulation and control of quantum systems [94]. This goal is non-trivial since microscopic quantum systems have many unique characteristics (e.g., coherence, entanglement and measurement back action) which do not occur in classical systems and the dynamics of quantum systems are governed by the laws of quantum mechanics. In this chapter, we first provide a brief introduction to the foundation of quantum mechanics, especially several quantum mechanics postulates, which will help achieve a basic understanding of the states, dynamics and measurement in quantum systems and may be helpful to readers who are not familiar with quantum mechanics. We briefly introduce several classes of quantum systems including atoms, molecules, spins, quantum optical systems and quantum superconducting systems, which are considered from time to time in quantum learning control and robust control problems in the subsequent chapters.

This monograph covers some topics in quantum control from a control systems perspective. This chapter also provides a brief introduction to quantum control. We first introduce five typical classes of quantum control models. Then we briefly dis-

D. Dong and I. R. Petersen, *Learning and Robust Control in Quantum Technology*, Communications and Control Engineering,
https://doi.org/10.1007/978-3-031-20245-2_2

cuss the controllability of quantum systems. The concept of controllability concerns whether one can drive a quantum system to a desired state [62]. For a controllable quantum system, it is desirable to develop a good control strategy to accomplish the required quantum control tasks [35]. An early paradigm of quantum control was open-loop coherent control [158], which achieved much success in the quantum control of chemical reactions [120, 127]. Optimal control theory has been successfully applied to designing open-loop coherent control in order to find the best way of achieving given control objectives in physical chemistry. Time-optimal control problems for spin systems have been solved to achieve specified control objectives in minimum time. Optimal control techniques have also been successfully applied to spin systems to improve the sensitivity of quantum sensors and achieve time-minimum control [11, 13, 70, 71, 73]. Lyapunov control is another useful approach for quantum control design [76, 99]. Besides these open-control strategies, closed-loop control approaches have also been widely investigated in the area of quantum control. Two paradigms for closed-loop control have been proposed: closed-loop learning control [116] and quantum feedback control. Closed-loop learning control involves a closed-loop operation where each cycle of the loop is executed with a new sample. Quantum feedback control has been used to improve the system performance in various tasks, and two typical feedback strategies including measurement-based quantum feedback and coherent quantum feedback have been developed. In this chapter, we also give a brief introduction to these control approaches. Then we discuss quantum learning control and quantum robust control which are two main research focuses of this monograph.

2.2 Quantum Mechanics Postulates

2.2.1 Four Fundamental Postulates

Quantum mechanics provides a mathematical framework for describing physical systems (referred to quantum systems) at the scale of atoms and subatomic particles. The theory of quantum mechanics is built on several axiomatic fundamental postulates. In different textbooks, these postulates have slightly different forms, and here we use a version similar to that in [101].

Postulate 1: Any closed quantum system is associated to a complex vector space with inner product (i.e., a Hilbert space), known as the state space. The system is described by a unit vector in the state space of the system.

A closed quantum system means that this system is isolated from its environment, i.e., it has no interaction with other systems in the environment. Mathematically, the quantum state of a closed quantum system is usually represented by a unit complex vector $|\psi\rangle$ (a column vector) in the underlying Hilbert space $\mathbb{H}$. The dimension of the underlying Hilbert space $\mathbb{H}$ can be infinite or finite, depending upon the specific physical system. The simplest non-trivial Hilbert space $\mathbb{H}$ for a quantum system is two-dimensional. A two-dimensional quantum system is usually called a quantum bit (qubit) in quantum technology. An orthonormal basis of a qubit is usually denoted as $|0\rangle$ and $|1\rangle$, which correspond to the classical bit states 0 and 1. A qubit is the basic information unit to carry information in quantum information technology. Two states $|\psi\rangle$ and $e^{i\theta}|\psi\rangle$ with difference of a global phase factor $e^{i\theta}$ can be regarded as the same state since they have no observable physical difference. The adjoint of $|\psi\rangle$ is denoted as $\langle\psi|$, which corresponds to the conjugate $(*)$ and transpose (T) of $|\psi\rangle$, i.e.,

$$\langle\psi| = ((|\psi\rangle)^*)^T = (|\psi\rangle)^\dagger.$$

The inner product between two states $|\psi_1\rangle$ and $|\psi_2\rangle$ is defined as

$$\langle\psi_2|\psi_1\rangle \triangleq \langle\psi_2|\cdot|\psi_1\rangle.$$

Since a quantum state is represented by a unit vector $|\psi\rangle$, we have $\langle\psi|\psi\rangle = 1$, which is often called the normalization condition. An important property in quantum mechanics is the superposition principle. This states that the superposition of quantum states $|\psi_i\rangle$ for a quantum system in the form of $\sum_i c_i|\psi_i\rangle$ is also a valid state, where the complex coefficients c_i satisfy $\sum_i |c_i|^2 = 1$.

Postulate 2: The time evolution of the state $|\psi\rangle$ of a closed quantum system is described by the Schrödinger equation

$$i\hbar\frac{d|\psi(t)\rangle}{dt} = H|\psi(t)\rangle, \tag{2.1}$$

where $i = \sqrt{-1}$, $\hbar$ is the reduced Planck constant, and the Hermitian operator H describes the system Hamiltonian.

Postulate 2 describes how the state of a quantum system will evolve with time. When the system Hamiltonian H is time-independent, the solution to the Schrödinger equation is thus

$$|\psi(t)\rangle = U(t)|\psi(0)\rangle, \tag{2.2}$$

where we define

$$U(t) \triangleq \exp[-\frac{i}{\hbar}Ht], \tag{2.3}$$

and $U(0) = I$. The operator $U(t)$ in the form of (2.3) is always unitary, i.e., $U^\dagger U = UU^\dagger = I$. It is also called a propagator in some applications or a quantum gate in quantum information. For example, the following are several common single-qubit gates (Pauli gates or Pauli operators X, Y, Z and Hadamard gate H):

$$X = \sigma_x = \begin{bmatrix} 0 & 1 \\ 1 & 0 \end{bmatrix},$$

$$Y = \sigma_y = \begin{bmatrix} 0 & -\mathrm{i} \\ \mathrm{i} & 0 \end{bmatrix},$$

$$Z = \sigma_z = \begin{bmatrix} 1 & 0 \\ 0 & -1 \end{bmatrix},$$

$$H = \frac{1}{\sqrt{2}} \begin{bmatrix} 1 & 1 \\ 1 & -1 \end{bmatrix}.$$

Postulate 3: A quantum measurement is associated with a collection $\{M_i\}$ of measurement operators. These measurement operators satisfy the completeness condition

$$\sum_i M_i^\dagger M_i = I. \tag{2.4}$$

The index i labels possible measurement outcomes. If the quantum system is in state $|\psi\rangle$ before the measurement, then the probability that the ith outcome occurs is given by

$$p_i = \langle\psi|M_i^\dagger M_i|\psi\rangle,$$

and after the measurement the state immediately collapses into

$$\frac{M_i|\psi\rangle}{\sqrt{\langle\psi|M_i^\dagger M_i|\psi\rangle}}.$$

The randomness in obtaining potentially different results for the same measurement on identical quantum systems is a non-classical phenomenon [57]. The unique property of quantum measurement brings new challenges for estimating quantum states or identifying parameters in quantum systems [106, 113, 114, 149, 150, 156, 157, 167, 170, 177, 189]. The completeness condition is required to guarantee that all of the measurement outcome probabilities sum to one. If $\{M_i\}$ further satisfies $M_i M_j = \delta_{ij} M_i$ and each M_i is Hermitian, then the measurement is called a projective measurement, and M_i is a projector. A more generally used form of quantum measurement is a positive operator-valued measure (POVM) measurement. If we define $P_i \triangleq M_i^\dagger M_i$, then a POVM measurement is associated with a set of positive operators $\{P_i\}$. The probability of the ith outcome can be calculated as

$$p_i = \langle\psi|P_i|\psi\rangle.$$

Postulate 4: The state space of a composite quantum system is described by the tensor product of the state spaces of its component quantum systems. Specifically, let $|\psi_j\rangle$ be the state of the jth subsystem for a composite system consisting of n subsystems, then the state of the composite system is described as $|\psi_1\rangle \otimes |\psi_2\rangle \otimes ... \otimes |\psi_n\rangle$ (often denoted as $|\psi_1\psi_2...\psi_n\rangle$).

Postulate 4 shows how to obtain the state of a composite system from the states of its subsystems. To obtain the state of a subsystem from the state of a composite system, we may use partial trace to calculate the state. For any $|a\rangle, |b\rangle \in \mathbb{H}_A$, $|c\rangle, |d\rangle \in \mathbb{H}_B$, the partial trace over $\mathbb{H}_A$ is defined by

$$\mathrm{Tr}_A(|a\rangle\langle b| \otimes |c\rangle\langle d|) = \mathrm{Tr}(|a\rangle\langle b|)|c\rangle\langle d|.$$

2.2.2 *Pure State Versus Mixed State*

In the above postulates, we only consider quantum pure states where a pure state can be represented by a single unit vector. In many practical applications, we often need to consider a statistical ensemble of pure states $|\psi_i\rangle$ which cannot be described with a single vector and we call such a state of the statistical ensemble a mixed state. We usually use a density matrix ρ to represent a quantum mixed state which can be described as

$$\rho = \sum_i p_i |\psi_i\rangle\langle\psi_i|$$

with $p_i > 0$ and $\sum_i p_i = 1$. A physical quantum state ρ should be Hermitian ($\rho = \rho^\dagger$), positive semidefinite ($\rho \geq 0$), and satisfy $\mathrm{Tr}(\rho) = 1$. For a pure state $|\psi\rangle$, we have $\rho = |\psi\rangle\langle\psi|$. It is clear that $\mathrm{Tr}(\rho^2) = 1$ for a pure state and $\mathrm{Tr}(\rho^2) < 1$ for a mixed state. For an open quantum system, a density matrix ρ is required to describe its quantum state since the state of an open quantum is usually a mixed state.

When the quantum state is represented by a density matrix ρ, the evolution of the state ρ in Postulate 2 follows the Liouville–von Neumann equation

$$i\hbar\dot{\rho} = [H, \rho], \tag{2.5}$$

where $[A, B] = AB - BA$ is the commutator. In terms of the density matrix, (2.2) becomes

$$\rho(t) = U(t)\rho(0)U^\dagger(t). \tag{2.6}$$

From Postulate 3, the probability of the ith outcome can be calculated as

$$p_i = \mathrm{Tr}(P_i\rho). \tag{2.7}$$

From Postulate 4 we know the composite of density matrices is also in the form of tensor product. For a composite system consisting of two subsystems A and B, we have $\rho^{AB} = \rho^A \otimes \rho^B$ which is a state on $\mathbb{H}_A \otimes \mathbb{H}_B$. The reduced density operator for subsystem $\mathbb{H}_B$ can be calculated using the partial trace as

$$\rho^B \triangleq \mathrm{Tr}_A(\rho^{AB}).$$

2.3 Several Classes of Quantum Systems

2.3.1 *Atomic, Molecular and Spin Systems*

Atomic systems are a class of typical quantum systems. An atom is a basic unit of ordinary matter and every atom is composed of a nucleus and one or more electrons. An example is a hydrogen atom which contains a single positively charged proton and a single negatively charged electron bound to its nucleus. There are many other natural and artificial atoms. Due to quantum effects, it is impossible to accurately predict the behavior and dynamics of an atom using classical physics. We usually use a Schrödinger equation to describe the dynamics of an atom where the state is represented by a wavefunction. The eigenvalues of the system Hamiltonian correspond to the energies of different energy levels. The lowest energy state of a bound electron in an atom is called the ground state of the atom, and when an electron transitions to a higher level, it is called as being in an excited state. The simplest example is a two-level atom which can be used as a qubit and the transition between the ground state and excited state can be realized by absorbing or emitting a photon with a specific frequency. Controlling the population transfer between different energy levels is a common target in quantum control.

Molecular systems are another class of quantum systems where a molecule is usually composed of two or more atoms held together by chemical bonds. In molecules, vibrational states and rotational states are two widely considered classes of quantum states [80]. A vibration in a molecule corresponds to a periodic motion of the atoms relative to each other when the molecular center of mass remains unchanged. A molecular vibration is excited when the molecule absorbs a specific amount of energy and its vibrational state changes accordingly. The nuclei in a molecule may have rotational motion about its center of mass. The transition between different rotational states may occur via absorbing or emitting photons with specific frequencies. Besides controlling vibrational states and rotational states, another two common quantum control targets for molecules are dissociation and association. A purpose of dissociation is to selectively break some chemical bonds to generate expected products or achieve the expected ratio of different products. Molecular association usually aims to form a larger molecule or complex.

Spin is an intrinsic angular momentum of a particle, different from the orbital angular momentum due to the motion of the particle in space [80]. For example, in the Stern–Gerlach experiment, silver atoms without orbital angular momentum were observed to possess two possible discrete angular momenta, which demonstrates the existence of electron spin angular momentum. Electron spin and nuclear spin in nuclear magnetic resonance (NMR) have found wide applications in medicine, imaging, quantum sensing and quantum computation. It is a significant aim to accurately or robustly control spin states in different spin systems for various applications.

2.3.2 *Quantum Optical Systems*

Quantum optical systems are another typical class of quantum systems where it is assumed that light of a certain frequency ν is made up of discrete units of energy $h\nu$ (with Planck constant h) [7]. Such individual quanta of light are called photons which have both wave and particle-like properties. Current technology has allowed the generation of individual photons in a controlled way [7]. A photon may lie in different polarization states. Let $|H\rangle$ denote horizontal polarization and $|V\rangle$ denote vertical polarization. The state of a photon may be described as

$$|\psi\rangle = \alpha|H\rangle + \beta|V\rangle.$$

A photon can be used as a qubit where $|H\rangle$ and $|V\rangle$ correspond to $|0\rangle$ and $|1\rangle$ of the qubit, respectively.

The quantum states of light are often represented by a creation operator $a^\dagger$ and an annihilation operator a in Fock space. The operators $a^\dagger$ and a raise and lower the excitation of a single mode by one photon, respectively [7]. The Hamiltonian operator of the radiation field of a single mode can be written as

$$H = \hbar\omega\left(a^\dagger a + \frac{1}{2}\right),$$

where $\hat{n} = a^\dagger a$ is defined as the photon number operator. The eigenstates of $\hat{n}$ are denoted as $|n\rangle$ and we have

$$a^\dagger a|n\rangle = \hat{n}|n\rangle \doteq n|n\rangle.$$

The ground state $|0\rangle$ is often called the vacuum state, satisfying

$$a|0\rangle = 0.$$

The operation of the creation operator $a^\dagger$ and the annihilation operator a on a number state $|n\rangle$ is described as follows:

$$a|n\rangle = \sqrt{n}|n-1\rangle,$$

$$a^\dagger|n\rangle = \sqrt{n+1}|n+1\rangle.$$

Besides number states, a widely considered state in quantum optics is the coherent state $|\alpha\rangle$, which is the eigenstate of the annihilation operator a satisfying

$$a|\alpha\rangle = \alpha|\alpha\rangle$$

with complex number α. A coherent state can be written as the superposition of number states [48]

$$|\alpha\rangle = \exp\left(-\frac{1}{2}|\alpha|^2\right)\sum_{n=0}^{\infty}\frac{\alpha^n}{\sqrt{n!}}|n\rangle.$$

A quantum optical system usually consists of many optical components which are combined to achieve various targets such as controlling polarization and preparing given coherent states. One of the most common optical components is a mirror that is usually referred to as a beamsplitter. When an input beam is injected to a beamsplitter, we usually obtain two output beams, one reflected and the other transmitted. Assume the intensity reflection coefficient of a beamsplitter is η. The beamsplitter can be simply described by a transformation from two input mode operators a_{i1} and a_{i2} into two output mode operators a_{o1} and a_{o2} as follows [7]:

$$\begin{bmatrix} a_{o1} \\ a_{o2} \end{bmatrix} = \begin{bmatrix} \sqrt{\eta} & \sqrt{1-\eta} \\ \sqrt{1-\eta} & -\sqrt{\eta} \end{bmatrix}\begin{bmatrix} a_{i1} \\ a_{i2} \end{bmatrix}. \tag{2.8}$$

Another widely used optical component is an optical cavity, which is composed of two or more mirrors. These mirrors are usually aligned to make the beam of light be reflected in a closed path and interfere with the incident wave after one round trip [7]. The dynamics of an optical cavity can usually be described by linear differential equations. For example, for an empty cavity with the annihilation operator $a(t)$ consisting of m mirrors, the dynamics of the optical field in the cavity can be described as

$$da(t) = -\frac{\kappa}{2}a(t)\mathrm{d}t + \sum_{j=1}^{m}\sqrt{\kappa_j}\mathrm{d}A_{in,j}(t), \tag{2.9}$$

where the decay rate of the jth mirror is κ_j and $\kappa = \sum_{j=1}^{m}\kappa_j$ represents the total decay rates. $A_{in,j}$ represents the input field of the jth mirror, and the output of each mirror is described as

$$\mathrm{d}A_{out,j}(t) = \sqrt{\kappa_j}a(t)\mathrm{d}t - \mathrm{d}A_{in,j}(t). \tag{2.10}$$

2.3.3 Quantum Superconducting Systems

With recent technological advances, artificial quantum systems such as quantum superconducting systems can also be constructed. Quantum superconducting systems have been widely used as qubits, and there has been great progress in developing quantum computers based on quantum superconducting systems. In quantum superconducting circuits involving Josephson junctions, the Josephson coupling energy E_J and the charging energy E_C are two significant quantities. Their ratio determines whether the phase or the charge dominates the behavior of the qubit [176]. The simplest flux qubit consists of a superconducting loop interrupted by one Josephson junction in the phase regime, $E_J \gg E_C$ [33]. The phase difference across the junction is related to the flux Φ in the loop, and an external flux Φ_x can be applied to bias the system. The Hamiltonian can be described as [95]

$$H = -E_J \cos\left(2\pi \frac{\Phi}{\Phi_0}\right) + \frac{(\Phi - \Phi_x)^2}{2L} + \frac{Q^2}{2C_J}, \tag{2.11}$$

where $\Phi_0 = \frac{h}{2e}$, L is the self-inductance of the loop, C_J is the capacitance of the junction and the charge

$$Q = -i\hbar \frac{\partial}{\partial \Phi}$$

is the canonically conjugate to the flux Φ.

In many practical application, one may introduce a DC SQUID to replace the junction. This replacement enables one to control the Josephson coupling energy. Moreover, the structure with three junctions is usually used to replace the one with one junction, allowing more design flexibility. At low temperatures and with appropriate approximation, the system can be used as a qubit with the following Hamiltonian:

$$H = f(V_g)\sigma_z - g(\Phi)\sigma_x, \tag{2.12}$$

where $f(V_g)$ is related to the charging energy E_C and this term can be adjusted through an external voltage V_g, and $g(\Phi)$ corresponds to a controllable term including different control parameters [33].

Superconducting quantum circuits based on Josephson junctions are macroscopic circuits but can behave quantum mechanically like artificial atoms, which allows for the observation of quantum entanglement and quantum coherence on a macroscopic scale [22, 168]. These artificial atoms can be used to test the laws of quantum mechanics on macroscopic systems and also offer one of the most promising systems for implementing quantum computers due to their advantages such as scalability, design flexibility and tunability. Superconducting qubits can be controlled by adjusting external parameters such as currents, voltages and microwave photons, and the coupling between two qubits can be turned on and off at will [22].

2.4 Introduction to Quantum Control

2.4.1 *Quantum Control Models*

There are various models which can be used to describe different classes of quantum control problems. Several commonly used models include bilinear models, Markovian master equations (MMEs), stochastic master equations (SMEs) and quantum stochastic differential equations (QSDEs). The first three classes of quantum control models are presented in the Schrödinger picture while QSDEs are considered under the Heisenberg picture.

The bilinear model is actually the Schrödinger equation that describes the controlled evolution

$$i\hbar\frac{\mathrm{d}}{\mathrm{d}t}|\psi(t)\rangle = [H_0 + \sum_k u_k(t)H_k]|\psi(t)\rangle, \tag{2.13}$$

where $|\psi(t)\rangle$ represents the state of the system, $|\psi(t=0)\rangle = |\psi_0\rangle$, H_0 is the free Hamiltonian of the system (i.e., a Hermitian operator on $\mathbb{H}$), and the initial state has unit norm $\|\psi_0\|^2 \equiv \langle\psi_0|\psi_0\rangle = 1$, the control of the system may be realized by a set of control functions $u_k(t) \in \mathbb{R}$ coupled to the system via time-independent Hermitian interaction Hamiltonians H_k $(k = 1, 2, \dots)$. For simplicity, we usually consider finite-dimensional (e.g., N-dimensional) quantum control systems, which is an appropriate approximation in many practical situations.

A typical goal in a quantum control problem defined for the system (2.13) is to find a final time $T > 0$ and a set of admissible controls $u_k(t) \in \mathbb{R}$ which drives the system from the initial state $|\psi_0\rangle$ into a given target state $|\psi_f\rangle$. In some applications, T is also often fixed in advance. The total Hamiltonian

$$H(t) = H_0 + \sum_k u_k(t)H_k$$

defines a unitary transformation $U(t)$ which can accomplish the transition from the pure state $|\psi_0\rangle$ to the pure state $|\psi(t)\rangle$, and $U(t)$ satisfies

$$\mathrm{i}\dot{U}(t) = [H_0 + \sum_k u_k(t)H_k]U(t),\ U(0) = I. \tag{2.14}$$

Bilinear models are widely used to describe closed quantum control systems such as spin systems in NMR and molecular systems in physical chemistry. In many practical applications, the quantum systems should be considered as open quantum systems. Since most quantum control systems unavoidably interact with their external environments (including control inputs, measurement devices and coupling between the systems and their environments), quantum control systems are often required to deal with open quantum systems. For an open quantum system, the evolution of its state cannot generally be described in terms of a unitary transformation. In many situations, a quantum master equation for $\rho(t)$ is a suitable way to describe the dynamics of an open quantum system. One of the simplest cases is when a Markovian approximation can be applied where a short environmental correlation time is supposed and memory effects may be neglected [15]. For an N-dimensional open quantum system with Markovian dynamics, its state $\rho(t)$ can be described by the following Markovian master equation (MME) (for details, see, e.g., [2, 15, 84]):

$$\dot{\rho}(t) = -\mathrm{i}[H(t), \rho(t)] + \frac{1}{2}\sum_{j,k=0}^{N^2-1} \alpha_{jk} \left\{ \left[F_j\rho(t), F_k^{\dagger} \right] + \left[F_j, \rho(t)F_k^{\dagger} \right] \right\}. \tag{2.15}$$

Here $\{F_j\}_{j=0}^{N^2-1}$ is a basis for the space of linear bounded operators on $\mathbb{H}$ with $F_0 = I$, the coefficient matrix $A = (\alpha_{jk})$ is positive semidefinite and physically specifies the relevant relaxation rates.

In feedback control, we generally need to continuously monitor a quantum system to obtain feedback information. The evolution of a quantum system under continuous measurements of an observable X can be described by the following master equation [63]:

$$\mathrm{d}\rho = -\mathrm{i}[H, \rho]\mathrm{d}t - \kappa[X, [X, \rho]]\mathrm{d}t + \sqrt{2\kappa}(X\rho + \rho X - 2\langle X\rangle\rho)\mathrm{d}W, \tag{2.16}$$

where κ is a parameter related to the measurement strength, $\langle X\rangle = \mathrm{Tr}(X\rho)$, $\mathrm{d}W$ is a Wiener increment with zero mean and variance equal to $\mathrm{d}t$ and satisfies the following relationship to the measurement output y:

$$\mathrm{d}W = \mathrm{d}y - 2\sqrt{\kappa}\mathrm{Tr}(X\rho)\mathrm{d}t. \tag{2.17}$$

Equation (2.16) is usually called a stochastic master equation (SME). Such a stochastic master equation can be obtained as a filtering equation from quantum filtering theory (for details, see [14, 41–45, 137, 138]). It should be pointed out that (2.16) is only a typical form of SME and there exist many different types of SMEs which depend on different measurement processes [111].

In the bilinear model (2.13), MME (2.15) and SME (2.16), we use the Schrödinger picture of quantum mechanics where equations describing the time dependence of quantum states are given. That is, the state vectors of quantum systems evolve in time while the observables or operators (excluding the system Hamiltonian operators) are usually constant with respect to time. In quantum mechanics, another representation is the Heisenberg picture where the states of quantum systems remain constant while the observables evolve in time. Although the two representations of the Schrödinger picture and the Heisenberg picture are equivalent, it may be more convenient in some cases to adopt the Heisenberg picture to describe the quantum dynamics. An interesting case is a class of non-commutative linear quantum stochastic systems which includes many examples of interest in quantum technology, especially in linear quantum optics [66, 97, 105, 144, 151, 152, 179, 182]. This class of systems can be described by the following quantum stochastic differential equations (QSDEs) [66]:

$$\begin{aligned} dx(t) &= Ax(t)dt + Bdw(t); \; x(0) = x_0 \\ dy(t) &= Cx(t)dt + Ddw(t), \end{aligned} \tag{2.18}$$

where A, B, C and D are, respectively, $\mathbb{R}^{n\times n}$, $\mathbb{R}^{n\times n_w}$, $\mathbb{R}^{n_y\times n}$ and $\mathbb{R}^{n_y\times n_w}$ matrices (n, n_w, n_y are positive integers), and $x(t) = [x_1(t)\dots x_n(t)]^T$ is a vector of self-adjoint possibly non-commutative system variables. The initial system variables $x(0) = x_0$ consist of operators (on an appropriate Hilbert space) satisfying the commutation relations

$$[x_j(0), x_k(0)] = 2\mathrm{i}\Theta_{jk}, \; j, k = 1, \dots, n, \tag{2.19}$$

where Θ_{jk} are the components of a real antisymmetric matrix Θ. For simplicity, we may take Θ to have one of the following forms: (i) Canonical if $\Theta = \mathrm{diag}(J, \dots, J)$, or (ii) degenerate canonical if $\Theta = \mathrm{diag}(0_{n'\times n'}, J, \dots, J)$, where $0 < n' \leq n$, n' is the number of classical variables, and

$$J = \begin{bmatrix} 0 & 1 \\ -1 & 0 \end{bmatrix}.$$

The vector quantity w describes the input signals and is assumed to admit the decomposition

$$dw(t) = \beta_w(t)dt + d\tilde{w}(t), \tag{2.20}$$

where $\tilde{w}(t)$ is the noise part of $w(t)$ and $\beta_w(t)$ is a self-adjoint, adapted process (see [107], [8]). The noise $\tilde{w}(t)$ is a vector of self-adjoint quantum noises with Itô table

$$d\tilde{w}(t)d\tilde{w}^T(t) = F_{\tilde{w}}dt, \tag{2.21}$$

where $F_{\tilde{w}}$ is a positive semidefinite Hermitian matrix (see [8, 107] for details). Equation (2.18) can describe quantum systems such as linear quantum optical systems. However, it does not necessarily represent the dynamics of a meaningful physical system. We may need to add some additional constraints to ensure that a system described by equations of the form (2.18) is physically realizable [66].

In [66], a notion of physical realizability has been developed based on the concept of an open quantum harmonic oscillator as the basic component of a physically realizable quantum system. Denote $\mathrm{diag}_m(J)$ as a $m \times m$ block diagonal matrix with m matrices J on the diagonal. Under several reasonable assumptions on a system of the form (2.18) (e.g., n_y is even, $n_w \geq n_y$, $F_w = I + \mathrm{i}\mathrm{diag}(J, \dots, J)$, etc.; see [66] for details), we have the following results on physical realizability [66]:

Theorem 2.1 *The system (2.18) is physically realizable if and only if:*

$$iA\Theta + i\Theta A^T + BT_wB^T = 0, \tag{2.22}$$

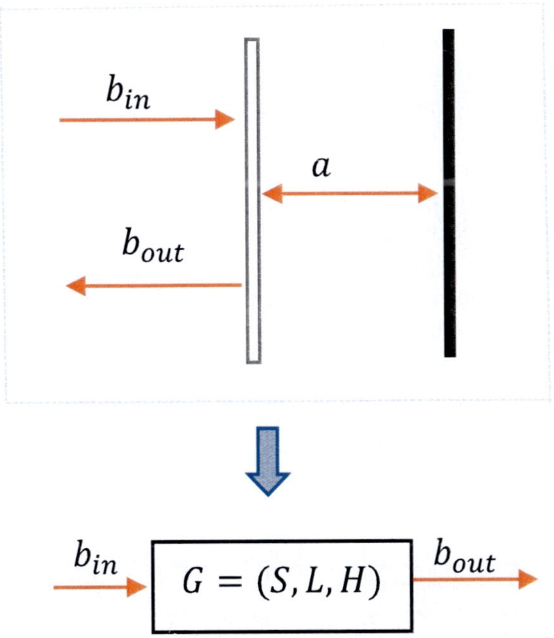

Fig. 2.1 [169] An illustrative example for an optical cavity and its $G = (S, L, H)$ description

$$B\begin{bmatrix} I_{n_y \times n_y} \\ 0_{(n_w - n_y)\times n_y} \end{bmatrix} = \Theta C^T diag_{N_y}(J), \tag{2.23}$$

$$D = \begin{bmatrix} I_{n_y \times n_y} & 0_{n_y \times (n_w - n_y)} \end{bmatrix}, \tag{2.24}$$

where $N_y = \frac{n_y}{2}$ *and* $T_w = \frac{1}{2}(F_w - F_w^T)$.

Another widely used modeling method is the (S, L, H) parametrization which provides a powerful framework for modeling input–output quantum networks [23]. In this framework, S is a scattering matrix, L is a vector of coupling operators, (S, L) specifies the interface of the system to external fields, and the Hamiltonian H describes the internal energy of the system [169]. The (S, L, H) framework provides a convenient tool for control theoretic analysis of quantum systems (especially quantum optical systems). For example, for a quantum optical system as shown in Fig. 2.1 consisting of a pair of mirrors where the right mirror is perfectly reflecting while the left mirror is partially transmitting. An electromagnetic (optical) mode is trapped between the two mirrors. This mode can be described by a quantum harmonic oscillator using annihilation operator a and creation operator $a^\dagger$. Assume the detuned parameter $\Delta = \omega_c - \omega_0$ where ω_c is the center frequency and ω_0 is the reference frequency. The energy decay rate of this cavity is η. This system can be described by (S, L, H) as

$$G = (S, L, H) = (I, \sqrt{\eta}a, \Delta a^\dagger a). \tag{2.25}$$

The (S, L, H) description of many other examples (e.g., phase shifter, beam splitter, one-sided cavity with a Kerr nonlinearity) can be found in [23, 50]. The (S, L, H) framework can be extended in a straightforward way to systems with multiple input–output modes, which is often encountered in quantum networks. The (S, L, H) framework has found wide application for analyzing input–output relations of quantum systems in physics and control communities. Various results have shown that the framework is useful and convenient for modeling fundamental properties of light-matter interactions, designing complex quantum networks and developing coherent feedback control theory [23, 65]. In this book, the (S, L, H) framework is used mainly in Chap. 7, and the QSDE approach is employed to model quantum control systems mainly in Chap. 8.

2.4.2 *Controllability*

Controllability is a fundamental theoretical notion for classical and quantum systems. In quantum control, controllability also has a close connection with the universality of quantum computation [118] and the possibility of attaining atomic or molecular scale transformations [119, 166]. A common research focus on controllability is on bilinear models for quantum systems in which controllability criteria may be expressed in terms of the structure and rank of the corresponding Lie groups and Lie algebras [25]. Different definitions of controllability such as pure state controllability, operator controllability [1], eigenstate controllability [180] and their corresponding testing criteria have been presented. In some works, pure state controllability is also called wavefunction controllability (e.g., [34, 135]). Operator controllability, in the unitary case, is also called complete controllability [124, 125]. Some algorithms have been developed for testing the controllability of specific quantum systems [3, 25, 119, 124]. The Lie algebra method allows for a straightforward mathematical treatment of closed quantum systems. However, the relevant criteria may be computationally difficult when the dimension of the controlled system is large. Hence, another method based on graph theory has been developed for pure state controllability for which the controllability criterion becomes easy to verify [135, 136].

The above discussion on controllability mainly focuses on finite dimensional systems. Many practical quantum systems, especially those with continuous spectra, are essentially infinite dimensional and their quantum states should be described on an infinite dimensional Hilbert space $\mathbb{H}$. The controllability of such infinite-dimensional quantum systems has been studied in [62, 79, 98, 166]. Another interesting focus is on the controllability of open quantum systems. It has been proven that a finite-dimensional open quantum system with Markovian dynamics (i.e., MME (2.15)) is not controllable when using only coherent control [4]. However, finite-dimensional open quantum systems with Kraus-map dynamics are complete kinematic state controllable [165], and a specific Kraus map can be constructed for transformation from an arbitrary initial state to a predefined target state if some incoherent resources are available as control tools [75, 90]. Another two interesting topics in controllability are the controllability of quantum ensembles and quantum networks, and some results have also been presented [9, 16, 82, 83, 153].

2.4.3 Quantum Optimal Control

In many quantum control problems with practical applications, it is expected to design a control law to achieve given objectives with optimal performance. Optimal control theory and methods have been a powerful tool in seeking an optimal control law [19, 54, 108, 159].

In quantum optimal control, the control problem is usually formulated as a problem of seeking a set of admissible controls satisfying the system dynamic equations and simultaneously minimizing a cost functional related to practical requirements. The cost functional may be different according to the practical requirements of the quantum control problem, such as minimizing the control time (time-optimal control problem) [70, 133], the control energy [26, 52], the error between the final state and target state, the sensitivity of a quantum sensor [110], the ratio of two products [37] or a combination of these requirements. Many useful tools in traditional optimal control, such as the variational method, the Pontryagin minimum principle, gradient algorithms and convergent iterative algorithms [191], can be adapted to quantum systems and applied to search for optimal controls. Optimal control techniques have been widely applied to control quantum phenomena in physical chemistry (for details, see, e.g., [18, 55, 116, 117, 120, 127–129, 131]) and NMR experiments [71, 73, 139]. Optimal control algorithms in both the time domain and the frequency domain have been developed to seek for control pulses [72, 130]. For example, the GRadient Ascent Pulse Engineering (GRAPE) algorithm has been developed and applied in many NMR applications [72]. Time-optimal control has been investigated for many spin systems theoretically as well as experimentally [46, 67, 148]. Optimal control is often combined with other control strategies, such as closed-loop learning control and quantum feedback control for achieving various tasks such as manipulating quantum entanglement [96], identifying system parameters [47, 88, 178], controlling chemical reactions [116, 127], enhancing sensitivity of quantum sensors [110] and tracking quantum states [20].

2.4.4 Quantum Lyapunov Control

Lyapunov control approaches are powerful tools for feedback controller design in classical control systems. In quantum control, the acquisition of feedback information through measurements usually changes the state being measured, which makes it difficult to directly apply Lyapunov control methods to quantum feedback design. However, we may first implement the feedback control design by simulation on a computer, which will provide a sequence of controls. Then we may apply the control sequence to the quantum system under consideration in an open-loop form [6, 40, 76, 99, 154, 155]. This is a "feedback design and open-loop control" strategy. This strategy is especially useful for some difficult quantum control tasks [5]. The most important aspects in Lyapunov-based control design include the construction of the Lyapunov function, the determination of the control law and the analysis of asymptotic convergence.

To design a control law for quantum systems, several classes of Lyapunov functions have been proposed. For example, if the target state is $|\psi_f\rangle$, we may select one of the following Lyapunov functions: $V_1(t) = \frac{1}{2}(1 - |\langle\psi_f|\psi(t)\rangle|^2)$ [140], $V_2(t) = \langle\psi(t) - \psi_f|\psi(t) - \psi_f\rangle$ [99], or $V_3(t) = \langle\psi(t)|P|\psi(t)\rangle$, where P is a positive semidefinite Hermitian operator [51]. It is clear that $V_j(t) \geq 0$ $(j = 1, 2, 3)$. We may select the control function to guarantee that the first-order time derivative of the Lyapunov function is negative semidefinite. That is, we may determine the control law using the condition $\dot{V}(t) \leq 0$ on the Lyapunov function $V(t)$. LaSalle's invariance principle is useful to analyze the asymptotic behavior of the system dynamics [25, 81]. Lyapunov control techniques have also been formulated using density operators for the control of a spin ensemble [5, 6] and various strategies have been developed for achieving rapid or finite-time convergence [60, 77, 78]. Moreover, the Lyapunov approach has been used in stochastic quantum control systems to help design feedback control laws [85–87, 100].

2.4.5 *Quantum Feedback Control*

Feedback is a fundamental concept in classical control theory and a major aim of introducing feedback is to compensate for the effects of unpredictable disturbances on a system under control, or to make automatic control possible when the initial state of the system is unknown. In the last thirty years, feedback control has also been widely investigated in quantum control [163, 186]. There are two main classes of quantum feedback control schemes including measurement-based feedback control and coherent feedback control. In measurement-based feedback control, it is usually necessary to obtain information about the state of system through measurement. However, measurements on a quantum system will unavoidably disturb the state of the system under measurement. Projective measurement and continuous weak measurement are the two main approaches of information acquisition in measurement-based feedback control and the controller is usually a classical system. Markovian quantum feedback [160, 162] and Bayesian quantum feedback [29–31] were the two main typical feedback strategies in the early development of quantum feedback control. In Markovian quantum feedback, any time delay is ignored and a memoryless controller is assumed. That is, the measurement record is immediately fed back onto the system to alter the system dynamics and may then be forgotten [161]. Hence, the equation describing the resulting evolution is a Markovian master equation. In Bayesian quantum feedback, the process is divided into two steps involving state estimation and feedback control. The best estimates of the dynamical variables are obtained continuously from the measurement record and fed back to control the system dynamics [30]. In Bayesian quantum feedback, the feedback Hamiltonian is a general function of the measurement record [122], which is used to control the system dynamics. Bayesian feedback is usually superior to Markovian feedback since it uses more information. However, it is more difficult to implement Bayesian feedback than Markovian feedback due to the existence of the estimation step [161]. Measurement-based quantum feedback has been used to improve the control performance, e.g., quantum entanglement [121, 172, 173], quantum state reduction [115, 138], quantum state preparation in many areas such as quantum optics [49, 123, 163] and superconducting quantum systems [141].

Another important feedback paradigm is quantum coherent feedback where the feedback controller itself is a quantum system, and it processes quantum information. This is greatly different from Markovian and Bayesian quantum feedback where the feedback information from measurement results is classical information and the feedback controller is a classical controller. Quantum coherent feedback has also been widely studied for achieving various performance requirements [24, 89, 143, 145, 183, 184, 187, 190]. For example, an H^∞ control method has been used to investigate the robust control problem of linear quantum stochastic systems via quantum coherent feedback [66]. Mabuchi [93] demonstrated an experimental realization of a coherent feedback control system and verified the theory of linear quantum stochastic control in [66]. An LQG optimal control approach has been used to analyze quantum coherent feedback problem for linear quantum stochastic systems [103]. Quantum coherent feedback has been used to develop controller design strategies in order to control a single qubit in diamond [58] and cool a quantum oscillator [56]. The performance of measurement-based feedback control and coherent feedback control have also compared for different scenarios [64]. A more comprehensive introduction to quantum feedback control can be found in [163] and [186].

2.5 Quantum Learning Control

Learning control has been recognized as a powerful approach for solving many complicated quantum control tasks and has achieved great success in laser control of molecules and various other applications since the method was presented in [32, 69, 91]. Many quantum control tasks may be formulated as optimization problems, and a learning algorithm can be employed to search for a good control field achieving desired performance. Gradient algorithms have been demonstrated to be an excellent candidate for numerically finding an optimal control field and achieved many successful applications in NMR systems due to their high efficiency [72]. In many other quantum optimal control problems, the gradient information may not be easy to obtain and some complex quantum control problems may have local optima. For these situations, stochastic search algorithms combined with a closed-loop strategy usually can achieve improved performance to find a good control field. Closed-loop learning control using genetic algorithms (GA) or differential evolution (DE) [39] has achieved great successes in the control of laboratory chemical reactions [69, 116]. The closed-loop learning control procedure generally involves three components [116]: (i) a trial laser control input design, (ii) the laboratory generation of the control that is applied to the sample and subsequently observed for its impact and (iii) a learning algorithm that considers the prior experiments and suggests the form of the next control input. The initial trial control input may be a well-designed laser pulse or a random control input. The control objective is usually formulated as an optimal control problem by converting the task into a problem of optimizing a functional of the quantum states, control inputs, control time, etc. In the learning process, one first applies a trial input to a sample to be controlled and observes the result. Second, a learning algorithm suggests a better control input based on the prior experiments. Third, one applies the "better" control input to a new sample. This process continues in order to achieve the control objective. It is often easy to produce many identical-state samples in laboratory chemical reactions. If the control objective is well defined, there is a capability to apply specified control inputs to the samples, there are sufficiently rich control inputs and a sufficiently intelligent learning algorithm is applied to adjust the control inputs, this process will converge to optimize the required objective and an optimal control can be found [116].

Learning control of quantum systems is one of the two main focuses of this monograph. In particular, we present sampling-based learning control in Chaps. 3 and 4 for achieving robust performance in various tasks including control and classification of inhomogeneous quantum ensembles, control of superconducting qubits and photoassociation, synchronization of laser and molecules for charge transfer, and generation of quantum gates. In these results, gradient-based algorithms are integrated in the SLC scheme. In Chap. 5, differential evolution is applied to the control of open quantum ensembles and the synchronization of quantum networks. Besides these open-loop learning control results, we also present experimental results on closed-loop learning control for fragmentation control of $Pr(hfac)_3$ and CH_2BrI molecules using femtosecond laser pulses, and for quantum autoencoders using quantum optical systems. Moreover, quantum control using reinforcement learning is also discussed.

2.6 Quantum Robust Control

In practical applications, it is unavoidable that quantum systems are subject to all kinds of disturbances and uncertainties [38, 112, 164, 188]. Many instances of incomplete knowledge and unknown errors can also be treated as uncertainties. For example, in NMR an ensemble consisting of around 10^{23} spins can be utilized to perform quantum information processing. The chemical shift of their spectrometers may not be known exactly [68]. In trapped ions, the bichromatic laser beams may slightly interfere with each other [102]. The temperature may influence the polarization control achieved by using liquid crystal variable retarders in the semiconductor quantum dots [74]. The operations of multiple superconducting qubits may be confronted with possible fluctuations in the coupling energy of a Josephson junction [10, 33]. Hence, it is both theoretically and practically important to develop systematic approaches for attaining robust control in quantum systems.

In the last twenty years, a number of results on quantum robust control have been presented. For example, the small gain theorem has been applied to the stability analysis of quantum feedback networks [27]. A noise filtering method has been presented to enhance robustness in quantum control [132]. A transfer function approach has been applied to feedback and robust control of single-input single-output quantum systems [174, 175]. An H^∞ controller synthesis problem has been formulated for a class of linear quantum stochastic systems [66]. An idea of sampling uncertainty parameters has been presented to design robust control pulses for electron shuttling [181]. Stimulated Raman adiabatic passage has been extensively studied due to its independence of the pulse shape, which makes it robust against fluctuations in the experimental parameters [12, 53, 134]. Dynamical decoupling has been widely used to provide robust schemes in the face of environmental noise in quantum control [17, 142]. A robust optimal control landscape for the generation of quantum unitary transformations has been investigated by analyzing the second-order Hessian and the topology of the critical regions [59, 61].

Quantum robust control is the other focus of this monograph. In particular, a sampling-based learning control method is introduced for robust control of inhomogeneous quantum ensembles, quantum state transfer and quantum gate generation in Chaps. 3 and 4. Robust control tasks such as control of open inhomogeneous ensembles, robust synchronization of quantum networks and robust photofragmentation using femtosecond laser pulses are solved using machine learning in Chap. 5. A sliding mode control approach is systematically presented in Chap. 6 to deal with uncertainties in two-level and multi-level quantum systems as well as uncertainties in the system–environment coupling for open quantum systems. Robust stability, performance analysis, H^∞ control [66, 92] and fault-tolerant control for several classes of stochastic quantum systems are discussed in Chaps. 7 and 8.

2.7 Summary and Further Reading

This chapter provides a brief introduction to quantum mechanics foundations and quantum control. The knowledge will facilitate the understanding of the relevant background of the topics and the following chapters in this monograph. In order to obtain a comprehensive understanding to quantum mechanics, readers may refer to many excellent textbooks (e.g., [28, 80]). Moreover, further reading may include [7, 126, 146] for a comprehensive introduction to quantum optics, [101] for the topics of quantum information and quantum computation, [25, 35, 36, 104, 163] for introductions to various topics in quantum control, [109] for a brief introduction to quantum robust control, [32] for a brief introduction to learning control of quantum systems and [23, 50] for comprehensive introductions to the (S, L, H) framework.

References

1. Albertini F, D'Alessandro D (2003) Notions of controllability for bilinear multilevel quantum systems. IEEE Trans Autom Control 48(8):1399–1403
2. Alicki R, Lendi K (2007) Quantum dynamical semigroups and applications, 2nd edn. Springer, Berlin
3. Altafini C (2002) Controllability of quantum mechanical systems by root space decomposition of su(N). J Math Phys 43:2051–2062
4. Altafini C (2003) Controllability properties for finite dimensional quantum Markovian master equations. J Math Phys 44:2357–2372
5. Altafini C (2007) Feedback control of spin systems. Quant Inf Process 6:9–36
6. Altafini C (2007) Feedback stabilization of isospectral control systems on complex flag manifolds: application to quantum ensembles. IEEE Trans Autom Control 52(11):2019–2028
7. Bachor HA, Ralph TC (2019) A guide to experiments in quantum optics, 3rd edn. Wiley-VCH
8. Belavkin VP (1992) Quantum stochastic calculus and quantum nonlinear filtering. J Multivar Anal 42:171–201
9. Belhadj M, Salomon J, Turinici G (2015) Ensemble controllability and discrimination of perturbed bilinear control systems on connected, simple, compact Lie groups. Eur J Control 22:23–29
10. Bialczak RC, Ansmann M, Hofheinz M, Lenander M, Lucero E, Neeley M, O'Connell AD, Sank D, Wang H, Weides M, Wenner J, Yamamoto T, Cleland AN, Martinis JM (2011) Fast tunable coupler for superconducting qubits. Phys Rev Lett 106(6):060501
11. Boscain U, Charlot G, Gauthier JP, Guérin S, Jauslin HR (2002) Optimal control in laser-induced population transfer for two- and three-level quantum systems. J Math Phys 43(5):2107–2132
12. Boscain U, Chittaro F, Mason P, Sigalotti M (2012) Adiabatic control of the Schrödinger equation via conical intersections of the eigenvalues. IEEE Trans Autom Control 57(8):1970–1983
13. Boscain U, Mason P (2006) Time minimal trajectories for a spin 1/2 particle in a magnetic field. J Math Phys 47:062101
14. Bouten L, van Handel R, James MR (2007) An introduction to quantum filtering. SIAM J Control Optim 46(6):2199–2241
15. Breuer H-P, Petruccione F (2002) The theory of open quantum systems, 1st edn. Oxford University Press
16. Burgarth D, Bose S, Bruder C, Giovannetti V (2009) Local controllability of quantum networks. Phys Rev A 79(6):060305
17. Bylander J, Gustavsson S, Yan F, Yoshihara F, Harrabi K, Fitch G, Cory DG, Nakamura Y, Tsai J-S, Oliver WD (2011) Noise spectroscopy through dynamical decoupling with a superconducting flux qubit. Nat Phys 7:565–570
18. Chakrabarti R, Rabitz H (2007) Quantum control landscapes. Int Rev Phys Chem 26(4):671–735
19. Chakrabarti R, Wu RB, Rabitz H (2008) Quantum Pareto optimal control. Phys Rev A 78:033414
20. Chen BS, Chen WH, Hsu F, Zhang W (2008) Optimal tracking control design of quantum systems via tensor formal power series method. Open Autom Control Syst J 1:50–64
21. Chu S (2002) Cold atoms and quantum control. Nature 416:206–210
22. Clarke J, Wilhelm FK (2008) Superconducting quantum bits. Nature 453:1031–1042
23. Combes J, Kerckhoff J, Sarovar M (2017) The H^∞ framework for modeling quantum input-output networks. Adv Phys X 2:784–888
24. Cui L, Dong Z, Zhang G, Lee HWJ (2017) Mixed LQG and H^∞ coherent feedback control for linear quantum systems. Int J Control 90:2575–2588
25. D'Alessandro D (2007) Introduction to quantum control and dynamics. Chapman & Hall/CRC
26. D'Alessandro D, Dahleh M (2001) Optimal control of two-level quantum systems. IEEE Trans Autom Control 46:866–876

27. D'Helon C, James MR (2006) Stability, gain, and robustness in quantum feedback networks. Phys Rev A 73:053803
28. Dirac PAM (1958) The principles of quantum mechanics, 4th edn. Oxford University Press
29. Doherty AC, Habib S, Jacobs K, Mabuchi H, Tan SM (2000) Quantum feedback control and classical control theory. Phys Rev A 62:012105
30. Doherty AC, Jacobs K (1999) Feedback control of quantum systems using continuous state estimation. Phys Rev A 60:2700–2711
31. Doherty AC, Jacobs K, Jungman G (2001) Information, disturbance, and Hamiltonian quantum feedback control. Phys Rev A 63:062306
32. Dong D (2021) Learning control of quantum systems. In: Baillieul J, Samad T (eds) Encyclopedia of systems and control. Springer-Verlag, London Ltd., London, pp 1090–1096
33. Dong D, Chen C, Qi B, Petersen IR, Nori F (2015) Robust manipulation of superconducting qubits in the presence of fluctuations. Sci Rep 5:7873 (Springer Nature)
34. Dong D, Petersen IR (2009) Variable structure control of uncontrollable quantum systems. IFAC Proc Volumes 42(6):237–242
35. Dong D, Petersen IR (2010) Quantum control theory and applications: a survey. IET Control Theor Appl 4:2651–2671
36. Dong D, Petersen IR (2022) Quantum estimation, control and learning: opportunities and challenges. Ann Rev Control 54:243–251
37. Dong D, Shu C-C, Chen J, Xing X, Ma H, Guo Y, Rabitz H (2021) Learning control of quantum systems using frequency-domain optimization algorithms. IEEE Trans Control Syst Technol 29(4):1791–1798
38. Dong D, Wang Y (2017) Several recent developments in estimation and robust control of quantum systems. In: Proceedings of the 2017 Australian and New Zealand control conference, Gold Coast, Australia, December 17–20, pp 190–195
39. Dong D, Xing X, Ma H, Chen C, Liu Z, Rabitz H (2020) Learning-based quantum robust control: algorithm, applications, and experiments. IEEE Trans Cybern 50:3581–3593
40. Ferrante A, Pavon M, Raccanelli G (2002) Driving the propagator of a spin system: a feedback approach. In: Proceedings of the 41st IEEE conference on decision and control, Las Vegas, Nevada USA, pp 46–50
41. Gao Q, Dong D, Petersen IR, Ding SX (2020) Design of a quantum projection filter. IEEE Trans Autom Control 65:3693–3700
42. Gao Q, Dong D, Petersen IR, Ding SX (2021) Design of a discrete-time fault-tolerant quantum filter and fault detector. IEEE Trans Cybern 51:889–899
43. Gao Q, Dong D, Petersen IR, Rabitz H (2016) Fault tolerant filtering and fault detection for quantum systems driven by fields in single photon states. J Math Phys 57(6):062201
44. Gao Q, Zhang G, Petersen IR (2019) An exponential quantum projection filter for open quantum systems. Automatica 99:59–68
45. Gao Q, Zhang G, Petersen IR (2020) An improved quantum projection filter. Automatica 112:108716
46. Geng J, Wu Y, Wang X, Xu K, Shi F, Xie Y, Rong X, Du J (2016) Experimental time-optimal universal control of spin qubits in solids. Phys Rev Lett 117:170501
47. Geremia JM, Rabitz H (2002) Optimal identification of Hamiltonian information by closed-loop laser control of quantum systems. Phys Rev Lett 89:263902
48. Gerry CC, Knight PL (2005) Introductory Quantum Optics. Cambridge University Press, Cambridge, UK
49. Gillett GG, Dalton RB, Lanyon BP, Almeida MP, Barbieri M, Pryde GJ, O'Brien JL, Resch KJ, Bartlett SD, White AG (2010) Experimental feedback control of quantum systems using weak measurements. Phys Rev Lett 104:080503
50. Gough JE, James MR (2009) The series product and its application to quantum feedforward and feedback networks. IEEE Trans Autom Control 54:2530–2544
51. Grivopoulos S, Bamieh B (2003, December) Lyapunov-based control of quantum systems. In: Proceedings of the 42nd IEEE conference on decision and control, Maui, Hawaii USA, pp 434-438

52. Grivopoulos S, Bamieh B (2008) Optimal population transfers in a quantum system for large transfer time. IEEE Trans Autom Control 53:980–992
53. Guérin S, Hakobyan V, Jauslin HR (2011) Optimal adiabatic passage by shaped pulses: efficiency and robustness. Phys Rev A 84:013417
54. Guo Y, Dong D, Shu C-C (2018) Optimal and robust control of quantum state transfer by shaping the spectral phase of ultrafast laser pulses. Phys Chem Chem Phys 20:9498–9506
55. Guo Y, Shu C-C, Dong D, Nori F (2019) Vanishing and revival of resonance Raman scattering. Phys Rev Lett 123(22):223202
56. Hamerly R, Mabuchi H (2012) Advantages of coherent feedback for cooling quantum oscillators. Phys Rev Lett 109:173602
57. Helstrom CW (1976) Quantum detection and estimation theory. Academic Press, USA, New York
58. Hirose M, Cappellaro P (2016) Coherent feedback control of a single qubit in diamond. Nature 532:77–80
59. Hocker D, Brif C, Grace MD, Donovan A, Ho T, Tibbetts KM, Wu RB, Rabitz H (2014) Characterization of control noise effects in optimal quantum unitary dynamics. Phys Rev A 90(6):062309
60. Hou SC, Khan MA, Yi XX, Dong D, Petersen IR (2012) Optimal Lyapunov-based quantum control for quantum systems. Phys Rev A 86(2):022321
61. Hsieh M, Rabitz H (2008) Optimal control landscape for the generation of unitaty tranformations. Phys Rev A 77:042306
62. Huang GM, Tarn TJ, Clark JW (1983) On the controllability of quantum-mechanical systems. J Math Phys 24:2608–2618
63. Jacobs K, Steck DA (2006) A straightforward introduction to continuous quantum measurement. Contemp Phys 47:279–303
64. Jacobs K, Wang X, Wiseman HM (2014) Coherent feedback that beats all measurement-based feedback protocols. New J Phys 16:073036
65. James MR, Gough JE (2010) Quantum dissipative systems and feedback control design by interconnection. IEEE Trans Autom Control 55:1806–1821
66. James MR, Nurdin HI, Petersen IR (2008) H^{∞} control of linear quantum stochastic systems. IEEE Trans Autom Control 53:1787–1803
67. Ji Y, Bian J, Jiang M, D'Alessandro D, Peng X (2018) Time-optimal control of independent spin-1/2 systems under simultaneous control. Phys Rev A 98:062108
68. Jones JA (2011) Quantum computing with NMR. Prog Nucl Magn Reson Spectro 59:91–120
69. Judson RS, Rabitz H (1992) Teaching lasers to control molecules. Phys Rev Lett 68:1500–1503
70. Khaneja N, Brockett R, Glaser SJ (2001) Time optimal control in spin systems. Phys Rev A 63:032308
71. Khaneja N, Kehlet C, Luy B, Glaser SJ (2003) Broadband relaxation optimized polarization transfer in magnetic resonance. Proc Natl Acad Sci U S A 101:14742–14747
72. Khaneja N, Reiss T, Kehlet C, Schulte-Herbrüggen T, Glaser SJ (2005) Optimal control of coupled spin dynamics: design of NMR pulse sequences by gradient ascent algorithms. J Magn Reson 172(2):296–305
73. Khaneja N, Reiss T, Luy B, Glaser SJ (2003) Optimal control of spin dynamics in the presence of relaxation. J Mag Reson 162:311–319
74. Kodriano Y, Schmidgall ER, Benny Y, Gershoni D (2014) Optical control of single excitons in semiconductor quantum dots. Semiconductor Science and Technology 29:053001
75. Kraus K (1983) States, effects, and operations, 1st edn. Springer, Berlin
76. Kuang S, Cong S (2008) Lyapunov control methods of closed quantum systems. Automatica 44:98–108
77. Kuang S, Dong D, Petersen IR (2017) Rapid Lyapunov control of finite-dimensional quantum systems. Automatica 81:164–175
78. Kuang S, Guan X, Dong D (2021) Finite-time stabilization control of quantum systems. Automatica 123:109327

79. Lan CH, Tarn TJ, Chi QS, Clark JW (2005) Analytic controllability of time-dependent quantum control systems. J Math Phys 46:052102
80. Landau LD, Lifshitz EM (1999) Quantum mechanics: non-relativistic theory, 3rd edn. Butterworth-Heinemann
81. LaSalle J, Lefschetz S (1961) Stability by Liapunov's direct method with applications. Academic Press, New York
82. Li J-S, Khaneja N (2009) Ensemble control of Bloch equations. IEEE Trans Autom Control 54(3):528–536
83. Li J-S, Zhang W, Tie L (2020) On separating points for ensemble controllability. SIAM J Control Optim 58:2740–2764
84. Lindblad G (1976) On the generators of quantum dynamical semigroups. Commun Math Phys 48(2):119–130
85. Liu Y, Dong D, Kuang S, Petersen IR, Yonezawa H (2021) Two-step feedback preparation of entanglement for qubit systems with time delay. Automatica 125:109174
86. Liu Y, Dong D, Petersen IR, Yonezawa H (2019) Filter-based feedback control for a class of Markovian open quantum systems. IEEE Control Syst Lett 3:565–570
87. Liu Y, Kuang S, Cong S (2017) Lyapunov-based feedback preparation of GHZ entanglement of N-qubit systems. IEEE Trans Cybern 47(11):3827–3839
88. Liu J, Yuan H (2017) Quantum parameter estimation with optimal control. Phys Rev A 96:012117
89. Lloyd S (2000) Coherent quantum feedback. Phys Rev A 62:022108
90. Lloyd S, Viola L (2001) Engineering quantum dynamics. Phys Rev A 65:010101(R)
91. Ma H, Chen C (2020) Several developments in learning control of quantum systems. In: IEEE International conference on systems, man and cybernetics, Toronto, Canada, October 11–14, pp 4165-4172
92. Maalouf AI, Petersen IR (2011) Coherent H^∞ control for a class of annihilation operator linear quantum systems. IEEE Trans Autom Control 56:309–319
93. Mabuchi H (2008) Coherent-feedback quantum control with a dynamic compensator. Phys Rev A 78:032323
94. Mabuchi H, Khaneja N (2005) Principles and applications of control in quantum systems. Int J Robust Nonlinear Control 15:647–667
95. Makhlin Y, Schön G, Shnirman A (2001) Quantum-state engineering with Josephson-junction devices. Rev Mod Phys 73:357–400
96. Mancini S, Wiseman HM (2007) Optimal control of entanglement via quantum feedback. Phys Rev A 75:012330
97. Miao Z, James MR, Petersen IR (2016) Coherent observers for linear quantum stochastic systems. Automatica 71:264–271
98. Mirrahimi M, Rouchon P (2004) Controllability of quantum harmonic oscillators. IEEE Trans Autom Control 49:745–747
99. Mirrahimi M, Rouchon P, Turinici G (2005) Lyapunov control of bilinear Schrödinger equations. Automatica 41:1987–1994
100. Mirrahimi M, van Handel R (2007) Stabilizing feedback controls for quantum systems. SIAM J Control Optim 46:445–467
101. Nielsen MA, Chuang IL (2010) Quantum computation and quantum information. Cambridge University Press, England, Cambridge
102. Noh C, Roddriguez-lara BM, Angelakis DG (2012) Quantum simulation of neutrino oscillations with trapped ions. New J Phys 14:033028
103. Nurdin HI, James MR, Petersen IR (2009) Coherent quantum LQG control. Automatica 45:1837–1846
104. Nurdin HI, Yamamoto N (2017) Linear dynamical quantum systems. Springer International Publishing
105. Pan Y, Dong D, Petersen IR (2017) Dark modes of quantum linear systems. IEEE Trans Autom Control 62:4180–4186

106. Paris M, Řeháček J (2004) Quantum state estimation (lecture notes in physics), vol 649. Springer, Berlin, Germany
107. Parthasarathy K (1992) An introduction to quantum stochastic calculus. Birkhauser, Berlin
108. Peirce AP, Dahleh M, Rabitz H (1988) Optimal control of quantum-mechanical systems: existence, numerical approximation, and applications. Phys Rev A 37:4950–4964
109. Petersen IR (2021) Robustness issues in quantum control. In: Baillieul J, Samad T (eds) Encyclopedia of systems and control. Springer-Verlag, London Ltd, London, pp 1986–1991
110. Poggiali F, Cappellaro P, Fabbri N (2018) Optimal control for one-qubit quantum sensing. Phys Rev X 8:021059
111. Qi B (2009) On the quantum master equation under feedback control. Sci China Ser F Inf Sci 52:2133–2139
112. Qi B (2013) A two-step strategy for stabilizing control of quantum systems with uncertainties. Automatica 49:834–839
113. Qi B, Hou Z, Li L, Dong D, Xiang G-Y, Guo G-C (2013) Quantum state tomography via linear regression estimation. Sci Rep 3:3496
114. Qi B, Hou Z, Wang Y, Dong D, Zhong H-S, Li L, Xiang G-Y, Wiseman HM, Li C-F, Guo G-C(2017) Adaptive quantum state tomography via linear regression estimation: theory and two-qubit experiment. npj Quant Inf 3:19
115. Qi B, Pan H, Guo L (2013) Further results on stabilizing control of quantum systems. IEEE Trans Autom Control 58:1349–1354
116. Rabitz H, De Vivie-Riedle R, Motzkus M, Kompa K (2000) Whither the future of controlling quantum phenomena? Science 288(5467):824–828
117. Rabitz H, Hsieh M, Rosenthal C (2004) Quantum optimally controlled transition landscapes. Science 303:1998–2001
118. Ramakrishna V, Rabitz H (1996) Relation between quantum computing and quantum controllability. Phys Rev A 54:1715–1716
119. Ramakrishna V, Salapaka MV, Dahleh M, Rabitz H, Peirce A (1995) Controllability of molecular systems. Phys Rev A 51:960–966
120. Rice SA, Zhao MS (2000) Optical control of molecular dynamics, 1st edn. Wiley
121. Rouchon P, Ralph JF (2015) Efficient quantum filtering for quantum feedback control. Phys Rev A 91:012118
122. Ruskov R, Korotkov AN (2002) Quantum feedback control of a solid-state qubit. Phys Rev B 66:041401(R)
123. Sayrin C, Dotsenko I, Zhou X, Peaudecerf B, Rybarczyk T, Gleyzes S, Rouchon P, Mirrahimi M, Amini H, Brune M, Raimond JM, Haroche S (2011) Real-time quantum feedback prepares and stabilizes photon number states. Nature 477:73–77
124. Schirmer SG, Fu H, Solomon AI (2001) Complete controllability of quantum systems. Phys Rev A 63:063410
125. Schirmer SG, Solomon AI, Leahy JV (2002) Degrees of controllability for quantum systems and application to atomic systems. J Phys A: Math General 35:4125–4141
126. Scully MO, Zubairy MS (1997) Quantum optics. Cambridge University Press
127. Shapiro M, Brumer P Principles of the quantum control of molecular processes, 1st edn. Wiley
128. Shu CC, Dong D, Petersen IR, Henriksen NE (2017) Complete elimination of nonlinear light-matter interactions with broadband ultrafast laser pulses. Phys Rev A 95:033809
129. Shu CC, Henriksen NE (2012) Phase-only shaped laser pulses in optimal control theory: application to indirect photofragmentation dynamics in the weak-field limit. J Chem Phys 136:044303
130. Shu CC, Ho TS, Xing X, Rabitz H (2016) Frequency domain quantum optimal control under multiple constraints. Phys Rev A 93:033417
131. Shu CC, Ho TS, Rabitz H (2016) Monotonic convergent quantum optimal control method with exact equality constraints on the optimized control fields. Phys Rev A 93:053418
132. Soare A, Ball H, Hayes D, Sastrawan J, Jarratt MC, McLoughlin JJ, Zhen Z, Green TG, Biercuk MJ (2014) Experimental noise filtering by quantum control. Nat Phys 10:825–829

133. Sugny D, Kontz C, Jauslin HR (2007) Time-optimal control of a two-level dissipative quantum system. Phys Rev A 76:023419
134. Torosov BT, Guérin S, Vitanov V (2011) High-fidelity adiabatic passage by composite sequences of chirped pulses. Phys Rev Lett 106(23):233001
135. Turinici G, Rabitz H (2001) Quantum wavefunction controllability. Chem Phys 267:1–9
136. Turinici G, Rabitz H (2003) Wavefunction controllability for finite-dimensional bilinear quantum systems. J Phys A: Math General 36:2565–2576
137. van Handel R, Stockton JK, Mabuchi H (2005) Modelling and feedback control design for quantum state preparation. J Opt B: Quant Semiclassical Opt 7:S179–S197
138. van Handel R, Stockton JK, Mabuchi H (2005) Feedback control of quantum state reduction. IEEE Trans Autom Control 50:768–780
139. Vandersypen LMK, Chuang IL (2004) NMR techniques for quantum control and computation. Rev Mod Phys 76:1037
140. Vettori P (2002) On the convergence of a feedback control strategy for multilevel quantum systems. In: Proceedings of the fifteenth international symposium on mathematical theory of networks and systems, South Bend, Indiana, USA
141. Vijay R, Macklin C, Slichter DH, Weber SJ, Murch KW, Naik R, Korotkov AN, Siddiqi I (2012) Stabilizing Rabi oscillations in a superconducting qubit using quantum feedback. Nature 490:77–80
142. Viola L, Knill E, Lloyd S (1999) Dynamical decoupling of open quantum systems. Phys Rev Lett 82:2417–2421
143. Vladimirov IG, Petersen IR (2013) A quasi-separation principle and Newton-like scheme for coherent quantum LQG control. Syst Control Lett 62:550–559
144. Vladimirov IG, Petersen IR, James MR (2021) Multi-point Gaussian states, quadratic exponential cost functionals, and large deviations estimates for linear quantum stochastic systems. Appl Math Optim 83:83–137
145. Vuglar SL, Petersen IR (2016) Quantum noise, physical realizability and coherent quantum feedback control. IEEE Trans Autom Control 62:998–1003
146. Walls DF, Milburn GJ (2008) Quantum optics, 2nd edn. Springer, Berlin
147. Wan KH, Dahlsten O, Kristjánsson H, Gardner R, Kim M (2017) Quantum generalisation of feedforward neural networks. npj Quant Inf 3:36
148. Wang X, Allegra M, Jacobs K, Lloyd S, Lupo C, Mohseni M (2015) Quantum brachistochrone curves as geodesics: obtaining accurate minimum-time protocols for the control of quantum systems. Phys Rev Lett 114:170501
149. Wang Y, Dong D, Qi B, Zhang J, Petersen IR, Yonezawa H (2018) A quantum Hamiltonian identification algorithm: computational complexity and error analysis. IEEE Trans Autom Control 63:1388–1403
150. Wang Y, Dong D, Sone A, Petersen IR, Yonezawa H, Cappellaro P (2020) Quantum Hamiltonian identifiability via a similarity transformation approach and beyond. IEEE Trans Autom Control 65(11):4632–4647
151. Wang S, Nurdin HI, Zhang G, James MR (2013) Quantum optical realization of classical linear stochastic systems. Automatica 49(10):3090–3096
152. Wang S, Nurdin HI, Zhang G, James MR (2018) Representation and network synthesis for a class of mixed quantum-classical linear stochastic systems. Automatica 96:84–97
153. Wang X, Pemberton-Ross P, Schirmer SG (2012) Symmetry and subspace controllability for spin networks with a single-node control. IEEE Trans Autom Control 57(8):1945–1956
154. Wang X, Schirmer SG (2010) Analysis of Lyapunov method for control of quantum states. IEEE Trans Autom Control 55:2259–2270
155. Wang X, Schirmer SG (2010) Analysis of effectiveness of Lyapunov control for non-generic quantum states. IEEE Trans Autom Control 55:1406–1411
156. Wang Y, Yin Q, Dong D, Qi B, Petersen IR, Hou Z, Yonezawa H, Xiang G-Y (2019) Quantum gate identification: error analysis, numerical results and optical experiment. Automatica 101:269–279

157. Wang Y, Yokoyama S, Dong D, Petersen IR, Huntington EH, Yonezawa H (2021) Two-stage estimation for quantum detector tomography: error analysis, numerical and experimental results. IEEE Trans Inf Theor 67(4):2293–2307
158. Warren WS, Rabitz H, Dahleh M (1993) Coherent control of quantum dynamics: the dream is alive. Science 259:1581–1589
159. Werschnik J, Gross EKU (2007) Quantum optimal control theory. J Phys B: At Mol Opt Phys 40:R175–R211
160. Wiseman HM (1994) Quantum theory of continuous feedback. Phys Rev A 49:2133–2150
161. Wiseman HM, Mancini S, Wang J (2002) Bayesian feedback versus Markovian feedback in a two-level atom. Phys Rev A 66:013807
162. Wiseman HM, Milburn GJ (1993) Quantum theory of optical feedback via homodyne detection. Phys Rev Lett 70:548–551
163. Wiseman HM, Milburn GJ (2010) Quantum measurement and control. Cambridge University Press, England, Cambridge
164. Wu RB, Ding HJ (2019) Robust quantum control against clock noises in multiqubit systems. Phys Rev A 100:022302
165. Wu R, Pechen A, Brif C, Rabitz H (2007) Controllability of open quantum systems with Kraus-map dynamics. J Phys A: Math Theor 40:5681–5693
166. Wu RB, Tarn TJ, Li CW (2006) Smooth controllability of infinite-dimensional quantum-mechanical systems. Phys Rev A 73:012719
167. Xiao S, Wang Y, Dong D, Zhang J (2022) Optimal and two-step adaptive quantum detector tomography. Automatica 141:110296
168. Xiang Z-L, Ashhab S, You JQ, Nori F (2013) Hybrid quantum circuits: superconducting circuits interacting with other quantum systems. Rev Mod Phys 85:623–653
169. Xiang C, Petersen IR, Dong D (2022) Guaranteed cost coherent control for quantum systems with non-quadratic perturbations in the system Hamiltonian. Automatica 139:110201
170. Xue S, Wu RB, Ma S, Li D, Jiang M (2021) Gradient algorithm for Hamiltonian identification of open quantum systems. Phys Rev A 103:022604
171. Yamamoto N (2006) Robust observer for uncertain linear quantum systems. Phys Rev A 74:032107
172. Yamamoto N, Nurdin HI, James MR, Petersen IR (2008) Avoiding entanglement sudden death via measurement feedback control in a quantum network. Phys Rev A 78:042339
173. Yanagisawa M (2006) Quantum feedback control for deterministic entangled photon generation. Phys Rev Lett 97:190201
174. Yanagisawa M, Kimura H (2003) Transfer function approach to quantum control-part I: dynamics of quantum feedback systems. IEEE Trans Autom Control 48:2107–2120
175. Yanagisawa M, Kimura H (2003) Transfer function approach to quantum control-part II: control concepts and applications. IEEE Trans Autom Control 48:2121–2132
176. You JQ, Nori F (2005) Superconducting circuits and quantum information. Phys Today 58:42–47
177. Yuan H (2016) Sequential feedback scheme outperforms the parallel scheme for Hamiltonian parameter estimation. Phys Rev Lett 117:160801
178. Yuan H, Fung C-H (2015) Optimal feedback scheme and universal time scaling for Hamiltonian parameter estimation. Phys Rev Lett 115:110401
179. Zhang G (2014) Analysis of quantum linear systems response to multi-photon states. Automatica 50(2):442–451
180. Zhang C-B, Dong D, Chen Z-H (2005) Control of non-controllable quantum systems: a quantum control algorithm based on Grover iteration. J Opt B: Quant Semiclassical Opt 7:S313–S317
181. Zhang J, Greenman L, Deng X, Whaley KB (2014) Robust control pulses design for electron shuttling in solid-state devices. IEEE Trans Control Syst Technol 22(6):2354–2359
182. Zhang G, Grivopoulos S, Petersen IR, Gough JE (2018) The Kalman decomposition for linear quantum systems. IEEE Trans Autom Control 63(2):331–346

183. Zhang G, James MR (2012) Quantum feedback networks and control: a brief survey. Chin Sci Bull 57(18):2200–2214
184. Zhang G, Joseph Lee HW, Huang B, Zhang H (2012) Coherent feedback control of linear quantum optical systems via squeezing and phase shift. SIAM J Control Optim 50(4):2130–2150
185. Zhang J, Li CW, Wu RB, Tarn TJ, Liu XS (2005) Maximal suppression of decoherence in Markovian quantum systems. J Phys A: Math Gen 38:6587–6601
186. Zhang J, Liu Y-X, Wu RB, Jacobs K, Nori F (2017) Quantum feedback: theory, experiments, and applications. Phys Rep 679:1–60
187. Zhang G, Pan Y (2020) On the dynamics of two photons interacting with a two-qubit coherent feedback network. Automatica 117:108978
188. Zhang H, Rabitz H (1994) Robust optimal control of quantum molecular systems in the presence of disturbances and uncertainties. Phys Rev A 49:2241–2254
189. Zhang J, Sarovar M (2014) Quantum Hamiltonian identification from measurement time traces. Phys Rev Lett 113(8):080401
190. Zhang J, Wu RB, Liu YX, Li CW, Tarn TJ (2012) Quantum coherent nonlinear feedback with applications to quantum optics on chip. IEEE Trans Autom Control 57:1997–2008
191. Zhu WS, Rabitz H (1998) A rapid monotonically convergent iteration algorithm for quantum optimal control over the expectation value of a positive definite operator. J Chem Phys 109:385–391

Chapter 3
Control and Classification of Inhomogeneous Quantum Ensembles

Abstract This chapter discusses control and classification of inhomogeneous quantum ensembles. A sampling-based learning control method is systematically developed for ensemble control and classification. Inhomogeneous quantum ensembles are introduced in Sect. 3.1, and a sampling-based learning control method is presented in Sect. 3.2. The results on control of inhomogeneous quantum ensembles are presented in Sect. 3.3[a]. Section 3.4 through Sect. 3.7 present the results on quantum discrimination and ensemble classification using the sampling-based learning control method[b]. Section 3.8 includes the conclusion and further reading.

3.1 Inhomogeneous Quantum Ensembles

A quantum ensemble consists of a large number of (e.g., up to $\sim 10^{23}$) single quantum systems (e.g., spin systems), and every quantum system is referred to as a member of the ensemble. In this chapter, we consider an ensemble in the sense of the individual systems slightly varying over a distribution of characteristics, rather than in the context of a mixed state. Quantum ensembles have wide applications in emerging quantum technology including quantum computation [14], long-distance quantum communication [18], quantum memory [8] and magnetic resonance imaging [29]. Several results on quantum ensemble control have been presented including unitary control in homogeneous quantum ensembles for maximizing signal intensity in coherent spectroscopy [19] and feedback stabilization of quantum ensembles [1].

In practical applications, the members of a quantum ensemble could have variations in the parameters that characterize the system dynamics [23, 24]. For example,

[a] Sections 3.2 and 3.3 contain materials reprinted, with permission, from Physical Review A [11]. Copyright (2014) by the American Physical Society.
[b] Section 3.4 through Sect. 3.7 contain materials reprinted, with permission, from IEEE Transactions on Neural Networks and Learning Systems [12] © 2017 IEEE.

D. Dong and I. R. Petersen, *Learning and Robust Control in Quantum Technology*, Communications and Control Engineering,
https://doi.org/10.1007/978-3-031-20245-2_3

the spins of an ensemble in NMR experiments may encounter a large dispersion in the strength of the applied radio frequency field (rf imhomogeneity) as well as members exhibiting variations in their natural frequencies (Larmor dispersion) [22, 47]. In this book, these situations are referred to as *inhomogeneous* quantum ensembles. It is generally impractical to employ different control inputs for individual members of a quantum ensemble in the laboratory. Hence, it is important to develop the means for designing control fields that can simultaneously steer the ensemble of systems from an initial state to a desired target state when variations exist in the system parameters [16]. Such controls are also called compensating pulse sequences in NMR spectroscopy [23, 52]. Other applications include control of a randomly oriented ensemble of molecules in physical chemistry [49], the design of slice selective excitation and inversion pulses in magnetic resonance imaging, and the correction of systematic errors in quantum information processing [24]. Theoretical results show that under some commonly arising conditions, there exist optimal laser fields to control all molecules in an inhomogeneous ensemble, regardless of their orientation or spatial location [41, 50]. Li and his coworkers [24–27, 29, 31, 45, 46, 53, 54] presented a series of theoretical results on controllability of inhomogeneous quantum ensembles and developed systematic algorithms to achieve optimal control of inhomogeneous ensembles. For example, the controllability of an inhomogeneous quantum ensemble was understood by the investigation of the algebra of polynomials defined on the vector fields characterizing the system dynamics where the role of Lie brackets and non-commutativity in designing control fields for a quantum ensemble was highlighted [25]. A sufficient and necessary controllability condition was derived for a finite-dimensional time-varying linear ensemble system in [26]. The notion of separating points and polynomial approximation was exploited to present sufficient and necessary controllability conditions for a class of time-invariant linear ensemble systems evolving on an infinite-dimensional space of continuous functions [31]. A multi-dimensional pseudospectral method has been employed for the optimal control of inhomogeneous quantum ensembles [45]. Iterative methods have been presented to construct optimal controls for inhomogeneous ensemble systems with fixed-endpoint and free-endpoint conditions where the convergence of the iterative algorithms and the optimality of the solutions have also been discussed [53]. Many other researchers have also investigated the controllability and control design of inhomogeneous quantum ensembles [5, 39, 48, 51]. For example, the adiabatic approximation theory has been used to analyze the controllability of quantum ensembles, and sufficient conditions were established for approximate ensemble controllability [4]. A Lyapunov control methodology has been developed to examine the stabilization of an inhomogeneous ensemble of non-interacting spin systems [6]. Control of an inhomogeneous spin ensemble coupled to a cavity has been investigated using optimal control theory [2]. The simultaneous control of an ensemble of springs with different frequencies has been investigated by employing an adiabatic shortcut to adiabaticity and optimal process [35]. In this chapter, we present a systematic method of sampling-based learning control (SLC) [11] for designing effective control fields that can simultaneously steer the members of an inhomogeneous quantum ensemble to the same target state, and for classification of inhomogeneous quantum ensembles.

3.2 Sampling-Based Learning Control

Sampling-based learning control (SLC) was first presented for the control of inhomogeneous quantum ensembles in [11], which includes two steps of training and testing. Here, we first use an example of closed quantum ensembles to demonstrate the idea. Consider a finite-dimensional closed quantum system with the evolution of its state $|\psi(t)\rangle$ described by the Schrödinger equation:

$$\frac{\mathrm{d}}{\mathrm{d}t}|\psi(t)\rangle = -\mathrm{i}H(t)|\psi(t)\rangle. \tag{3.1}$$

The solution of (3.1) is given by $|\psi(t)\rangle = U(t)|\psi_0\rangle$, where $|\psi(0)\rangle = |\psi_0\rangle$ and the propagator $U(t)$ satisfies

$$\frac{\mathrm{d}}{\mathrm{d}t}U(t) = -\mathrm{i}H(t)U(t) \tag{3.2}$$

with $U(0) = I$.

We consider an inhomogeneous ensemble in which the Hamiltonian of each member has the following form:

$$H_{\omega,\theta}(t) = f(\omega)H_0 + \sum_{m=1}^{M} g(\theta_m)u_m(t)H_m. \tag{3.3}$$

The functions $f(\omega)$ and $g(\theta)$ characterize the inhomogeneous distribution in the free Hamiltonian and control Hamiltonian, respectively. For simplicity, we assume that $f(\omega) = \omega$ and $g(\theta_m) = \theta$, and the parameters ω and θ are time independent and uniformly distributed over $[1 - \bar{\Omega}, 1 + \bar{\Omega}]$ and $[1 - \Theta, 1 + \Theta]$, respectively. The constants $\bar{\Omega} \in [0, 1]$ and $\Theta \in [0, 1]$ represent the bounds on the parameter dispersion. The objective is to design the controls $\{u_m(t), m = 1, 2, \ldots, M\}$ to simultaneously drive the members (with different ω and θ) of the quantum ensemble from an initial state $|\psi_0\rangle$ to the same target state $|\psi_{\text{target}}\rangle$ with high fidelity. The control outcome is described by a *performance function* $J(u)$ for each control strategy $u = \{u_m(t), m = 1, 2, \ldots, M\}$. The control problem can then be formulated as a maximization problem as follows:

$$\begin{aligned}
&\max_u J(u) := \max_u \mathbb{A}_v[J_{\omega,\theta}(u)] \\
&\text{s.t. } \frac{\mathrm{d}}{\mathrm{d}t}|\psi_{\omega,\theta}(t)\rangle = -\mathrm{i}H_{\omega,\theta}(t)|\psi(t)\rangle, \ t \in [0, T], \\
&|\psi_{\omega,\theta}(0)\rangle = |\psi_0\rangle \\
&H_{\omega,\theta}(t) = \omega H_0 + \theta \sum_{m=1}^{M} u_m(t)H_m, \\
&\omega \in [1 - \bar{\Omega}, 1 + \bar{\Omega}], \ \ \theta \in [1 - \Theta, 1 + \Theta],
\end{aligned} \tag{3.4}$$

where $J_{\omega,\theta}(u)$ is the fidelity measure of each member of the ensemble and $\mathbb{A}_v[J_{\omega,\theta}(u)]$ denotes the average value of $J_{\omega,\theta}$ over the ensemble. The fidelity between the final state $|\psi_{\omega,\theta}(T)\rangle$ and the target state $|\psi_{\text{target}}\rangle$ is defined as follows [38]:

$$F(|\psi_{\omega,\theta}(T)\rangle, |\psi_{\text{target}}\rangle) = |\langle\psi_{\omega,\theta}(T)|\psi_{\text{target}}\rangle|. \tag{3.5}$$

The fidelity F is used to evaluate the performance of a designed control in the testing step. However, for the convenient calculation of a gradient flow in the training step, we consider the performance function $J(u) = F^2$, i.e.,

$$J_{\omega,\theta}(u) := |\langle\psi_{\omega,\theta}(T)|\psi_{\text{target}}\rangle|^2.$$

Note that $J_{\omega,\theta}$ depends implicitly on the control u through the Schrödinger equation.

In the SLC method, first some samples are drawn from the distribution of inhomogeneous parameters to design the control. Then the resultant control is applied to additional ensemble members to test the control performance.

In the training step, we select N sampled members from the quantum ensemble according to the distribution (e.g., uniform distribution) of the inhomogeneous parameters and then construct an augmented system as follows:

$$\frac{\mathrm{d}}{\mathrm{d}t}\begin{bmatrix} |\psi_{\omega_1,\theta_1}(t)\rangle \\ |\psi_{\omega_2,\theta_2}(t)\rangle \\ \vdots \\ |\psi_{\omega_N,\theta_N}(t)\rangle \end{bmatrix} = -\mathrm{i}\begin{bmatrix} H_{\omega_1,\theta_1}(t)|\psi_{\omega_1,\theta_1}(t)\rangle \\ H_{\omega_2,\theta_2}(t)|\psi_{\omega_2,\theta_2}(t)\rangle \\ \vdots \\ H_{\omega_N,\theta_N}(t)|\psi_{\omega_N,\theta_N}(t)\rangle \end{bmatrix}, \tag{3.6}$$

where

$$H_{\omega_n,\theta_n} = \omega_n H_0 + \theta_n \sum_m u_m(t) H_m$$

with $n = 1, 2, \ldots, N$. The performance function for the augmented system is defined by

$$J_N(u) := \frac{1}{N}\sum_{n=1}^{N} J_{\omega_n,\theta_n}(u) = \frac{1}{N}\sum_{n=1}^{N} |\langle\psi_{\omega_n,\theta_n}(T)|\psi_{\text{target}}\rangle|^2. \tag{3.7}$$

The goal of the training step is to find a control u^* that maximizes the performance function defined in (3.7). The performance function is denoted as $J_N(u^0)$ with an initial control $u^0 = \{u_m^0(t)\}$.

The motivation behind SLC is to design the control using a minimal number of sampled members. Therefore, it is necessary to choose a representative set of samples. For example, when the distributions of both ω and θ are uniform, we may choose equally spaced samples in the $\omega - \theta$ space. In this case, the intervals of $[1 - \bar{\Omega}, 1 + \bar{\Omega}]$ and $[1 - \Theta, 1 + \Theta]$ are divided into $N_{\bar{\Omega}} + 1$ and $N_\Theta + 1$ subintervals, respectively, where $N_{\bar{\Omega}}$ and N_Θ are conveniently chosen positive odd integers. Then the total number of samples is $N = N_{\bar{\Omega}} N_\Theta$, where ω_n and θ_n are chosen from all combinations of $(\omega_{n_{\bar{\Omega}}}, \theta_{n_{\Theta}})$ as follows:

$$\begin{cases} \omega_n \in \{\omega_{n_{\bar{\Omega}}} = 1 - \bar{\Omega} + \frac{(2n_{\bar{\Omega}}-1)\bar{\Omega}}{N_{\bar{\Omega}}}, \ n_{\bar{\Omega}} = 1, 2, \ldots, N_{\bar{\Omega}}\}, \\ \theta_n \in \{\theta_{n_{\Theta}} = 1 - \Theta + \frac{(2n_{\Theta}-1)\Theta}{N_{\Theta}}, \ n_{\Theta} = 1, 2, \ldots, N_{\Theta}\}. \end{cases} \tag{3.8}$$

For testing, we apply the optimal control u^* obtained in the training step to additional samples randomly selected from the inhomogeneous quantum ensemble and evaluate the control performance of each sample in terms of the fidelity $F(|\psi(T)\rangle, |\psi_{\text{target}}\rangle)$ between the final state achieved by each sample $|\psi(T)\rangle$ and the target state $|\psi_{\text{target}}\rangle$. If both the average value and the minimum value of the fidelity $F(|\psi(T)\rangle, |\psi_{\text{target}}\rangle)$ for all the tested samples are satisfactory, we accept the designed control law and end the control design process. Otherwise, we return to the training step and generate another optimized control strategy (e.g., restarting the training step with a new initial control strategy or a new set of samples guided by the performance of the tested members).

3.3 Sampling-Based Learning Control of Inhomogeneous Quantum Ensembles

3.3.1 Gradient-Based Optimal Algorithm

The SLC method provides a systematic methodology for control design of inhomogeneous quantum ensembles for the state-to-state transition probability. In the training step, we sample several members according to the distribution of inhomogeneous parameters from the ensemble and construct an augmented system using these collective samples. Then, we may employ different algorithms to find the control providing good performance for the augmented system. In particular, here we use a gradient flow-based learning and optimization algorithm to achieve this task [34]. In the process of testing the deduced controls, we randomly select a number of sampling members to evaluate the control performance.

To obtain an optimal control strategy

$$u^* = \{u_m^*(t), (t \in [0, T]), m = 1, 2, \ldots, M\}$$

for the augmented system (3.6), one technique is to follow the gradient of $J_N(u)$ in the steepest ascent direction. For ease of notation, we present the method for $M = 1$. We introduce a time-like variable s to characterize different control strategies $u^{(s)}(t)$. Then a gradient flow in the control space is defined as

$$\frac{du^{(s)}}{ds} = \nabla J_N(u^{(s)}), \tag{3.9}$$

where $\nabla J_N(u)$ denotes the gradient of J_N with respect to the control u. If $u^{(s)}$ is the solution to (3.9) starting from an arbitrary initial condition $u^{(0)}$, then the value of J_N will increase along $u^{(s)}$, i.e., $\frac{\mathrm{d}}{\mathrm{d}s}J_N(u^{(s)}) \geq 0$. Starting from a trial guess u^0, we solve the following initial value problem:

$$\begin{cases} \dfrac{\mathrm{d}u^{(s)}}{\mathrm{d}s} = \nabla J_N(u^{(s)}), \\ u^{(0)} = u^0 \end{cases} \tag{3.10}$$

in order to find a control strategy which maximizes J_N. This initial value problem can then be solved numerically by using a forward Euler method over the s-domain, i.e.,

$$u(s + \Delta s, t) = u(s, t) + \Delta s \nabla J_N(u^{(s)}). \tag{3.11}$$

Algorithm 3.1 Gradient flow-based iterative learning

1: Set the iteration index $k = 0$
2: Choose a set of arbitrary controls $u^{k=0}(t) = \{u_m^0(t),\ m = 1, 2, \ldots, M\}, t \in [0, T]$
3: **repeat** steps 4-6 (corresponding to one iteration)
4: Compute the propagator $U_{\omega_n,\theta_n}(t)$ with $u^k(t)$ for all the training sample members $(n = 1, 2, \ldots, N)$
5: Update each control

$$u_m^{k+1}(t) = u_m^k(t) + \eta_k \delta_m^k(t)$$

with $m = 1, 2, \ldots, M$ and

$$\delta_m^k(t) = \frac{2}{N}\sum_{n=1}^{N} \Im\left(\langle\psi_{\omega_n,\theta_n}(T)|\psi_{\text{target}}\rangle\langle\psi_{\text{target}}|U_{\omega_n,\theta_n}(T)U^{\dagger}_{\omega_n,\theta_n}(t)\theta_n H_m U_{\omega_n,\theta_n}(t)|\psi_0\rangle\right)$$

6: $k = k + 1$
7: **until** the learning process ends (i.e., the algorithm converges)
8: The optimal control strategy $u^* = \{u_m^*\} = \{u_m^k\},\ m = 1, 2, \ldots, M$

In practical applications, we use the following discrete iteration rule to update the control field (for details relating to gradient algorithms, see, e.g., [9, 21, 34, 43, 47]):

$$u^{k+1}(t) = u^k(t) + \eta_k \nabla J_N(u^k), \tag{3.12}$$

where η_k is the updating step size (learning rate) for the kth iteration. By (3.7), we also obtain that

$$\nabla J_N(u) = \frac{1}{N}\sum_{n=1}^{N} \nabla J_{\omega_n,\theta_n}(u). \tag{3.13}$$

In addition, we have

$$\nabla J_{\omega_n,\theta_n}(u) = 2\Im\left(\langle\psi_{\omega_n,\theta_n}(T)|\psi_{\text{target}}\rangle\langle\psi_{\text{target}}|U_{\omega_n,\theta_n}(T)U^{\dagger}_{\omega_n,\theta_n}(t)\theta_n H_1 U_{\omega_n,\theta_n}(t)|\psi_0\rangle\right), \tag{3.14}$$

where $\Im(\cdot)$ denotes the imaginary part of a complex number, and the propagator $U_{\omega_n,\theta_n}(t)$ satisfies

$$\frac{\mathrm{d}}{\mathrm{d}t}U_{\omega_n,\theta_n}(t) = -\mathrm{i}H_{\omega_n,\theta_n}(t)U_{\omega_n,\theta_n}(t), \quad U(0) = I.$$

The gradient flow method can be generalized to the case with $M > 1$ as shown in Algorithm 3.1.

3.3.2 Numerical Examples of Inhomogeneous Ensemble Control

Example 3.1 Consider a quantum ensemble consisting of two-level quantum systems (e.g., spins). We let the free Hamiltonian be $H_0 = \frac{1}{2}\sigma_z$. The control Hamiltonian is

$$H_u = \frac{1}{2}u_1(t)\sigma_x + \frac{1}{2}u_2(t)\sigma_y.$$

For the inhomogeneous ensemble, the Hamiltonian of each member is described as

$$H_{\omega,\theta}(t) = \omega H_0 + \theta H_u(t). \tag{3.15}$$

The state of the quantum system can be represented as

$$|\psi(t)\rangle = c_0(t)|0\rangle + c_1(t)|1\rangle.$$

Denote $C(t) = \begin{bmatrix} c_0(t) \\ c_1(t) \end{bmatrix}$, where $c_0(t)$ and $c_1(t)$ are complex amplitudes. We have

$$\mathrm{i}\dot{C}(t) = (H_0 + H_u(t))C(t). \tag{3.16}$$

To construct an augmented system for the training step, we select N members ($n = 1, 2, \ldots, N$) from the ensemble and each satisfies

$$\begin{bmatrix} \dot{c}_{0,n}(t) \\ \dot{c}_{1,n}(t) \end{bmatrix} = \begin{bmatrix} 0.5\omega_n\mathrm{i} & \theta_n u_2(t) - 0.5\mathrm{i}u_1(t) \\ -\theta_n u_2(t) + 0.5\mathrm{i}u_1(t) & -0.5\omega_n\mathrm{i} \end{bmatrix}\begin{bmatrix} c_{0,n}(t) \\ c_{1,n}(t) \end{bmatrix}, \tag{3.17}$$

where $\omega_n \in [1 - \bar{\Omega}, 1 + \bar{\Omega}]$ and $\theta_n \in [1 - \Theta, 1 + \Theta]$ have uniform distributions. The objective is to find a control $u(t) = \{u_m(t), m = 1, 2\}$ to drive all the inhomogeneous members from an initial state $|\psi_0\rangle = |0\rangle$, i.e., $C_0 = [1, 0]^T$, to the target state $|\psi_{\text{target}}\rangle = |1\rangle$, i.e., $C_{\text{target}} = [0, 1]^T$. We construct an augmented system for the training samples using (3.6) with the performance function $J_N(u)$ in (3.7).

The task is to find the control $u(t)$ to maximize the performance function $J_N(u)$. For a given small threshold $\bar{\delta} > 0$, if $J_N(u^*) > 1 - \bar{\delta}$, then we have found a suitable optimal control candidate for the augmented system. We employ Algorithm 3.1 to find the optimal control $u^*(t) = \{u_m^*(t), m = 1, 2\}$ for this augmented system. This optimal control is then applied to other randomly selected members to test its performance.

Several groups of numerical experiments are carried out on an inhomogeneous spin ensemble to demonstrate the performance of SLC. The parameter settings are as follows: $\bar{\Omega} = 0.2$ and $\Theta = 0.2$; the target time is $T = 2$, and the total time interval $[0, T]$ is divided equally into $Q = 200$ time steps, $\Delta t = \frac{T}{Q} = 0.01$; the learning rate is set as $\eta_k = 0.2$; the control strategy is initialized as $u^{k=0}(t) = \{u_1^0(t) = \sin t, u_2^0(t) = \sin t\}$.

To construct an augmented system for the inhomogeneous ensemble with parameter dispersion on both ω and θ, we choose $N_{\bar{\Omega}} = 5$ and $N_\Theta = 5$ such that $N = N_{\bar{\Omega}} N_\Theta = 25$ samples are employed in the learning phase. Using (3.8), we have

$$\begin{cases} \omega_n = 1 - 0.2 + \dfrac{0.2(2\text{fix}(n/5) - 1)}{5}, \\ \theta_n = 1 - 0.2 + \dfrac{0.2(2(\text{mod}(n, 5) - 1)}{5}, \end{cases} \tag{3.18}$$

where $n = 1, 2, \ldots, 25$, $\text{fix}(x) = \max\{z \in \mathbf{Z} | z \leq x\}$, $\text{mod}(n, 5) = n - 5z$ $(z \in \mathbf{Z}$ and $\frac{n}{5} - 1 < z \leq \frac{n}{5})$ and $\mathbf{Z}$ is the set of integers. We set $\varepsilon = 5 \times 10^{-5}$. The algorithm converges after around 380 iterations. The learned optimal control strategy is given as in Fig. 3.1, and the testing performance in Fig. 3.2 shows that the fidelities for the state transition lie in the interval of $[0.9985, 1]$ with a mean value of 0.9997. For comparison, if we use only one sample $(\omega = 1, \theta = 1)$ for training to obtain a control law, the testing performance gives fidelities that lie in $[0.9436, 1]$ with a mean value of 0.9808. The numerical results show that SLC is effective for control design of the two-level inhomogeneous ensemble. The fidelities of the controlled state for the randomly selected members approach very near to 1 even with $\pm 20\%$ parameter dispersion over a uniform distribution.

Using the optimal control strategy in Fig. 3.1, we randomly select several thousand members and present the state transition trajectories of the two-level ensemble on the Bloch sphere. As shown in Fig. 3.3, although the trajectories of these randomly selected members considerably differ from each other due to the inhomogeneity of the ensemble, they are all successfully driven from the initial state $|\psi_0\rangle = |0\rangle$ to the same target state $|\psi_{\text{target}}\rangle = |1\rangle$ with the high average fidelity of 0.9997.

Example 3.2 We consider a Λ-type atomic ensemble and demonstrate the SLC design process. For a Λ-type atomic system [20, 56], we assume that the initial state is

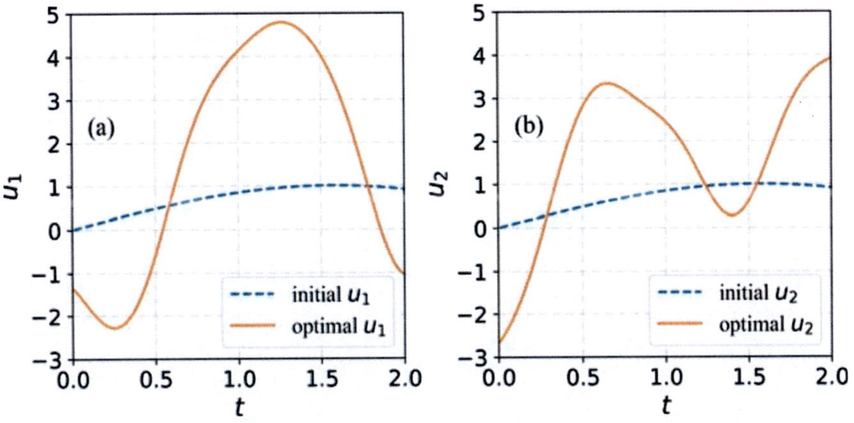

Fig. 3.1 Chen et al. [11] Learned optimal control strategy that maximizes $J_N(u)$ for the two-level ensemble

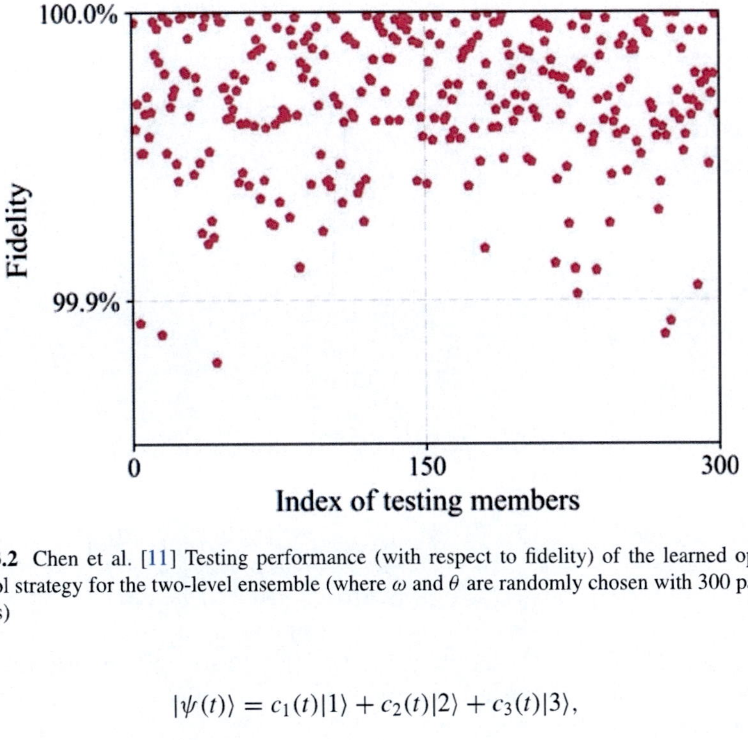

Fig. 3.2 Chen et al. [11] Testing performance (with respect to fidelity) of the learned optimal control strategy for the two-level ensemble (where ω and θ are randomly chosen with 300 pairs of values)

$$|\psi(t)\rangle = c_1(t)|1\rangle + c_2(t)|2\rangle + c_3(t)|3\rangle,$$

and denote $C(t) = [c_1(t)\ c_2(t)\ c_3(t)]^T$. We have

$$\mathrm{i}\dot{C}(t) = (H_0 + H_u(t))C(t). \tag{3.19}$$

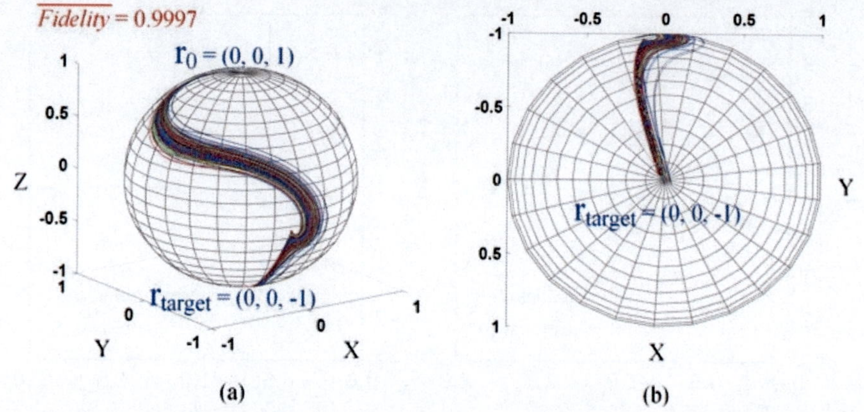

Fig. 3.3 Chen et al. [11] State transition trajectories of the two-level ensemble with inhomogeneities due to both ω and θ using the learned optimal control strategy as shown in Fig. 3.1 with initial state $|\psi_0\rangle = |0\rangle$ (corresponding to $r_0 = (0, 0, 1)$) and the target state $|\psi_{\text{target}}\rangle = |1\rangle$) (corresponding to $r_{\text{target}} = (0, 0, -1)$)

We take $H_0 = \text{diag}(1.5, 1, 0)$ and choose H_1 and H_2 in the control Hamiltonian of (3.3) as follows [20]:

$$H_1 = \begin{bmatrix} 0 & 0 & 0 \\ 0 & 0 & 1 \\ 0 & 1 & 0 \end{bmatrix}, \tag{3.20}$$

$$H_2 = \begin{bmatrix} 0 & 0 & 1 \\ 0 & 0 & 0 \\ 1 & 0 & 0 \end{bmatrix}. \tag{3.21}$$

To construct an augmented system for the SLC training step, we choose N samples from the ensemble to form:

$$\begin{bmatrix} \dot{c}_{1,n}(t) \\ \dot{c}_{2,n}(t) \\ \dot{c}_{3,n}(t) \end{bmatrix} = \begin{bmatrix} -1.5\omega_n \mathrm{i} & 0 & -\mathrm{i}\theta_n u_2(t) \\ 0 & -\omega_n \mathrm{i} & -\mathrm{i}\theta_n u_1(t) \\ -\mathrm{i}\theta_n u_2(t) & -\mathrm{i}\theta_n u_1(t) & 0 \end{bmatrix} \begin{bmatrix} c_{1,n}(t) \\ c_{2,n}(t) \\ c_{3,n}(t) \end{bmatrix}, \tag{3.22}$$

where $\omega_n \in [1 - \bar{\Omega}, 1 + \bar{\Omega}]$ and $\theta_n \in [1 - \Theta, 1 + \Theta]$ have uniform distributions. The objective is to find a control strategy $u(t) = \{u_m(t), m = 1, 2\}$ to drive all the inhomogeneous members from

$$|\psi_0\rangle = \frac{1}{\sqrt{3}}(|1\rangle + |2\rangle + |3\rangle)$$

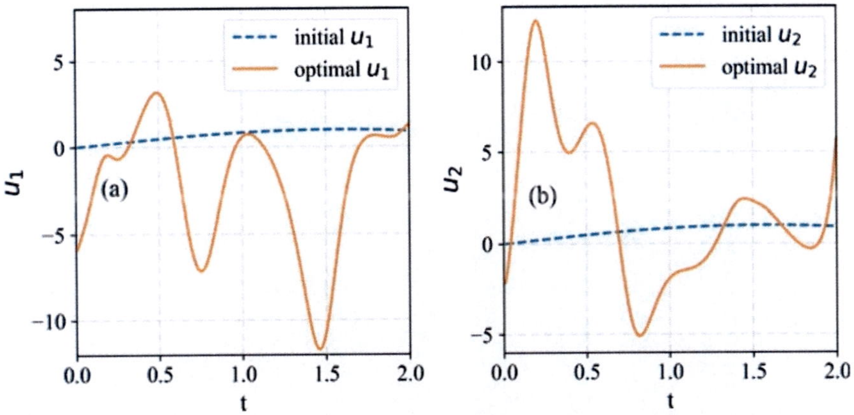

Fig. 3.4 Learned optimal control strategy to maximize $J_N(u)$ for the Λ-type atomic ensemble

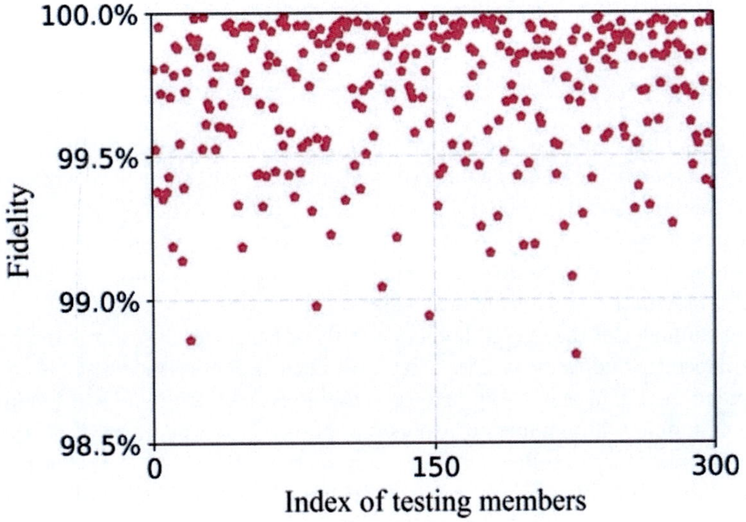

Fig. 3.5 Testing performance (with respect to fidelity) of the learned optimal control strategy for the Λ-type atomic ensemble (where ω and θ are randomly chosen with 300 pairs of values)

to $|\psi_{\text{target}}\rangle = |3\rangle$. We aim to maximize the performance function $J_N(u)$ in (3.7) and employ Algorithm 3.1 to find the optimal control $u^*(t) = \{u_m^*(t), m = 1, 2\}$ for this augmented system. Then the optimal control strategy is applied to other randomly selected members to test its performance.

We choose the parameter settings as follows: the control strategy is initialized with $u^{k=0}(t) = \{u_m^0(t) = \sin t, m = 1, 2\}$; $\bar{\delta} = 10^{-4}$; the other parameter settings are the same as those in Example 3.1. To construct an augmented system for the training step, we have the training samples selected according to (3.18).

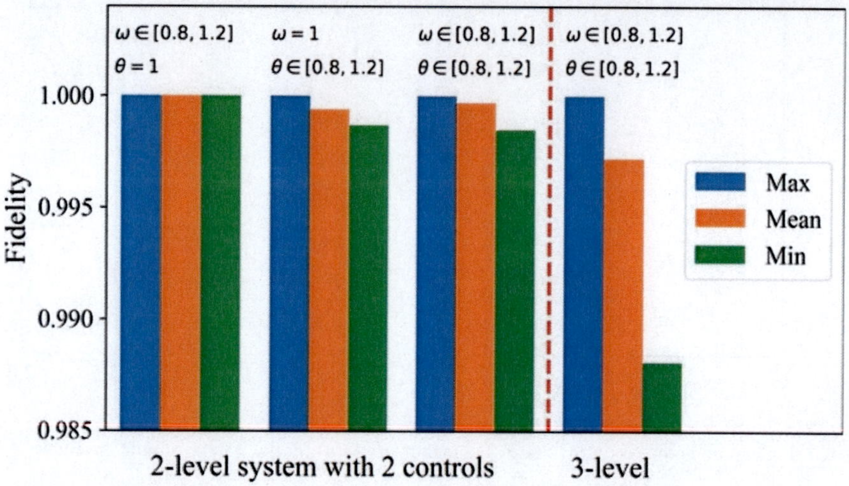

Fig. 3.6 Control performance with respect to fidelity for the two-level and three-level inhomogeneous ensembles

The algorithm converges after around 2000 iterations. The learned optimal control strategy is given in Fig. 3.4, and the testing results are given in Fig. 3.5, which shows that the fidelities for all the 300 testing members lie in the interval of $[0.9881, 1]$ with the mean value of 0.9972. For comparison, if we use only one sample ($\omega = 1, \theta = 1$) for training to obtain a control, the testing performance gives fidelities that lie in $[0.8279, 1]$ with the mean value of 0.9449.

As a summary of the overall numerical tests of SLC, Fig. 3.6 shows the control performance (including some cases that have been explicitly shown above) for the aforementioned spin and Λ-type atomic ensembles. For the two-level inhomogeneous ensemble with parameter dispersion only in ω, the fidelities of all the 300 testing members are excellent and lie in the interval of $[1 - 10^{-6}, 1]$. For the case with parameter dispersion only in θ, the fidelities lie in the interval of $[0.9987, 1]$ with the mean value 0.9994. The collective numerical results show that the SLC method has potential for practical control design of various inhomogeneous quantum ensembles. These findings support the previous theoretical analysis suggesting that control of inhomogeneous ensembles should generally be feasible [25, 49, 50].

3.4 Quantum Discrimination and Ensemble Classification

The classification of inhomogeneous quantum ensembles is a significant issue and has great potential applications in the discrimination of atoms (or molecules), the separation of isotopes and quantum information extraction. However, quantum mechanics forbids deterministic discrimination among non-orthogonal states [17, 37]. A useful idea is to first drive the members of a quantum ensemble from an initial state to different orthogonal states corresponding to different classes (e.g., eigenstates) before classifying them. It often becomes easy to discriminate (or separate) different orthogonal states.

Here, we recast the quantum ensemble classification task as a supervised quantum learning problem and present a systematic classification methodology by using an SLC method [11, 12, 15]. In this proposed method, we first learn an optimal control strategy to steer the members in a quantum ensemble belonging to different classes into their corresponding target states, and then employ a physical read-out process (e.g., projective measurement, fluorescence images of molecules [10], Stern–Gerlach experiments for spin systems [3, 38]) to classify these classes. For example, the states of single molecules could be read out using a visualization technique, where the highly photostable chromophore dinaphtoquaterrylenebis (dicarboximide) (DNQDI) is embedded in thin polymer films in concentrations sufficiently low to allow individual DNQDI molecules to be spatially resolved in an epifluorescence confocal microscope (for details, see [10]). It is feasible to read out the intensity of the single-molecule fluorescence after they are excited with laser pulses. For a spin ensemble, when some members are driven to spin up and the others are driven to spin down (orthogonal to spin up), it is feasible to physically separate the two classes of members using Stern–Gerlach experiments [38]. We first develop an approach for the discrimination of two similar quantum systems and the binary classification of quantum ensembles. Then we apply the proposed approach to multi-class classification of multi-level quantum ensembles.

We consider the classification problem for a quantum ensemble of similar members with different Hamiltonians, which is referred to as *quantum ensemble classification* (QEC). QEC can be taken as a generalization of quantum dynamic discrimination and has similar characteristics to selective excitation problems in NMR [13]. Suppose that for an inhomogeneous quantum ensemble, given an unknown member belonging to a certain class, we are interested in how well we can predict the class that the unknown member belongs to. In classical machine learning, this problem can be solved using typical supervised learning algorithms with a training set. However, this problem is much more difficult for quantum systems because we cannot achieve deterministic discrimination for given quantum systems unless they lie in mutually orthogonal states. We have to drive the members from different classes to appropriate orthogonal states (e.g., eigenstates) before we can discriminate them with high accuracy. The SLC approach can be combined with supervised learning for QEC. We define the sampling set in the training step for the QEC problem as follows:

Definition 3.1 A sampling set for the training step consists of N quantum systems (each of them labeled with an associated class) that are chosen from the quantum ensemble, and the set is denoted as

$$D_N = \{(H^1(t), y_1), (H^2(t), y_2), \dots, (H^N(t), y_N)\}, \tag{3.23}$$

where $H^n(t)$ $(n = 1, 2, \ldots, N)$ describes the nth quantum system in the sampling set and y_n is the associated class that this quantum system belongs to.

For ease of presentation, we first consider an inhomogeneous ensemble consisting of two classes of members (i.e., classes A and B) and propose an SLC approach for this binary quantum ensemble classification problem using a two-level quantum ensemble example. We further extend the proposed approach to the classification problem with multi-classes and multi-level quantum ensembles. For the binary quantum ensemble classification problem, the Hamiltonian of each member has the following form:

$$\begin{cases} H^A_{\varepsilon_0,\varepsilon_u}(t) = f^A(\varepsilon_0)H_0 + g^A_u(\varepsilon_u)\sum_{m=1}^{M} u_m(t)H_m \\ H^B_{\varepsilon_0,\varepsilon_u}(t) = f^B(\varepsilon_0)H_0 + g^B_u(\varepsilon_u)\sum_{m=1}^{M} u_m(t)H_m. \end{cases} \tag{3.24}$$

$f^A(\cdot)$ and $g^A_u(\cdot)$ are known functions, while the inhomogeneity parameters ε_0 and ε_u in the Hamiltonian $H^A_{\varepsilon_0,\varepsilon_u}(t)$ for class A are characterized by the distribution functions $d^A_0(\varepsilon_0)$ and $d^A_u(\varepsilon_u)$, respectively. We assume that the parameters ε_0 and ε_u are time independent. A similar explanation is applicable to the Hamiltonian $H^B_{\varepsilon_0,\varepsilon_u}(t)$ for class B. We assume that these parameter variations in ε_0 and ε_u originate from different inhomogeneities (i.e., ε_0 and ε_u are independent from each other). It is straightforward to extend the proposed method to the case where ε_u is dependent on ε_0 if an approximate dependency relationship is known.

For a binary quantum ensemble classification task, the objective is to design a control strategy $u(t) = \{u_m(t), m = 1, 2, \ldots, M\}$ to simultaneously stabilize the members in class A (with different ε_0 and ε_u) from an initial state $|\psi_0\rangle$ to the same target state $|\psi_{\text{targetA}}\rangle$, and at the same time to stabilize the members in class B (with different ε_0 and ε_u) from $|\psi_0\rangle$ to another target state $|\psi_{\text{targetB}}\rangle$. In many practical situations, it is natural to assume that the inhomogeneity parameters in different classes have a specific distribution (e.g., Gaussian distribution [47]). In particular, we consider the following binary quantum ensemble classification problem: A binary quantum ensemble classification (binary QEC) task is to construct a binary quantum classifier to maximize the classification accuracy, where this binary quantum classifier consists of three steps:

1. *Training step:* Learn an optimal control strategy $u(t)$ with the sampling set

$$D_N = \{(H^1(t), y_1), (H^2(t), y_2), \ldots, (H^N(t), y_N)\},$$

where $y_n \in \{A, B\}$ (A and B are symbolic constants) and $H^n(t)$ $(n = 1, 2, \ldots, N)$ is the time-dependent Hamiltonian describing the nth member in the sampling set.

2. *Coherent control step:* Apply the learned optimal control strategy $u(t)$ to all the members of the quantum ensemble.

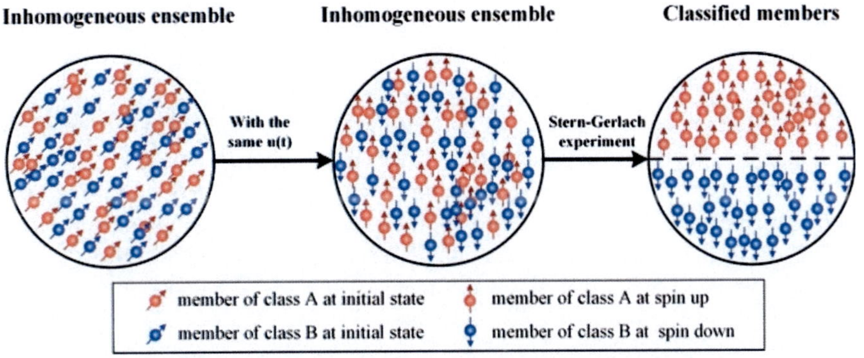

Fig. 3.7 Illustration of the binary classification for an inhomogeneous spin quantum ensemble [12]

3. *Classification step:* Predict the class y_j of an unknown quantum system in the quantum ensemble using a corresponding physical read-out process, where $j = 1, 2, \ldots, N_e$ and N_e is the number of members in the quantum ensemble.

For example, a schematic of the classification process for a spin-$\frac{1}{2}$ quantum ensemble is demonstrated in Fig. 3.7. As shown in Fig. 3.7, an ensemble of inhomogeneous spin-$\frac{1}{2}$ systems is prepared with an initial state. After learning using a training set from the quantum ensemble, we can find an optimal control strategy to simultaneously drive all the members of class A to the target state (spin up) and all the members of class B to another target state (spin down). Then we can use a Stern–Gerlach experiment to physically separate the two classes.

It is clear that the key task of the classification problem is to learn an optimal control strategy in the training step for the binary quantum classifier. The training performance is described by a *performance function* $J(u)$ for each learned control strategy $u = \{u_m(t), m = 1, 2, \ldots, M\}$. The binary QEC problem can then be formulated as a maximization problem as follows:

$$
\begin{aligned}
&\max_u J(u) := \max_u \{w_A \mathbb{A}_v[J^A_{\varepsilon_0,\varepsilon_u}(u)] + w_B \mathbb{A}_v[J^B_{\varepsilon_0,\varepsilon_u}(u)]\} \\
\text{s.t.}\ \ & t \in [0, T] \\
& |\psi^A_{\varepsilon_0,\varepsilon_u}(0)\rangle = |\psi^B_{\varepsilon_0,\varepsilon_u}(0)\rangle = |\psi_0\rangle \\
& \begin{cases} \dfrac{\mathrm{d}}{\mathrm{d}t}|\psi^A_{\varepsilon_0,\varepsilon_u}(t)\rangle = -\mathrm{i}[f^A(\varepsilon_0)H_0 + g^A_u(\varepsilon_u)\displaystyle\sum_{m=1}^{M} u_m(t)H_m(t)]|\psi^A_{\varepsilon_0,\varepsilon_u}(t)\rangle \\ J^A_{\varepsilon_0,\varepsilon_u}(u) := |\langle\psi^A_{\varepsilon_0,\varepsilon_u}(T)|\psi_{\mathrm{targetA}}\rangle|^2 \end{cases} \\
& \begin{cases} \dfrac{\mathrm{d}}{\mathrm{d}t}|\psi^B_{\varepsilon_0,\varepsilon_u}(t)\rangle = -\mathrm{i}[f^B(\varepsilon_0)H_0 + g^B_u(\varepsilon_u)\displaystyle\sum_{m=1}^{M} u_m(t)H_m]|\psi^B_{\varepsilon_0,\varepsilon_u}(t)\rangle \\ J^B_{\varepsilon_0,\varepsilon_u}(u) := |\langle\psi^B_{\varepsilon_0,\varepsilon_u}(T)|\psi_{\mathrm{targetB}}\rangle|^2 \end{cases}
\end{aligned}
, \quad (3.25)
$$

where $w_A, w_B \in [0, 1]$ are the weights assigned to classes A and B, respectively, satisfying $w_A + w_B = 1$. $J^A_{\varepsilon_0,\varepsilon_u}(u)$ is a measure of classification accuracy for each member in class A regarding the target state $|\psi_{\text{targetA}}\rangle$ and $\mathbb{A}_v[J^A_{\varepsilon_0,\varepsilon_u}(u)]$ denotes the average value of $J^A_{\varepsilon_0,\varepsilon_u}(u)$ over class A regarding all the members in this class. A similar expression holds for class B. It is clear that $J^A_{\varepsilon_0,\varepsilon_u}$ and $J^B_{\varepsilon_0,\varepsilon_u}$ depend implicitly on the control strategy $u(t)$ through the Schrödinger equation. The performance $J(u)$ represents the weighted accuracy of classification with an upper bound of 1. The larger $J(u)$ is, the better the classification performance is.

3.5 Discrimination of Two Similar Quantum Systems

Optimal dynamic discrimination between two similar quantum systems has been investigated using different techniques such as multi-objective optimization [7] and closed-loop learning control [30]. The quantum discrimination problem can be taken as a special case of the binary QEC problem with the number of members in an ensemble $N_e = 2$. In this sense, control design for quantum discrimination is the foundation of QEC. We develop a gradient-based learning control method for quantum discrimination of two similar quantum systems and then extend the method to control design for binary QEC.

3.5.1 *Learning Control for Quantum Discrimination*

Suppose two similar quantum systems to be discriminated (separated) a and b have the following Hamiltonians:

$$\begin{cases} H^a(t) = f(\varepsilon_0^a)H_0 + g_u(\varepsilon_u^a)\sum_{m=1}^{M} u_m(t)H_m \\ H^b(t) = f(\varepsilon_0^b)H_0 + g_u(\varepsilon_u^b)\sum_{m=1}^{M} u_m(t)H_m, \end{cases} \tag{3.26}$$

where ε_0^a, ε_u^a, ε_0^b and ε_u^b are predefined constants for functions $f(\cdot)$ and $g_u(\cdot)$. a and b are prepared in the same initial state $|\psi_0\rangle$. The objective is to find an optimal control strategy $u(t)$ ($t \in [0, T]$) to drive the state of system a to the target state $|\psi^a_{\text{target}}\rangle$ and the state of system b to the target state $|\psi^b_{\text{target}}\rangle$ at the same time. Usually, we let $\langle\psi^a_{\text{target}}|\psi^b_{\text{target}}\rangle = 0$ so that we can completely discriminate system a from system b. The control performance $J(u)$ is redefined for the discrimination problem as

$$J(u) := w_a J^a(u) + w_b J^b(u), \tag{3.27}$$

where $w_a, w_b \in [0, 1]$ are the weights assigned to the associated systems, respectively, and

$$\begin{aligned} J^a(u) &:= |\langle \psi^a(T)|\psi^a_{\text{target}}\rangle|^2, \\ J^b(u) &:= |\langle \psi^b(T)|\psi^b_{\text{target}}\rangle|^2. \end{aligned} \tag{3.28}$$

Here we set $w_a = w_b = 0.5$ for the discrimination problem. Using the gradient-based method, for each control $u_m(t)$ $(m = 1, 2, \ldots, M)$ of the control strategy $u(t)$, we have

$$\begin{aligned} \nabla J(u^k_m) =& 2w_a \Im\left(\langle \psi^a(T)|\psi^a_{\text{target}}\rangle\langle \psi^a_{\text{target}}|U_a(T)U^\dagger_a(t)g_u(\varepsilon^a_u)H_m U_a(t)|\psi_0\rangle\right) \\ &+ 2w_b \Im\left(\langle \psi^b(T)|\psi^b_{\text{target}}\rangle\langle \psi^b_{\text{target}}|U_b(T)U^\dagger_b(t)g_u(\varepsilon^b_u)H_m U_b(t)|\psi_0\rangle\right). \end{aligned} \tag{3.29}$$

A gradient-based iterative learning algorithm for the discrimination of quantum systems is shown in Algorithm 3.2.

Algorithm 3.2 Gradient-based iterative learning for quantum discrimination

1: Set the index of iterations $k = 0$
2: Choose a set of arbitrary controls $u^{k=0}(t) = \{u^0_m(t),\ m = 1, 2, \ldots, M\}, t \in [0, T]$
3: **repeat** (for each iterative process)
4: Compute the propagator $U^k_a(t)$ and $U^k_b(t)$ for systems a and b, respectively, with the control strategy $u^k(t)$
5: **repeat** (for each control $u_m(t)$ $(m = 1, 2, \ldots, M)$ of the control vector $u^k(t)$)
6: compute $\nabla J(u^k_m)$ using (3.29)
7: $u^{k+1}_m(t) = u^k_m(t) + \eta_k \nabla J(u^k_m)$
8: **until** $m = M$
9: $k = k + 1$
10: **until** the learning process ends
11: The optimal control strategy $u^*(t) = \{u^*_m(t)\} = \{u^k_m(t)\},\ m = 1, 2, \ldots, M$

3.5.2 *Numerical Examples*

Example 3.3 To demonstrate this learning control method for discrimination of two similar quantum systems, we consider two-level (e.g., spin-$\frac{1}{2}$) systems. We assume the free Hamiltonian $H_0 = \frac{1}{2}\sigma_z$. Its two eigenstates are denoted as $|0\rangle$ (e.g., spin up) and $|1\rangle$ (e.g., spin down). We use the control Hamiltonian

$$H_u = \frac{1}{2}u_1(t)\sigma_x + \frac{1}{2}u_2(t)\sigma_y.$$

Hence,

$$H(t) = H_0 + H_u(t) = \frac{1}{2}\sigma_z + \frac{1}{2}u_1(t)\sigma_x + \frac{1}{2}u_2(t)\sigma_y. \tag{3.30}$$

For two similar two-level systems, the Hamiltonian of each system can be described as

$$\begin{aligned} H(t) &= f(\varepsilon_0)H_0 + g_u(\varepsilon_u)H_u(t) \\ &= \frac{1}{2}f(\varepsilon_0)\sigma_z + \frac{1}{2}g_u(\varepsilon_u)(u_1(t)\sigma_x + u_2(t)\sigma_y). \end{aligned} \tag{3.31}$$

We assume $f(\varepsilon_0) = \varepsilon_0$ and $g_u(\varepsilon_u) = \varepsilon_u$. The state of the two quantum systems can be represented in the eigenbasis of H_0 by $|\psi(t)\rangle = c_0(t)|0\rangle + c_1(t)|1\rangle$. Denote

$$C(t) = \begin{bmatrix} c_0(t) \\ c_1(t) \end{bmatrix}.$$

We have

$$\begin{bmatrix} \dot{c}_0(t) \\ \dot{c}_1(t) \end{bmatrix} = \begin{bmatrix} 0.5\varepsilon_0 \mathrm{i} & \varepsilon_u(u_2(t) - 0.5\mathrm{i}u_1(t)) \\ -\varepsilon_u(u_2(t) + 0.5\mathrm{i}u_1(t)) & -0.5\varepsilon_0\mathrm{i} \end{bmatrix} \begin{bmatrix} c_0(t) \\ c_1(t) \end{bmatrix}, \tag{3.32}$$

where $(\varepsilon_0, \varepsilon_u) = (\varepsilon_0^a, \varepsilon_u^a)$ for system a and $(\varepsilon_0, \varepsilon_u) = (\varepsilon_0^b, \varepsilon_u^b)$ for system b.

Define the performance function as

$$J(u) = \frac{1}{2}J^a(u) + \frac{1}{2}J^b(u) = \frac{1}{2}|\langle\psi^a(T)|\psi^a_{\text{target}}\rangle|^2 + \frac{1}{2}|\langle\psi^b(T)|\psi^b_{\text{target}}\rangle|^2. \tag{3.33}$$

The task is to find a control $u(t)$ to maximize the performance function in (3.33). For a given small threshold $\bar{\delta} > 0$, if $|J(u^{k+1}) - J(u^k)| < \bar{\delta}$ for uninterrupted n_e steps, a suitable control law for the problem has then been found. We set $\bar{\delta} = 10^{-4}$, and $n_e = 100$.

Now we employ Algorithm 3.2 to find an optimal control $u^*(t) = \{u_m^*(t), m = 1, 2\}$ and then apply the optimal control strategy for discriminating system a from system b. The parameter settings are listed as follows: the initial state $|\psi_0\rangle = |0\rangle$, and the target state for system a $|\psi^a_{\text{target}}\rangle = |0\rangle$; the target state for system b $|\psi^b_{\text{target}}\rangle = |1\rangle$; the ending time $T = 5$ (in atomic units) and the total time duration $[0, T]$ is equally discretized into $Q = 500$ time slices with each time slice

$$\Delta t = (t_q - t_{q-1})|_{q=1,2,\ldots,Q} = \frac{T}{Q} = 0.01;$$

the learning rate $\eta_k = 0.2$; the control strategy is initialized as

$$u^{k=0}(t) = \{u_1^0(t) = \sin t, u_2^0(t) = \sin t\}.$$

We first consider two similar systems a and b which are characterized with parameters $(\varepsilon_0^a, \varepsilon_u^a) = (0.9, 0.9)$ and $(\varepsilon_0^b, \varepsilon_u^b) = (1.1, 1.1)$, respectively. The numerical results are shown in Fig. 3.8. As shown in Fig. 3.8a, the learning process converges very quickly and the performance function $J(u)$ converges to 0.999 after about 2000

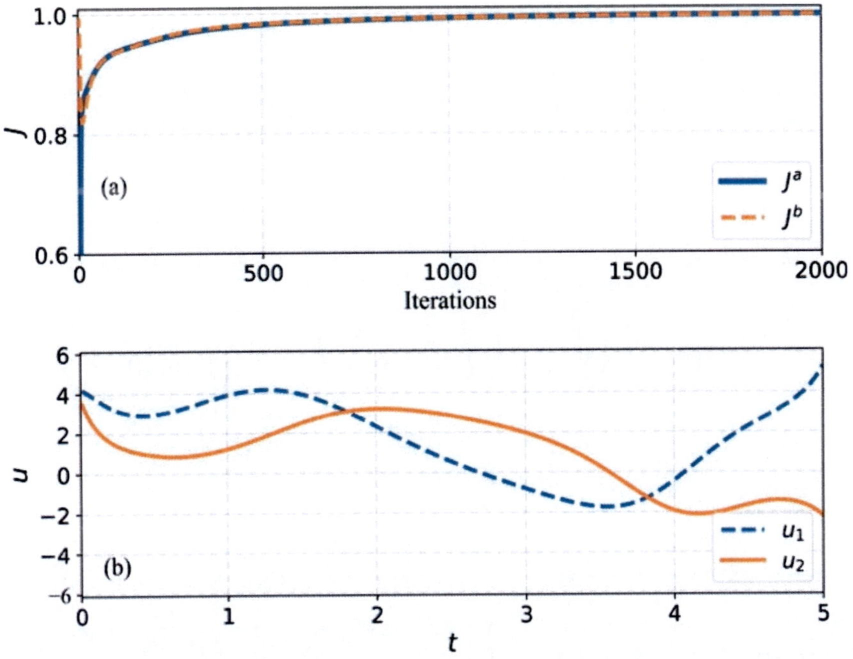

Fig. 3.8 Learning performance of discrimination between system a $((\varepsilon_0^a, \varepsilon_u^a) = (0.9, 0.9))$ and system b $((\varepsilon_0^b, \varepsilon_u^b) = (1.1, 1.1))$. **a** evolution of performance functions $J^a(u)$ and $J^b(u)$; **b** the learned optimal control strategy $u(t)$

steps of iterative learning with an optimized control strategy $u(t) = \{u_1(t), u_2(t)\}$ in Fig. 3.8b. Then we apply the learned optimal control strategy to systems a and b. At time $t = T = 5$, $|c_0^a(T)|^2 = 1.0000$ and $|c_0^b(T)|^2 = 0.0000$, which indicates that, after the coherent control step, we can discriminate system a from system b using a projective measurement and the success probability is almost 100%.

Now we consider two similar systems a and b which are characterized with parameters $(\varepsilon_0^a, \varepsilon_u^a) = (0.95, 0.95)$ and $(\varepsilon_0^b, \varepsilon_u^b) = (1.05, 1.05)$. Similar to the first case, we can also successfully learn an optimal control strategy to drive systems a and b to different target eigenstates from the same initial state. The evolution of their states is shown in Fig. 3.9 regarding their populations at the state $|0\rangle$, respectively. By comparing the two cases, it is clear that the difference lies in the similarity between Hamiltonians. For the second case, the increasing of the similarity of Hamiltonians makes it more difficult to discriminate system a from system b. More learning steps (about 15000 steps) are needed for the second case than the first one (around 2000 steps). This phenomenon is comprehensively tested through further numerical experiments with varied parameters. All of the results show that the gradient-based iterative learning method is successful for discrimination of similar quantum systems and also support the previous findings that optimal dynamic discrimination is fea-

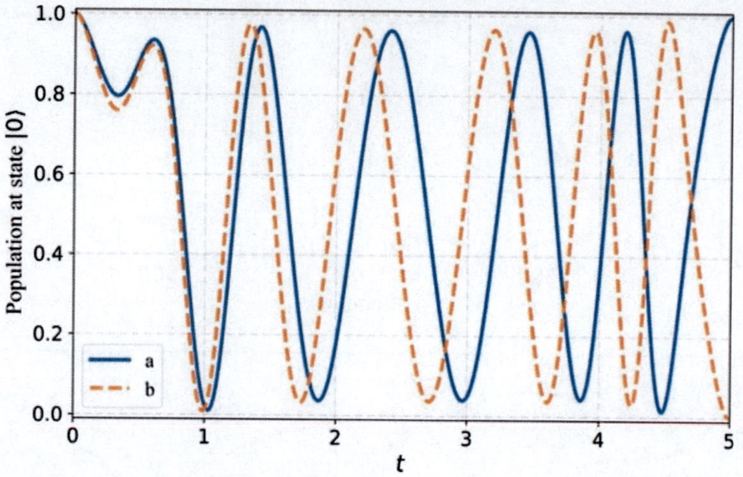

Fig. 3.9 Evolution of the states of system a $((\varepsilon_0^a, \varepsilon_u^a) = (0.95, 0.95))$ and system b $((\varepsilon_0^b, \varepsilon_u^b) = (1.05, 1.05))$ regarding the population at the state $|0\rangle$, respectively

sible for many similar quantum systems in the physics and chemistry communities [7, 28, 30, 32, 36, 44, 50, 55].

3.6 Binary Quantum Ensemble Classification via SLC

Binary classification is to classify the members of a given set of objects into two classes on the basis of whether they have certain properties or not. For a binary QEC problem, we have to learn from a sampling set as defined in Definition 3.1 and find an optimal control strategy for all the members in the quantum ensemble. Here, we combine an SLC approach with the quantum discrimination method introduced above to solve the QEC problem (i.e., the maximization problem formulated as (3.25)).

3.6.1 *Binary Ensemble Classification Algorithm*

The first key issue for QEC is how to obtain the sampling set for the training step. Generally there are two ways to construct a sampling set for QEC: (a) the data are provided initially and the sampling set can be constructed directly, but we do not know the distribution of parameters that characterize the members belonging to different classes; (b) no initial sampling data are provided but we know the distribution of parameters, and we can choose samples using the distribution information. The first way is very common in classical machine learning problems, while the second way is more suitable for the classification of quantum systems. In the quantum domain, it is difficult to obtain a specific description for a single system in a quantum ensemble, while we can characterize an ensemble of similar systems with a distribution of parameters (e.g., Gaussian distribution, Boltzmann distribution and uniform distribution). According to the distribution of parameters for a quantum ensemble, we can choose sample members to construct the sampling set for the classification task. To find an optimal control strategy using the sampling set, we adopt a modified SLC approach to solve the supervised quantum learning problem of QEC via constructing an augmented system using the sampling set.

Suppose we have obtained a sampling set $D_N = \{(H^n(t), y_n)\}$ $(n = 1, 2, \ldots, N)$ for the binary QEC problem, where $y_n \in \{A, B\}$ and $H^n(t)$ is the time-dependent Hamiltonian that describes the nth member of the quantum ensemble. Now we split D_N into two subsets according to the value of y_n ($n = p$ or q) and rewrite the sampling set for training as follows:

$$\begin{aligned} D_N &= D_{N_A} \cup D_{N_B},\ N = N_A + N_B, \\ D_{N_A} &= \{(H^A_{\varepsilon_0^p,\varepsilon_u^p}(t), y_p = A)\},\ p = 1, 2, \ldots, N_A, \\ D_{N_B} &= \{(H^B_{\varepsilon_0^q,\varepsilon_u^q}(t), y_q = B)\},\ q = N_A + 1, N_A + 2, \ldots, N, \end{aligned} \tag{3.34}$$

where

$$H^A_{\varepsilon_0^p,\varepsilon_u^p}(t) = f^A(\varepsilon_0^p)H_0 + g_u^A(\varepsilon_u^p)\sum_{m=1}^{M} u_m(t)H_m,$$

$$H^B_{\varepsilon_0^q,\varepsilon_u^q}(t) = f^B(\varepsilon_0^q)H_0 + g_u^B(\varepsilon_u^q)\sum_{m=1}^{M} u_m(t)H_m.$$

Using the sampling set (3.34), we can construct an augmented system as follows:

$$\frac{\mathrm{d}}{\mathrm{d}t}\begin{bmatrix} |\psi^A_{\varepsilon_0^1,\varepsilon_u^1}(t)\rangle \\ \vdots \\ |\psi^A_{\varepsilon_0^{N_A},\varepsilon_u^{N_A}}(t)\rangle \\ |\psi^B_{\varepsilon_0^{N_A+1},\varepsilon_u^{N_A+1}}(t)\rangle \\ \vdots \\ |\psi^B_{\varepsilon_0^N,\varepsilon_u^N}(t)\rangle \end{bmatrix} = -\mathrm{i}\begin{bmatrix} H^A_{\varepsilon_0^1,\varepsilon_u^1}(t)|\psi^A_{\varepsilon_0^1,\varepsilon_u^1}(t)\rangle \\ \vdots \\ H^A_{\varepsilon_0^{N_A},\varepsilon_u^{N_A}}(t)|\psi^A_{\varepsilon_0^{N_A},\varepsilon_u^{N_A}}(t)\rangle \\ H^B_{\varepsilon_0^{N_A+1},\varepsilon_u^{N_A+1}}(t)|\psi^B_{\varepsilon_0^{N_A+1},\varepsilon_u^{N_A+1}}(t)\rangle \\ \vdots \\ H^B_{\varepsilon_0^N,\varepsilon_u^N}(t)|\psi^B_{\varepsilon_0^N,\varepsilon_u^N}(t)\rangle \end{bmatrix}. \tag{3.35}$$

The performance function for this augmented system is defined by

$$J_N(u) := w_A J^A + w_B J^B, \tag{3.36}$$

where

$$
\begin{aligned}
J^A &= \frac{1}{N_A}\sum_{p=1}^{N_A} J^A_{\varepsilon_0^p,\varepsilon_u^p}(u) = \frac{1}{N_A}\sum_{p=1}^{N_A} |\langle \psi^A_{\varepsilon_0^p,\varepsilon_u^p}(T)|\psi_{\text{targetA}}\rangle|^2, \\
J^B &= \frac{1}{N_B}\sum_{q=N_A+1}^{N} J^B_{\varepsilon_0^q,\varepsilon_u^q}(u) = \frac{1}{N_B}\sum_{q=N_A+1}^{N} |\langle \psi^B_{\varepsilon_0^q,\varepsilon_u^q}(T)|\psi_{\text{targetB}}\rangle|^2.
\end{aligned}
\tag{3.37}
$$

The task of the training step is to find a control strategy that maximizes the performance function defined in (3.36). From (3.29), (3.36) and (3.37), we have

$$
\begin{aligned}
\nabla J_N(u_m^k) =& \frac{2w_a}{N_A}\sum_{p=1}^{N_A} \Im\Big(\big\langle \psi^A_{\varepsilon_0^p,\varepsilon_u^p}(T)\big|\psi_{\text{targetA}}\big\rangle \langle \psi_{\text{targetA}}|U_{\varepsilon_0^p,\varepsilon_u^p}(T)U^{\dagger}_{\varepsilon_0^p,\varepsilon_u^p}(t) g_u^A(\varepsilon_u^p) H_m U_{\varepsilon_0^p,\varepsilon_u^p}(t)|\psi_0\rangle\Big) \\
&+\frac{2w_b}{N_B}\sum_{q=N_A+1}^{N} \Im\Big(\langle \psi^B_{\varepsilon_0^q,\varepsilon_u^q}(T)|\psi_{\text{targetB}}\rangle\langle \psi_{\text{targetB}}|U_{\varepsilon_0^q,\varepsilon_u^q}(T)U^{\dagger}_{\varepsilon_0^q,\varepsilon_u^q}(t) g_u^B(\varepsilon_u^q) H_m U_{\varepsilon_0^q,\varepsilon_u^q}(t)|\psi_0\rangle\Big).
\end{aligned}
\tag{3.38}
$$

Then we design the SLC algorithm (Algorithm 3.3) for binary QEC using the gradient-based method to approximate an optimal control strategy $u^* = \{u_m^*(t)\}$.

Algorithm 3.3 SLC for binary QEC

Set the index of iterations $k = 0$
Choose a set of arbitrary controls $u^{k=0}(t) = \{u_m^0(t),\ m = 1, 2, \ldots, M\}, t \in [0, T]$
repeat (for each iterative process)
  repeat (for each member in training subset D_{N_A}, $p = 1, 2, \ldots, N_A$)
    Compute the propagator $U^k_{\varepsilon_0^p,\varepsilon_u^p}(t)$ with the control strategy $u^k(t)$
  until $p = N_A$
  repeat (for each member in training subset D_{N_B}, $q = N_A + 1, N_A + 2, \ldots, N$)
    Compute the propagator $U^k_{\varepsilon_0^q,\varepsilon_u^q}(t)$ with the control strategy $u^k(t)$
  until $q = N$
  repeat (for each control $u_m(t)$ ($m = 1, 2, \ldots, M$) of the control vector $u^k(t)$)
    compute $\nabla J_N(u_m^k)$ using (3.38)
    $u_m^{k+1}(t) = u_m^k(t) + \eta_k \nabla J_N(u_m^k)$
  until $m = M$
  $k = k + 1$
until the learning process ends
The optimal control strategy $u^*(t) = \{u_m^*(t)\} = \{u_m^k(t)\},\ m = 1, 2, \ldots, M$

3.6.2 Numerical Examples

Example 3.4 We consider two-level quantum systems. For two similar classes of members in an inhomogeneous quantum ensemble, the Hamiltonians can be described as

$$
\begin{aligned}
H^{A}_{\varepsilon_0,\varepsilon_u}(t) &= \frac{1}{2}f^{A}(\varepsilon_0)\sigma_z + \frac{1}{2}g^{A}_{u}(\varepsilon_u)(u_1(t)\sigma_x + u_2(t)\sigma_y), \\
H^{B}_{\varepsilon_0,\varepsilon_u}(t) &= \frac{1}{2}f^{B}(\varepsilon_0)\sigma_z + \frac{1}{2}g^{B}_{u}(\varepsilon_u)(u_1(t)\sigma_x + u_2(t)\sigma_y).
\end{aligned}
\tag{3.39}
$$

Assume $f^A(\varepsilon_0) = \varepsilon_0$ with distribution $d^A_0(\varepsilon_0)$, $g^A_u(\varepsilon_u) = \varepsilon_u$ with distribution $d^A_u(\varepsilon_u)$, $g^B_0(\varepsilon_0) = \varepsilon_0$ with distribution $d^B_0(\varepsilon_0)$ and $g^B_u(\varepsilon_u) = \varepsilon_u$ with distribution $d^B_u(\varepsilon_u)$.

Suppose the distributions of ε_0 and ε_u for class A are

$$
\mathrm{d}^A_0(\varepsilon_0) = \Phi\left(\frac{\varepsilon_0 - \mu^A_0}{\sigma^A_0}\right) \quad \text{and} \quad \mathrm{d}^A_u(\varepsilon_u) = \Phi\left(\frac{\varepsilon_u - \mu^A_u}{\sigma^A_u}\right),
$$

where

$$
\Phi(x) = \int_{-\infty}^{x} \frac{1}{\sqrt{2\pi}}\exp(-\frac{1}{2}\nu^2)\mathrm{d}\nu
$$

is the distribution function of the standard normal distribution. We may choose some equally spaced samples in the $\varepsilon_0 - \varepsilon_u$ space. For example, we may choose the intervals of $[\mu^A_0 - 3\sigma^A_0, \mu^A_0 + 3\sigma^A_0]$ and $[\mu^A_u - 3\sigma^A_u, \mu^A_u + 3\sigma^A_u]$ and divide them into $N^A_{\varepsilon_0} + 1$ and $N^A_{\varepsilon_u} + 1$ subintervals, respectively, where $N^A_{\varepsilon_0}$ and $N^A_{\varepsilon_u}$ are usually positive odd numbers. Then the number of samples for class A is $N_A = N^A_{\varepsilon_0}N^A_{\varepsilon_u}$, where ε^p_0 and ε^p_u can be chosen from the combination of $(\varepsilon^{p^0}_0, \varepsilon^{p^u}_u)$ as follows:

$$
\begin{cases}
\varepsilon^p_0 \in \left\{\varepsilon^{p^0}_0 = \mu^A_0 - 3\sigma^A_0 + \frac{(2p^0-1)3\sigma^A_0}{N^A_{\varepsilon_0}}, \; p^0 = 1, 2, \ldots, N^A_{\varepsilon_0}\right\}, \\
\varepsilon^p_u \in \left\{\varepsilon^{p^u}_u = \mu^A_u - 3\sigma^A_u + \frac{(2p^u-1)3\sigma^A_u}{N^A_{\varepsilon_u}}, \; p^u = 1, 2, \ldots, N^A_{\varepsilon_u}\right\}.
\end{cases}
\tag{3.40}
$$

In practical applications, the numbers of $N^A_{\varepsilon_0}$ and $N^A_{\varepsilon_u}$ can be chosen by experience or be tried through numerical computation. As long as the augmented system can model the quantum ensemble and is effective to find the optimal control strategy, we prefer smaller numbers $N^A_{\varepsilon_0}$ and $N^A_{\varepsilon_u}$ to speed up the training process and simplify the augmented system. A similar expression to (3.40) defines the samples for class B. We use the performance function as defined in (3.36) with $w_A = w_B = 0.5$. Now we use Algorithm 3.3 to find the optimal control strategy.

The parameter settings are listed as follows: $w_A = w_B = 0.5$, the initial state for each member of the quantum ensemble $|\psi_0\rangle = |0\rangle$, and the target state for members belonging to class A $|\psi_{\text{targetA}}\rangle = |0\rangle$; the target state for elements belonging to class B $|\psi_{\text{targetB}}\rangle = |1\rangle$; the ending time $T = 8$ (in atomic units) and the total time duration $[0, T]$ is equally discretized into $Q = 800$ time slices with each time slice $\Delta t = (t_q - t_{q-1})|_{q=1,2,\ldots,Q} = \frac{T}{Q} = 0.01$; $N^A_{\varepsilon_0} = N^A_{\varepsilon_u} = N^B_{\varepsilon_0} = N^B_{\varepsilon_u} = 5$; the learning rate $\eta_k = 0.2$; the control strategy is initialized as

$$u^{k=0}(t) = \{u_1^0(t) = \sin t, u_2^0(t) = \sin t\}.$$

In the training step, we use $J(u)$ as the performance function which represents the measure of weighted accuracy for QEC. After we apply the optimized control u^* to the inhomogeneous quantum ensemble, we use fidelity to characterize how well every member is classified. It is clear that the accuracy of QEC can be calculated with

$$\begin{aligned}\zeta &= J(u) = \tfrac{1}{2}(\mathbb{A}_\nu[J^A] + \mathbb{A}_\nu[J^B]) \\ &= \tfrac{1}{2}(\mathbb{A}_\nu[F^2(|\psi^A_{\varepsilon_0,\varepsilon_u}(T)\rangle, |\psi_{\text{targetA}}\rangle)] + \mathbb{A}_\nu[F^2(|\psi^B_{\varepsilon_0,\varepsilon_u}(T)\rangle, |\psi_{\text{targetB}}\rangle)]).\end{aligned} \tag{3.41}$$

The parameters are set as $(\mu_0^A, 3\sigma_0^A) = (0.85, 0.05)$, $(\mu_u^A, 3\sigma_u^A) = (0.85, 0.05)$, $(\mu_0^B, 3\sigma_0^B) = (1.15, 0.05)$ and $(\mu_u^B, 3\sigma_u^B) = (1.15, 0.05)$. The learning control performance is shown in Fig. 3.10. It is clear that the learning algorithm converges quickly after about 8000 iterations and finds an optimized control for the coherent control step of binary QEC. Applying the learned control to 300 randomly selected testing samples (150 for *class A* and 150 for *class B*), the mean value of fidelity for the testing of *class A* is 0.9976 and for *class B* is 0.9985. With additional 10^4 testing samples for both *class A* and *class B*, the classification accuracy is estimated as $\zeta = 99.62\%$. The result demonstrates that the SLC approach is effective for the binary QEC problem and can achieve a high level of classification accuracy. More numerical experiments show that the classification performance deteriorates with larger dispersion on the Hamiltonian and smaller difference of Hamiltonians between class A and class B. The tradeoff relationship between performance and dispersion (or relevant parameters) has been widely investigated in many contexts such as NMR [40].

3.7 Multi-class Classification of Multi-level Quantum Ensembles

In machine learning, multi-class classification involves classifying instances into more than two classes. Some classification algorithms naturally permit the use of more than two classes [33]. A useful strategy is the one-vs-all strategy, where a single classifier is trained per class to distinguish that class from all other classes [42].

The SLC-based QEC approach proposed for binary QEC can be extended to multi-class classification of multi-level quantum ensembles using the one-vs-all strategy. For example, suppose an inhomogeneous quantum ensemble consists of three classes of members (i.e., classes A, B and C). First, by applying the binary QEC approach, we can classify them into two classes (one for members belonging to class A and the other for all the members belonging to classes B and C). By a proper physical operation, we can separate the members in class A from classes B and C. Then we

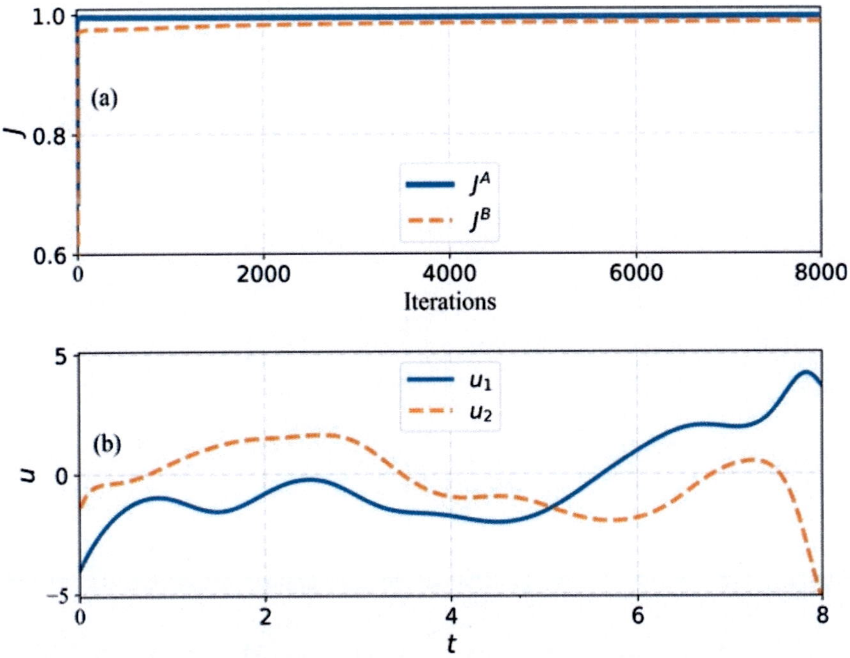

Fig. 3.10 Learning performance of binary QEC: **a** evolution of performance function J^A and J^B; **b** the learned optimal control for QEC

use the binary QEC approach again to classify the members belonging to class B from the members belonging to class C. However, the one-vs-all strategy may cause additional cost since the process involves multiple cases of binary classification and multiple learning control procedures.

Here, we use a different strategy from the one-vs-all strategy to extend the proposed SLC-based classification method to multi-class classification of multi-level quantum ensembles. In this strategy, only one quantum coherent control procedure is needed to implement the multi-class QEC.

We consider an inhomogeneous quantum ensemble of three-level Λ-type atomic systems [11]. The evolving state $|\psi(t)\rangle$ of the Λ-type system can be expanded in terms of the eigenstates as follows:

$$|\psi(t)\rangle = c_1(t)|1\rangle + c_2(t)|2\rangle + c_3(t)|3\rangle, \tag{3.42}$$

where $|1\rangle$, $|2\rangle$ and $|3\rangle$ are the basis states of the lower, middle and upper atomic states, respectively, corresponding to the free Hamiltonian

$$H_0 = \begin{bmatrix} 1.5 & 0 & 0 \\ 0 & 1 & 0 \\ 0 & 0 & 0 \end{bmatrix}. \tag{3.43}$$

Let

$$C(t) = \begin{bmatrix} c_1(t) \\ c_2(t) \\ c_3(t) \end{bmatrix}.$$

To control such a three-level system, we use the control Hamiltonian

$$H_u = u_1(t)H_1 + u_2(t)H_2,$$

where

$$H_1 = \begin{bmatrix} 0\,0\,0 \\ 0\,0\,1 \\ 0\,1\,0 \end{bmatrix}, \tag{3.44}$$

$$H_2 = \begin{bmatrix} 0\,0\,1 \\ 0\,0\,0 \\ 1\,0\,0 \end{bmatrix}. \tag{3.45}$$

Similarly, we describe the inhomogeneous three-level quantum ensemble with the Hamiltonian of each member as

$$H_{\varepsilon_0,\varepsilon_u}(t) = \varepsilon_0 H_0 + \varepsilon_u((u_1(t)H_1 + u_2(t)H_2). \tag{3.46}$$

Suppose that the inhomogeneous quantum ensemble consists of three classes of members labeled with classes A, B and C, respectively. For this multi-class QEC problem, we first use the same control field to drive the members belonging to classes A, B and C from an initial state $|\psi_0\rangle$ to three different target eigenstates ($|1\rangle$, $|2\rangle$ and $|3\rangle$), respectively, so that we can classify them with an additional physical operation (e.g., projective measurement).

We can modify Algorithm 3.3 into its multi-class version and then apply it to the three-level inhomogeneous quantum ensemble for finding an optimal control strategy $u^*(t) = \{u^*_m(t), m = 1, 2\}$ to maximize the performance function

$$\begin{aligned} J(u) =& \frac{1}{3}(\mathbb{A}_v[J^A] + \mathbb{A}_v[J^B] + \mathbb{A}_v[J^C]) \\ =& \frac{1}{3}(\mathbb{A}_v[F^2(|\psi^A_{\varepsilon_0,\varepsilon_u}(T)\rangle, |1\rangle)] + \mathbb{A}_v[F^2(|\psi^B_{\varepsilon_0,\varepsilon_u}(T)\rangle, |2\rangle)] \\ &+ \mathbb{A}_v[F^2(|\psi^C_{\varepsilon_0,\varepsilon_u}(T)\rangle, |3\rangle)]). \end{aligned} \tag{3.47}$$

The parameter settings are listed as follows: the initial state

$$C_0 = \begin{bmatrix} \frac{1}{\sqrt{3}} \\ \frac{1}{\sqrt{3}} \\ \frac{1}{\sqrt{3}} \end{bmatrix};$$

the three target eigenstates for classes A, B and C are $|1\rangle$, $|2\rangle$ and $|3\rangle$, respectively, i.e.,

$$C_{\text{targetA}} = \begin{bmatrix} 1 \\ 0 \\ 0 \end{bmatrix}, \quad C_{\text{targetB}} = \begin{bmatrix} 0 \\ 1 \\ 0 \end{bmatrix}, \quad C_{\text{targetC}} = \begin{bmatrix} 0 \\ 0 \\ 1 \end{bmatrix};$$

the ending time $T = 10$ and the total time duration $[0, T]$ is equally discretized into $Q = 1000$ time slices; the learning rate is $\eta_k = 0.2$; the control is initialized as

$$u^0(t) = \{u_1^0(t) = \sin t, u_2^0(t) = \sin t\}.$$

The parameters ε_0 and ε_u characterize the inhomogeneity of the quantum ensemble, and they have different normal distributions that are described with the distribution functions

$$\mathrm{d}_0(\varepsilon_0) = \Phi(\frac{\varepsilon_0 - \mu_0}{\sigma_0})$$

and

$$\mathrm{d}_u(\varepsilon_u) = \Phi(\frac{\varepsilon_u - \mu_u}{\sigma_u}),$$

where for class A ($\mu_0^A = 1, 3\sigma_0^A = 0.05$) and ($\mu_u^A = 0.8, 3\sigma_u^A = 0.05$), for class B ($\mu_0^B = 0.8, 3\sigma_0^B = 0.05$) and ($\mu_u^B = 1, 3\sigma_u^B = 0.05$), for class C ($\mu_0^C = 1.2, 3\sigma_0^C = 0.05$) and ($\mu_u^C = 1.2, 3\sigma_u^C = 0.05$). To construct the augmented system for learning the optimal control, the sampling method as described in (3.40) is adopted with setting

$$N_{\varepsilon_0}^A = N_{\varepsilon_u}^A = N_{\varepsilon_0}^B = N_{\varepsilon_u}^B = N_{\varepsilon_0}^C = N_{\varepsilon_u}^C = 3.$$

After we find an optimal control using the SLC method, 300 randomly selected samples for each class of members are tested and all of them are controlled to their corresponding target eigenstates, respectively, with high fidelity. The mean value of fidelity for testing of *class A* is 0.9897, for *class B* is 0.9953 and for *class C* is 0.9976. With additional 10^4 testing samples for each class, we have the classification accuracy $\zeta = 98.80\%$, which verifies the effectiveness of the proposed SLC-based approach for QEC.

3.8 Summary and Further Reading

In this chapter, we presented a systematic SLC method for control and classification of inhomogeneous quantum ensembles. For control of inhomogeneous quantum ensembles, we used two-level and three-level examples to demonstrate the performance. Future work will extend the control method to open inhomogeneous quantum ensembles. For the classification problem, we considered binary classification and multi-class classification of inhomogeneous ensembles, where there is no overlapping between the distribution of similar parameters. In practice, the parameter distributions between different classes may overlap. Usually, there is not an effective method to distinguish the members in the overlapping region between classes. Hence, there is a theoretical upper bound of classification precision (less than 100%) for this class of classification problem. Several results have also been presented in [12]. Future work will focus on more complex quantum classification problems.

Further reading includes [48] for stochastic sampling-based learning control of inhomogeneous quantum ensembles, [12] for more results on classification of inhomogeneous quantum ensembles, [25, 26, 31] for the controllability of quantum ensembles and [45, 53, 54] for optimal control design of inhomogeneous quantum ensembles.

References

1. Altafini C (2007) Feedback stabilization of isospectral control systems on complex flag manifolds: application to quantum ensembles. IEEE Trans Autom Control 52(11):2019–2028
2. Ansel Q, Probst S, Bertet P, Glaser SJ, Sugny D (2018) Optimal control of an inhomogeneous spin ensemble coupled to a cavity. Phys Rev A 98:023425
3. Appleby DM (2000) Optimal measurement of spin direction. Int J Theoret Phys 39(9):2231–2252
4. Augier N, Boscain U, Sigalotti M (2018) Adiabatic ensemble control of a continuum of quantum systems. SIAM J Control Optim 56(6):4045–4068
5. Beauchard K, Coron JM, Rouchon P (2010) Controllability issues for continuous-spectrum systems and ensemble controllability of Bloch equations. Commun Math Phys 296:525–557
6. Beauchard K, da Silva PSP, Rouchon P (2012) Stabilization for an ensemble of half-spin systems. Automatica 48:68–76
7. Beltrani V, Ghosh P, Rabitz H (2009) Exploring the capabilities of quantum optimal dynamic discrimination. J Chem Phys 130:164112
8. Bensky G, Petrosyan D, Majer J, Schmiedmayer J, Kurizki G (2012) Optimizing inhomogeneous spin ensembles for quantum memory. Phys Rev A 86:012310
9. Brif C, Chakrabarti R, Rabitz H (2010) Control of quantum phenomena: past, present and future. New J Phys 12:075008
10. Brinks D, Stefani FD, Kulzer F, Hildner R, Taminiau TH, Avlasevich Y, Müllen K, van Hulst NF (2010) Visualizing and controlling vibrational wave packets of single molecules. Nature 465(17):905–908
11. Chen C, Dong D, Long R, Petersen IR, Rabitz H (2014) Sampling-based learning control of inhomogeneous quantum ensembles. Phys Rev A 89:023402
12. Chen C, Dong D, Qi B, Petersen IR, Rabitz H (2017) Quantum ensemble classification: a sampling-based learning control approach. IEEE Trans Neural Netw Learn Syst 28:1345–1359
13. Conolly S, Nishimura D, Macovski A (1986) Optimal control solutions to the magnetic resonance selective excitation problem. IEEE Trans Med Imaging 5(2):106–115
14. Cory DG, Fahmy AF, Havel TF (1997) Ensemble quantum computing by NMR spectroscopy. Proc Natl Acad Sci U S A 94:1634–1639
15. Dong D, Chen C, Long R, Qi B, Petersen IR (2013) Sampling-based learning control for quantum systems with Hamiltonian uncertainties. In: Proceedings of the 52th IEEE conference on decision and control, December 10–13, Firenze, Italy, pp 1924–1929
16. Dong D, Petersen IR (2022) Quantum estimation, control and learning: opportunities and challenges. Ann Rev Control 54:243–251

17. Duan R, Feng Y, Ying M (2007) Entanglement is not necessary for perfect discrimination between unitary operations. Phys Rev Lett 98:100503
18. Duan LM, Lukin MD, Cirac JI, Zoller P (2001) Long-distance quantum communication with atomic ensembles and linear optics. Nature 414:413–418
19. Glaser SJ, Schulte-Herbrüggen T, Sieveking M, Schedletzky O, Nielsen NC, Sørensen OW, Griesinger C (1998) Unitary control in quantum ensembles: maximizing signal intensity in coherent spectroscopy. Science 280(5362):421–424
20. Hou SC, Khan MA, Yi XX, Dong D, Petersen IR (2012) Optimal Lyapunov-based quantum control for quantum systems. Phys Rev A 86(2):022321
21. Jirari H, Pötz W (2005) Optimal coherent control of dissipative N-level systems. Phys Rev A 72:013409
22. Khaneja N, Reiss T, Kehlet C, Schulte-Herbrüggen T, Glaser SJ (2005) Optimal control of coupled spin dynamics: design of NMR pulse sequences by gradient ascent algorithms. J Magn Reson 172(2):296–305
23. Levitt MH (1986) Composite pulses. Prog Nucl Magn Reson Spectrosc 18:61–122
24. Li J-S, Khaneja N (2006) Control of inhomogeneous quantum ensembles. Phys Rev A 73:030302
25. Li J-S, Khaneja N (2009) Ensemble control of Bloch equations. IEEE Trans Autom Control 54(3):528–536
26. Li J-S, Khaneja N (2011) Ensemble control of finite-dimensional time-varying linear systems. IEEE Trans Autom Control 56(2):345–357
27. Li J-S, Qi J (2016) Ensemble control of time-invariant linear systems with linear parameter variation. IEEE Trans Autom Control 61(10):2808–2820
28. Li B, Rabitz H, Wolf JP (2005) Optimal dynamic discrimination of similar quantum systems with time series data. J Chem Phys 122:154103
29. Li J-S, Ruths J, Yu TY, Arthanari H, Wagner G (2011) Optimal pulse design in quantum control: a unified computational method. Proc Natl Acade Sci U S A 108:1879–1884
30. Li B, Turinici G, Ramakrishna V, Rabitz H (2002) Optimal dynamic discrimination of similar molecules through quantum learning control. J Phys Chem B 106(33):8125–8131
31. Li J-S, Zhang W, Tie L (2020) On separating points for ensemble controllability. SIAM J Control Optim 58:2740–2764
32. Li B, Zhu W, Rabitz H (2006) Optimal dynamic discrimination of similar quantum systems in the presence of decoherence. J Chem Phys 124:024101
33. Lin M, Tang K, Yao X (2013) Dynamic sampling approach to training neural networks for multiclass imbalance classification. IEEE Trans Neural Netw Learn Syst 24(4):647–660
34. Long R, Riviello G, Rabitz H (2013) The gradient flow for control of closed quantum systems. IEEE Trans Autom Control 58(10):2665–2669
35. Martikyan V, Devra A, Guéry-Odelin D, Glaser SJ, Sugny D, (2020) Robust control of an ensemble of springs: application to ion cyclotron resonance and two-level quantum systems. Phys Rev A 102(5):053104
36. Mitra A, Rabitz H (2004) Mechanistic analysis of optimal dynamic discrimination of similar quantum systems. J Phys Chem A 108(21):4778–4785
37. Mohseni M, Steinberg AM, Bergou JA (2004) Optical realization of optimal unambiguous discrimination for pure and mixed quantum states. Phys Rev Lett 93(20):200403
38. Nielsen MA, Chuang IL (2010) Quantum computation and quantum information. Cambridge University Press, Cambridge
39. Owrutsky P, Khaneja N (2012) Control of inhomogeneous ensembles on the Bloch sphere. Phys Rev A 86:022315
40. Pauly J, Le Roux P, Nishimura D, Macovski A (1991) Parameter relations for the Shinnar-Le Roux selective excitation pulse design algorithm. IEEE Trans Med Imaging 10(1):53–65
41. Rabitz H, Turinici G (2007) Controlling quantum dynamics regardless of laser beam spatial profile and molecular orientation. Phys Rev A 75:043409
42. Rifkin R, Klautau A (2004) In defence of one-vs-all classification. J Mach Learn Res 5:101–141

43. Roslund J, Rabitz H (2009) Gradient algorithm applied to laboratory quantum control. Phys Rev A 79:053417
44. Roslund J, Roth M, Guyon L, Boutou V, Courvoisier F, Wolf J, Rabitz H (2011) Resolution of strongly competitive product channels with optimal dynamic discrimination: application to flavins. J Chem Phys 134:034511
45. Ruths J, Li J-S (2011) A multidimensional pseudospectral method for optimal control of quantum ensembles. J Chem Phys 134:044128
46. Ruths J, Li J-S (2012) Optimal control of inhomogeneous ensembles. IEEE Trans Autom Control 57(8):2021–2032
47. Skinner TE, Reiss TO, Luy B, Khaneja N, Glaser SJ (2003) Application of optimal control theory to the design of broadband excitation pulses for high-resolution NMR. J Magn Reson 163:8–15
48. Turinici G (2019) Stochastic learning control of inhomogeneous quantum ensembles. Phys Rev A 100:053403
49. Turinici G, Rabitz H (2004) Optimally controlling the internal dynamics of a randomly oriented ensemble of molecules. Phys Rev A 70:063412
50. Turinici G, Ramakrishna V, Li B, Rabitz H (2004) Optimal discrimination of multiple quantum systems: controllability analysis. J Phys A: Math Gen 37:273–282
51. van Damme L, Ansel Q, Glaser SJ, Sugny D (2017) Robust optimal control of two-level quantum systems. Phys Rev A 95(6):063403
52. Vandersypen LMK, Chuang IL (2004) NMR techniques for quantum control and computation. Rev Mod Phys 76:1037
53. Wang S, Li J-S (2017) Fixed-endpoint optimal control of bilinear ensemble systems. SIAM J Control Optim 55(5):3039–3065
54. Wang S, Li J-S (2018) Free-endpoint optimal control of inhomogeneous bilinear ensemble systems. Automatica 95:306–315
55. Yoshida B (2011) Classification of quantum phases and topology of logical operators in an exactly solved model of quantum codes. Ann Phys 326:15–95
56. You JQ, Nori F (2011) Atomic physics and quantum optics using superconducting circuits. Nature 474:589–597

Chapter 4
Sampling-Based Learning Control of Quantum Systems with Uncertainties

Abstract This chapter presents results on enhancing robustness of quantum systems with uncertainties using the sampling-based learning control (SLC) method. In Sect. 4.2, the SLC method is applied to state control of superconducting qubits, robust control of photoassociation of O+H and synchronizing a laser with molecules for charge transfer[a],[b],[c]. Robust control based on SLC and b-GRAPE is presented for generating quantum gates in Sect. 4.3 and SLC of open quantum systems is investigated in Sect. 4.4[d].

4.1 Introduction

In this chapter we present several results on the application of sampling-based learning control (SLC) to quantum robust control problems. In Chap.3, we developed an effective SLC method to guide the design for control and classification of inhomogeneous quantum ensembles. In this method, artificial samples are generated through sampling possible inhomogeneity parameters, and an augmented model is established using these artificial samples. Optimal control fields are learned by a learning and optimization algorithm and then they are applied to additional samples to test performance. SLC provides a "smart" open-loop control method for enhancing the robustness performance of inhomogeneous quantum ensembles. In this chapter, we further employ the SLC method for guiding the design of robust control fields for various

[a] Section 4.2 partially reproduced from [11], originally published under a Creative Commons Attribution-NonCommercial-NoDerivs 4.0 International License; doi:10.1038/srep07873.
[b] Section 4.2 contains materials reprinted, with permission, from Chemical Physics [69] © 2017 Elsevier.
[c] Section 4.2 contains materials reprinted, with permission, from RSC Advances [68]. Copyright (2016) by the Royal Society of Chemistry.
[d] Sections 4.3 and 4.4 partially reproduced from [14], originally published under a Creative Commons Attribution 4.0 International License; doi:10.1038/srep36090.

D. Dong and I. R. Petersen, *Learning and Robust Control in Quantum Technology*, Communications and Control Engineering,
https://doi.org/10.1007/978-3-031-20245-2_4

applications [13]. In particular, we consider the robust state control of superconducting qubits where uncertainties exist in the Hamiltonian. Robust control results are presented for the photoassociation of O+H where uncertainties are present in the potential curves or control fields. The SLC approach is also employed to enhance the robustness and reliability in synchronizing laser fields with molecules for achieving charge transfer. We also investigate robust control problems for quantum gates and open quantum systems using the SLC method. A so-called b-GRAPE algorithm is introduced to investigate robust control problems for quantum systems where techniques in training deep neural networks can be borrowed to search for robust control fields.

4.2 SLC for Robust State Control

4.2.1 State Control of Superconducting Qubits

Superconducting qubits have been widely investigated theoretically and implemented experimentally in the last thirty years due to their advantages, such as scalability, design flexibility and tunability, for solid-state quantum computation and quantum simulations [4, 6, 26, 38, 45, 49, 52, 54, 57, 62–64]. In practical applications, the existence of extrinsic and intrinsic noises, inaccuracies (e.g., inaccurate operation in the coupling between qubits) and fluctuations (e.g., fluctuations in magnetic fields and electric fields) in superconducting quantum circuits is unavoidable [53, 56]. For simplicity, here we use uncertainties to represent these noises, inaccuracies and uncertainties. These uncertainties degrade the performance of robustness and reliability in superconducting quantum circuits. Hence, it is highly desirable to design control fields that are robust against uncertainties for the development of practical quantum technologies [17, 34, 42, 47]. Here, we employ the SLC method for guiding the design of robust control fields for superconducting quantum systems with uncertainties.

Consider a quantum system with Hamiltonian $H(u, \varepsilon, t)$ and the evolution of its state $|\psi(t)\rangle$ is described by the following Schrödinger equation:

$$\mathrm{i}|\dot{\psi}(t)\rangle = H(u, \varepsilon, t)|\psi(t)\rangle, \tag{4.1}$$

where u represents the control field and ε characterizes possible uncertainties. In the SLC method, we first generate N artificial samples by selecting different values of ε (e.g., the N samples correspond to $\varepsilon^{(1)}$, $\varepsilon^{(2)}$,..., $\varepsilon^{(N)}$). We use these samples to construct an augmented system and define the performance function $J(u)$ for the augmented system as

$$J(u) = \frac{1}{N}\sum_{n=1}^{N} J_{\varepsilon^{(n)}}(u) = \frac{1}{N}\sum_{n=1}^{N} |\langle\psi_{\varepsilon^{(n)}}(T)|\psi_{\text{target}}\rangle|^2, \tag{4.2}$$

where $|\psi_{\text{target}}\rangle$ is the target state and $|\psi_{\varepsilon^{(n)}}(T)\rangle$ is the final state for one sample (corresponding to $\varepsilon^{(n)}$) at the time T. The task in the training step is to find an optimal control field u^* that maximizes the performance function defined in (4.2).

In order to construct an augmented system, we need to generate N artificial samples. We assume there are two uncertainty parameters ε^x and ε^z which have bounds $\mathcal{E}^x$ and $\mathcal{E}^z$, respectively; i.e., $|\varepsilon^x| \leq \mathcal{E}^x$ and $|\varepsilon^z| \leq \mathcal{E}^z$. We may choose some equally spaced samples in the ε^x–ε^z space. For example, the intervals $[-\mathcal{E}^x, \mathcal{E}^x]$ and $[-\mathcal{E}^z, \mathcal{E}^z]$ are divided into $N_x + 1$ and $N_z + 1$ subintervals, respectively, where N_x and N_z are usually positive odd numbers. Then, the number of samples $N = N_x N_z$, where ε^x_m and ε^z_n can be chosen from the combination of $(\varepsilon^x_m, \varepsilon^z_n)$ as follows:

$$\begin{cases} \varepsilon^x_m \in \{\varepsilon^x_m = -\mathcal{E}^x + \frac{(2m-1)\mathcal{E}^x}{N_x}, \ m = 1, 2, \ldots, N_x\}, \\ \varepsilon^z_n \in \{\varepsilon^z_n = -\mathcal{E}^z + \frac{(2n-1)\mathcal{E}^z}{N_z}, \ n = 1, 2, \ldots, N_z\}. \end{cases} \tag{4.3}$$

In order to find an optimal control field u^* for the augmented system, we employ the same gradient algorithm as that in Chap. 3 (Algorithm 3.1). Assume that the performance function is $J(u_0)$ with an initial field u_0. We can apply the gradient flow method to approximate an optimal control field u^*. This can be achieved by iterative learning using the following updating (for details, see, e.g., [3]):

$$u_{k+1}(t) = u_k(t) + \eta_k \nabla J(u_k), \tag{4.4}$$

where η_k is the updating step size for the kth iteration and $\nabla J(u)$ denotes the gradient of $J(u)$ with respect to the control u. For practical implementations, we usually divide the time interval $[0, T]$ equally into a number of smaller time intervals Δt and assume that the control fields are constant within each time interval Δt. In the algorithm, we assume $u(t) \in [V_-, V_+]$. If $u_{k+1} \leq V_-$, we let $u_{k+1} = V_-$. If $u_{k+1} \geq V_+$, we let $u_{k+1} = V_+$. In numerical calculation, if the change of the performance function for 100 consecutive training steps is less than a small threshold $\bar{\delta}$ (i.e., $|J(u_{k+100}) - J(u_k)| < \bar{\delta}$ for some k), then we declare that the algorithm has converged and we end the training step. In this section, we let $\bar{\delta} = 10^{-4}$ for all numerical results.

As an illustrative example, we first consider the robust control problem of single charge qubits with uncertainties. The simplest charge qubit is based on a small superconducting island (called a Cooper-pair box) coupled to the outside world through a weak Josephson junction and driven by a voltage source through a gate capacitance within the charge regime (i.e., $E_C \gg E_J$) [63]. The Hamiltonian of the Cooper-pair box can be described as [63]

$$H = E_C(n - n_g)^2 - E_J \cos\phi, \tag{4.5}$$

where the phase drop ϕ across the Josephson junction is conjugate to the number n of extra Cooper-pairs in the box, $n_g = \frac{C_g V_g}{2e}$ is controlled by the gate voltage V_g (C_g is the gate capacitance and $2e$ is the charge of each Cooper-pair). In practical applications, the Josephson junction in the charge qubit is usually replaced by a DC superconducting quantum interference device (SQUID) to make it easier to control the qubit. In a voltage range near a degeneracy point, the system can be approximated as a qubit with the following Hamiltonian:

$$H = f(V_g)\sigma_z - g(\Phi)\sigma_x. \tag{4.6}$$

For a practical superconducting qubit, the existence of uncertainties is unavoidable (e.g., fluctuations in the Josephson coupling energy and the charging energy, or uncertainties in the magnetic flux). We assume that possible uncertainties exist in both $f(V_g)$ and $g(\Phi)$. Suppose that the factors $f(V_g)$ and $g(\Phi)$ can be controlled by adjusting external parameters. Since E_J could be around ten GHz and E_C could be around one hundred GHz (e.g., the experiment in [45] used $E_{J1} = 13.4$ GHz, $E_{J2} = 9.1$ GHz, $E_{C1} = 117$ GHz and $E_{C2} = 152$ GHz), we assume $f(V_g) \in [0, 40]$ GHz and $g(\Phi) \in [0, 9.1]$ GHz. The practical control terms are $\bar{f}(V_g) = \varepsilon^z f(V_g)$ and $\bar{g}(\Phi) = \varepsilon^x g(\Phi)$ (where the uncertain parameters $\varepsilon^z \in [-\mathcal{E}^z, \mathcal{E}^z]$ and $\varepsilon^x \in [-\mathcal{E}^x, \mathcal{E}^x]$) due to possible multiplicative noises. Here the bounds of uncertainties $\mathcal{E}^z$ and $\mathcal{E}^x$ characterize the maximum ranges of uncertainties in ε^z and ε^x, respectively. The uncertainties can originate from the fluctuations in the magnetic flux Φ, the voltage V_g, the Josephson coupling energy E_J and the charging energy E_C.

As an example, we assume that the initial state is $|\psi_0\rangle = |g\rangle$, and the target state is either $|\psi_{\text{target}}\rangle = |e\rangle$ or $|\psi_{\text{target}}\rangle = \frac{1}{\sqrt{2}}(|g\rangle + |e\rangle)$. Let the operation time be $T = 1$ ns. Now we employ the SLC method to learn an optimal control field for manipulating the charge qubit system from the initial state to a target state. The time interval $t \in$

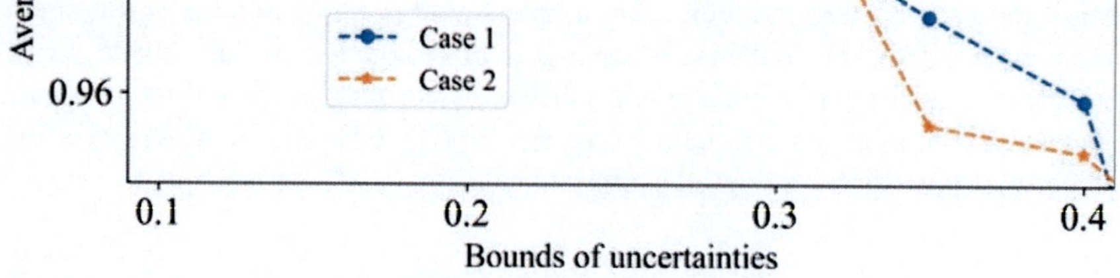

Fig. 4.1 Average fidelity versus the bounds of uncertainties $\mathcal{E}^z$ and $\mathcal{E}^x$ for charge qubits. The uncertainty parameters ε^z and ε^x have uniform distributions in $[-\mathcal{E}, \mathcal{E}]$ (i.e., we assume $\mathcal{E}^z = \mathcal{E}^x = \mathcal{E}$). Case 1: $|\psi_0\rangle = |g\rangle$ and $|\psi_{\text{target}}\rangle = |e\rangle$; Case 2: $|\psi_0\rangle = |g\rangle$ and $|\psi_{\text{target}}\rangle = \frac{1}{\sqrt{2}}(|g\rangle + |e\rangle)$. Every average fidelity is calculated using 5000 tested samples

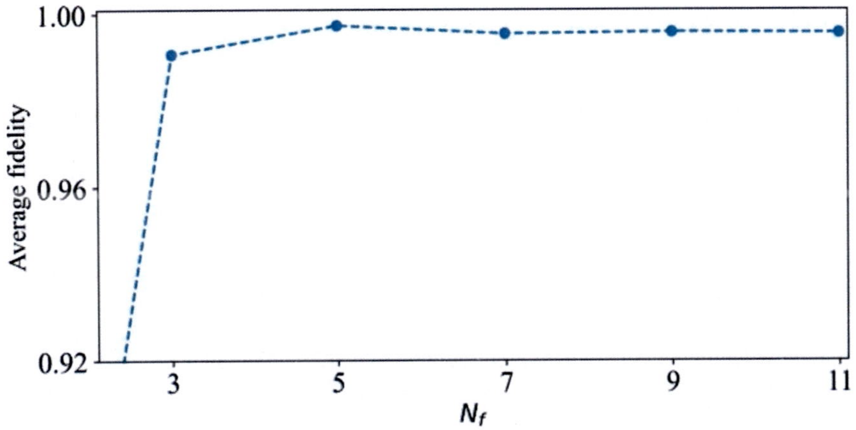

Fig. 4.2 Average fidelity versus the number N_f of values for ε^x and ε^z ($N_f = N_x = N_z$). Here, $\mathcal{E}^z = \mathcal{E}^x = 0.15$ (i.e., 30% fluctuations), $|\psi_0\rangle = |g\rangle$ and $|\psi_{\text{target}}\rangle = \frac{1}{\sqrt{2}}(|g\rangle + |e\rangle)$. Every average fidelity is calculated using 5000 samples

[0, 1] ns is equally divided into 100 smaller time intervals. Without loss of generality, we assume ε^x and ε^z to have uniform distributions and have the same bound of uncertainties (i.e., $\mathcal{E}^x = \mathcal{E}^z$). An augmented system is constructed by selecting $N_x = 5$ values for ε^x and $N_z = 5$ values for ε^z. The initial control fields are $f(V_g) = \sin t + \cos t + 20$ and $g(\Phi) = \sin t + \cos t + 5$. The learning algorithm runs for about 7000 iterations for $|\psi_{\text{target}}\rangle = |e\rangle$ (4000 iterations for $|\psi_{\text{target}}\rangle = \frac{1}{\sqrt{2}}(|g\rangle + |e\rangle)$) before it converges to the optimal control fields. After the optimal control fields are learned for the augmented system, they are applied to 5000 samples generated by stochastically selecting the values of the uncertainty parameters for evaluating the performance. The relationship between the average fidelity and the bounds of the uncertainties is shown in Fig. 4.1. Although the performance decreases when increasing the bounds on the uncertainties, the "smart" fields can still drive the system from the initial state $|\psi_0\rangle = |g\rangle$ to a given target state with high fidelity (the average fidelity is $\bar{F} = 0.9909$ for $|\psi_{\text{target}}\rangle = |e\rangle$, and $\bar{F} = 0.9952$ for $|\psi_{\text{target}}\rangle = \frac{1}{\sqrt{2}}(|g\rangle + |e\rangle)$) even though the bound on the uncertainties is 0.25 (i.e., 50% fluctuations relative to the nominal value).

We also test the relationship between the number of values N_f for ε^x and ε^z ($N_x = N_z = N_f$) and the average fidelity for bounds on the uncertainties $\mathcal{E}^z = \mathcal{E}^x = 0.15$. The performance is shown in Fig. 4.2. It is clear that the performance is excellent for $N_f = 5$ or 7. Although it is possible to improve the performance through using more samples, too many samples will cost more computation resources and spend too much time for learning a set of optimal fields. When increasing the number of uncertainty parameters, the advantage using a smaller N_f is more predominant. Hence, we choose $N_f = 5$ for each uncertainty parameter in all of the following numerical results in this section.

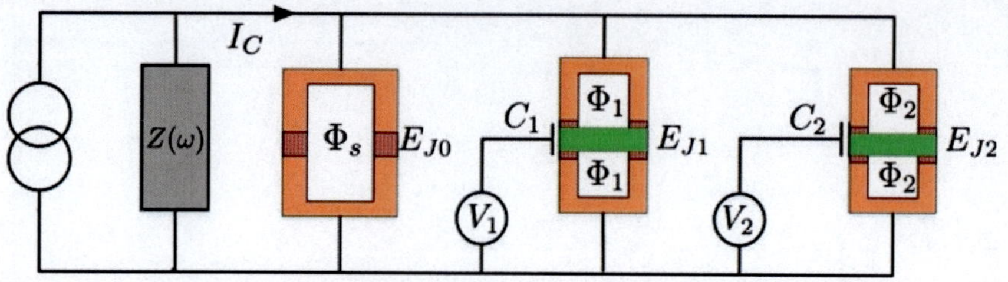

Fig. 4.3 A coupled-qubit circuit with a biased-current source of impedance $Z(\omega)$ [65]. Two charge qubits are coupled by the DC SQUID with two junctions with large E_{J0}

Now, we consider the coupled-qubit circuit in [65] where a symmetric DC SQUID with two sufficiently large junctions is used to couple two charge qubits (see Fig. 4.3). Each qubit is realized by a Cooper-pair box with Josephson coupling energy E_{Jj} and capacitance C_{Jj} $(j = 1, 2)$. Each Cooper-pair box is biased by an applied voltage V_j through the gate capacitance C_j $(j = 1, 2)$. We apply a flux Φ_s inside the large-junction DC SQUID loop with two junctions of large E_{J0}. The Hamiltonian of the coupled charge qubits can be described as

$$H = f(V_1)\sigma_z^{(1)} + f(V_2)\sigma_z^{(2)} - g(\Phi_1)\sigma_x^{(1)} - g(\Phi_2)\sigma_x^{(2)} - \chi(t)\sigma_x^{(1)}\sigma_x^{(2)}, \quad (4.7)$$

where $\sigma_\beta^{(i)}$ with $\beta = x, z$ and $i = 1, 2$ is Pauli matrix on the ith qubit. Due to possible uncertainties, we assume that the Hamiltonian for practical systems is

$$\begin{aligned} H = &(1+\varepsilon_1)f(V_1)\sigma_z^{(1)} + (1+\varepsilon_2)f(V_2)\sigma_z^{(2)} - (1+\varepsilon_3)g(\Phi_1)\sigma_x^{(1)} \\ &-(1+\varepsilon_4)g(\Phi_2)\sigma_x^{(2)} - (1+\varepsilon_5)\chi(t)\sigma_x^{(1)}\sigma_x^{(2)}, \end{aligned} \quad (4.8)$$

where the uncertainty parameters $\varepsilon_j \in [-\mathcal{E}_j, \mathcal{E}_j]$ $(j = 1, 2, 3, 4, 5)$.

We first let $g(\Phi_1) = g(\Phi_2) = 9.1$ GHz, the control terms $f(V_1) \in [0, 40]$ GHz, $f(V_2) \in [0, 40]$ GHz, $|\chi(t)| \leq 0.5$ GHz and $\varepsilon_5(t) \equiv 1$. The operation time is $T = 2$ ns. We assume that the uncertainty parameters ε_j $(j = 1, 2, 3, 4)$ may be time varying. Hence, ε_3 and ε_4 may correspond to time-varying additive uncertainties. As an illustrative example, we let $\varepsilon_j = 1 - \vartheta_j \cos t$, where each ϑ_j has a uniform distribution in the interval $[-\mathcal{E}_j, \mathcal{E}_j]$. For simplicity, we assume $\varepsilon_1 = \varepsilon_2$, $\varepsilon_3 = \varepsilon_4$ and $\mathcal{E}_1 = \mathcal{E}_2 = \mathcal{E}_3 = \mathcal{E}_4 = \mathcal{E}$. We now consider a controlled-phase-shift gate operation on an initial state

$$|\psi_0\rangle = \alpha_1|g, g\rangle + \alpha_2|g, e\rangle + \alpha_3|e, g\rangle + \alpha_4|e, e\rangle;$$

i.e., the target state is

$$|\psi_{\text{target}}\rangle = \alpha_1|g, g\rangle + \alpha_2|g, e\rangle + \alpha_3|e, g\rangle - \alpha_4|e, e\rangle.$$

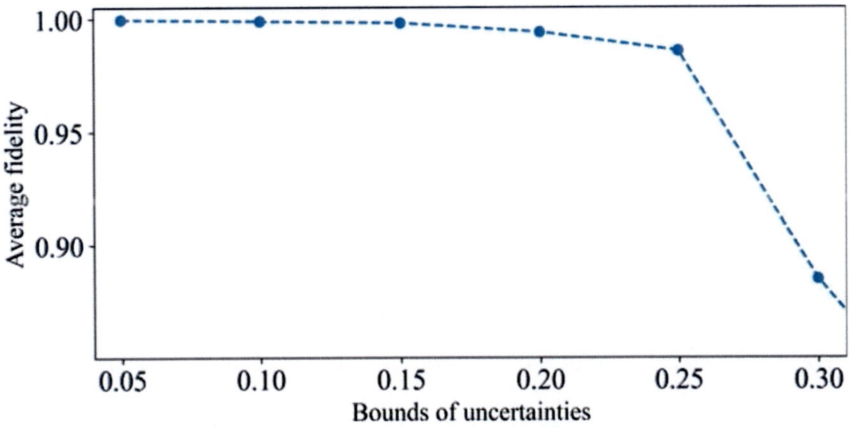

Fig. 4.4 Average fidelity versus the bound on the uncertainty $\mathcal{E}$ for two coupled charge qubits with a biased-current source. The uncertainty parameters $\varepsilon_j = 1 - \vartheta_j \cos t$ $(j = 1, 2, 3, 4)$, where each ϑ_j has a uniform distribution in $[-\mathcal{E}, \mathcal{E}]$. Here we assume $\varepsilon_1 = \varepsilon_2$ and $\varepsilon_3 = \varepsilon_4$. The initial state $|\psi_0\rangle = 0.7|g, g\rangle + 0.1|g, e\rangle + 0.7\mathrm{i}|e, g\rangle + 0.1\mathrm{i}|e, e\rangle$ and the target state $|\psi_{\text{target}}\rangle = 0.7|g, g\rangle + 0.1|g, e\rangle + 0.7\mathrm{i}|e, g\rangle - 0.1\mathrm{i}|e, e\rangle$. Each average fidelity is calculated using 5000 samples

In particular, we let $\alpha_1 = 0.7$, $\alpha_2 = 0.1$, $\alpha_3 = 0.7\mathrm{i}$ and $\alpha_4 = 0.1\mathrm{i}$. The time interval $t \in [0, 2]$ ns is equally divided into 200 smaller time intervals. The control fields are initialized as: $f(V_1) = f(V_2) = \sin t + \cos t + 5$, $\chi(t) = 0.25 \sin t$. The learning algorithm runs for about 5000 iterations before the optimal control fields are found. Then the learned fields are applied to 5000 samples that are generated through selecting the values of fluctuation parameters according to a uniform distribution. The performance is shown in Fig. 4.4. Although the performance decreases when increasing the bounds on the uncertainties, the "smart" fields can still drive the system from the initial state

$$|\psi_0\rangle = 0.7\,|g, g\rangle + 0.1\,|g, e\rangle + 0.7\mathrm{i}|e, g\rangle + 0.1\mathrm{i}|e, e\rangle$$

to the target state

$$|\psi_{\text{target}}\rangle = 0.7\,|g, g\rangle + 0.1\,|g, e\rangle + 0.7\mathrm{i}|e, g\rangle - 0.1\mathrm{i}|e, e\rangle$$

with high fidelity (average fidelity 0.9941) even with up to 40% uncertainties.

We also consider the control terms $f(V_1) \in [0, 40]$ GHz, $f(V_2) \in [0, 40]$ GHz, $g(\Phi_1) \in [0, 9.1]$ GHz, $g(\Phi_2) \in [0, 9.1]$ GHz, $|\chi(t)| \leq 0.5$ GHz and each uncertainty parameter ε_j $(j = 1, 2, \ldots, 5)$ in (4.8) has a truncated Gaussian distribution in $[-\mathcal{E}, \mathcal{E}]$. Assume that the probability density function of the truncated Gaussian distribution is

$$p(x, \mu, \sigma, l, r) = \frac{\phi(\frac{x-\mu}{\sigma})}{\sigma\left[\Phi\left(\frac{r-\mu}{\sigma}\right) - \Phi(\frac{l-\mu}{\sigma})\right]},$$

where $\mu = 0, \sigma = \frac{\mathcal{E}}{3}, l = -\mathcal{E}, r = \mathcal{E}$,

$$\phi(x) = (2\pi)^{-1/2}\exp\left(-\frac{1}{2}x^2\right)$$

is the probability density function of the standard normal distribution, and $\Phi(x)$ is its cumulative distribution function.

We now consider a CNOT operation. In particular, we let the initial state be

$$|\psi_0\rangle = \frac{1}{\sqrt{2}}(|g, g\rangle + |e, g\rangle)$$

and the target state be a maximum entangled state

$$|\psi_{\text{target}}\rangle = \frac{1}{\sqrt{2}}(|g, g\rangle + |e, e\rangle).$$

The operation time chosen here is $T = 2$ ns. In the training step, the uncertainties are uniformly sampled. However, in the testing step the samples are selected by sampling the uncertainty parameters with a truncated Gaussian distribution. For simplicity, we let $\varepsilon_1 = \varepsilon_2$ and $\varepsilon_3 = \varepsilon_4$. The performance is shown in Fig. 4.5. Compared with Fig. 4.4, better performance is achieved in this example since more control degrees of freedom are available. A set of "smart" fields is shown in Fig. 4.6 for $\mathcal{E} = 0.25$ (i.e., 50% fluctuations).

Remark 4.1 As examples, we only consider that each possible uncertainty parameter has a uniform distribution or a truncated Gaussian distribution in the testing step. However, the proposed method also works well for other time-varying or time-invariant distributions. Numerical results show that, in the training step, sampling uncertainty parameters according to simple uniform distributions can achieve excellent performance for these cases where the uncertainty parameters have other distributions.

4.2.2 *Robust Control of Photoassociation*

This section investigates a robust control problem for molecular systems and we apply the SLC approach to the photoassociation of O+H. Photoassociation is a fundamental task in the area of laser control of chemical reactions [66] and has wide applications in forming ultracold molecules and achieving ultralow temperature [31, 32, 51]. A basic setting in controlling photoassociation processes is to let two collid-

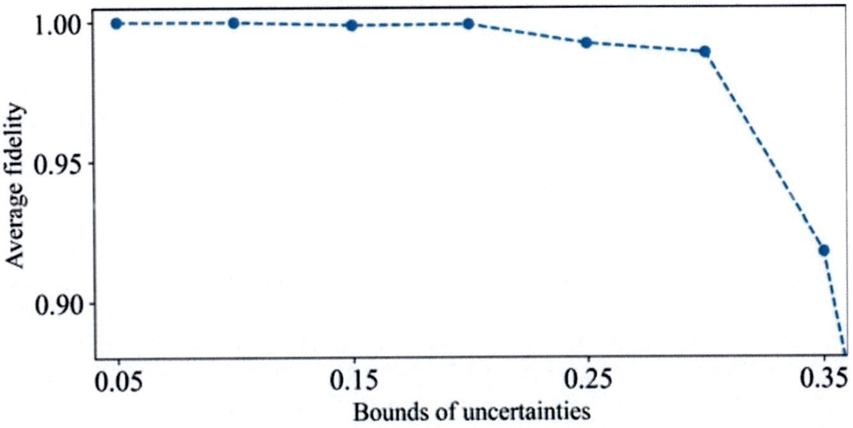

Fig. 4.5 Average fidelity versus the bound on the uncertainties $\mathcal{E}$ for two coupled charge qubits with a biased-current source. Each ε_j $(j = 1, 2, 3, 4, 5)$ has a truncated Gaussian distribution in $[-\mathcal{E}, \mathcal{E}]$, and we assume $\varepsilon_1 = \varepsilon_2$ and $\varepsilon_3 = \varepsilon_4$. The initial state is $|\psi_0\rangle = \frac{1}{\sqrt{2}}(|g, g\rangle + |e, g\rangle)$ and the target state is $|\psi_{\text{target}}\rangle = \frac{1}{\sqrt{2}}(|g, g\rangle + |e, e\rangle)$. Each average fidelity is calculated using 5000 samples

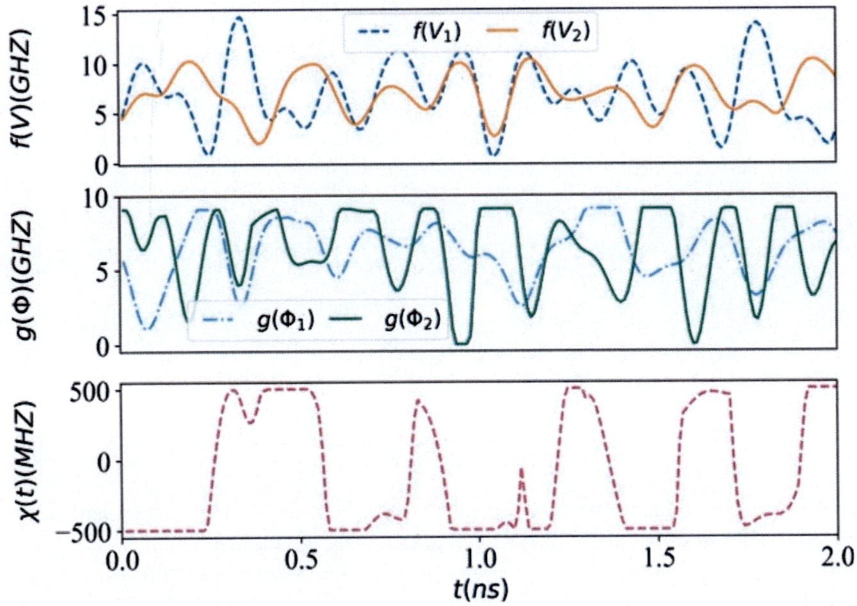

Fig. 4.6 A set of "smart" fields $f(V)$, $g(\Phi)$ and $\chi(t)$ for the problem of coupled-qubit circuit with a biased-current source when the bound on the uncertainty is very large with $\mathcal{E} = 0.25$ (i.e., 50% fluctuations)

ing atoms interact with an applied field to create a molecule [29]. Here, we consider photoassociation in the presence of infrared radiation to produce heteronuclear OH diatomic molecules directly in the electronic ground state [9]. This process is driven by a laser interacting with the colliding atoms via the permanent dipole moment associated with the collision. Optimal laser control of molecular photoassociation along with vibrational levels has been investigated previously using the Morse potential as the interatomic potential [9]. We consider the existence of uncertainties in the depth of the Morse potential in this model and aim to find a robust shaped laser pulse to associate the collisional O and H to the ground vibrational state. Such a laser pulse is searched for by employing a number of samples with different potential curves. In the numerical simulation, we adopt the fast-kick-off algorithm [36] based on the two-point boundary-value quantum control paradigm (TBQCP) method [24, 25]. For the photoassociation O+H→OH in the one-state model represented by a potential energy curve, the zero point of energy is chosen to be the asymptote of the potential curve (the detailed model can be found in [69]). We aim to find a robust control field driving photoassociation efficiently via discretizing the uncertainties to some samples. The dynamics can be expressed in terms of a one-dimensional nuclear wave packet whose evolution is governed by the time-dependent Schrödinger equation.

$$\mathrm{i}\frac{\partial}{\partial t}\psi(R,t) = H\psi(R,t). \tag{4.9}$$

Here, R denotes the internuclear distance between oxygen and hydrogen, $H = \mathrm{diag}(H_1, \ldots, H_N)$ denotes the Hamiltonian of N samples with H_j the Hamiltonian of the jth sample, and

$$\psi(R,t) = \begin{bmatrix} \psi_1(R,t) \\ \cdots \\ \psi_N(R,t) \end{bmatrix}$$

denotes the nuclear wave packet of the system with $\psi_j(R,t)$ the wave packet of the jth sample. The Hamiltonian matrix of each sample is given by

$$H_j = -\frac{1}{2m}\frac{\partial^2}{\partial R^2} + V_j(R) - \mu_j(R)u(t),\ j = 1, \ldots, N, \tag{4.10}$$

where m is the reduced mass of the OH system, $u(t)$ is the time-dependent laser pulse which is assumed to be linearly polarized along the molecular axis, and $V_j(R)$ and $\mu_j(R)$ are the potential energy and permanent dipole moment of the jth sample, respectively.

We first consider robust control with the uncertainties in the potential curve. The potential energy $V_j(R)$ is chosen to be a Morse function, and we consider the uncertainties of well depth $D_e \in [D_e^0 - \mathcal{E}, D_e^0 + \mathcal{E}]$ with the reference well depth D_e^0 and the maximum uncertainty $\mathcal{E}$. Hence, the potential curve for each sample is

$$V_j(R) = \frac{D_e^j}{D_e^0} V_0(R), \; j = 1, \ldots, N, \tag{4.11}$$

where

$$D_e^j = D_e^0 - \mathcal{E} + \frac{2(j-1)\mathcal{E}}{N-1}$$

and the reference potential

$$V_0(R) = D_e^0[e^{-2\alpha(R-R_e)} - 2e^{-\alpha(R-R_e)}], \tag{4.12}$$

where R_e is the equilibrium position, and α^{-1} is the parameter to depict potential range. Here we assume there are no uncertainties in the dipole moment, i.e., $\mu_j(R) = \mu(R)$ for $j = 1, \ldots, N$, and

$$\mu(R) = QR_e^{-\frac{R}{R_d}}, \tag{4.13}$$

where Q denotes the effective charge, and R_d denotes the range of the dipole interaction.

The initial state of each sample is taken to be a Gaussian wave packet

$$\Phi_j^i(R) = \left(\frac{2}{\pi\sigma_0}\right)^{\frac{1}{4}} \exp\left[\mathrm{i}k_0 R - \left(\frac{R-R_0}{\sigma_0}\right)^2\right], \; j = 1, \ldots, N, \tag{4.14}$$

where σ_0 denotes the width, R_0 the central position, and k_0 the momentum. The collision energy E_c associated with this incoming Gaussian wave packet is

$$E_c = \frac{1}{2m}(k_0^2 + \frac{1}{\sigma_0^2}). \tag{4.15}$$

In simulation, the wave packets are discretized into 5120 grid points extending to 150 a_0 in order to prevent the wave packet from reflecting to the boundary. The eigenvalues and eigenvectors of the bound states for the $1s\sigma$ potential are calculated by the Fourier-grid-Hamiltonian (FGH) method [41], and we also use the same spatial discretization for the time-dependent calculations. The Schrödinger Eq. (4.9) is solved using the second-order split-operator method [18, 19, 33]. The time step is chosen as 0.01 fs to ensure the required accuracy.

Consider the photoassociation from an initial state $\Phi^i(R) = [\Phi_1^i(R), \ldots, \Phi_N^i(R)]$ to be driven to a target state $\Phi^f(R) = [\Phi_1^f(R), \ldots, \Phi_N^f(R)]$ (usually a bound state for each sample) at some final time $T > 0$. We aim to find a control field to maximize the average photoassociation probability over the selected samples

$$K[T] = \frac{\sum_{j=1}^N |\langle \Phi_j^f(R)|\psi_j(T)\rangle|^2}{N}. \tag{4.16}$$

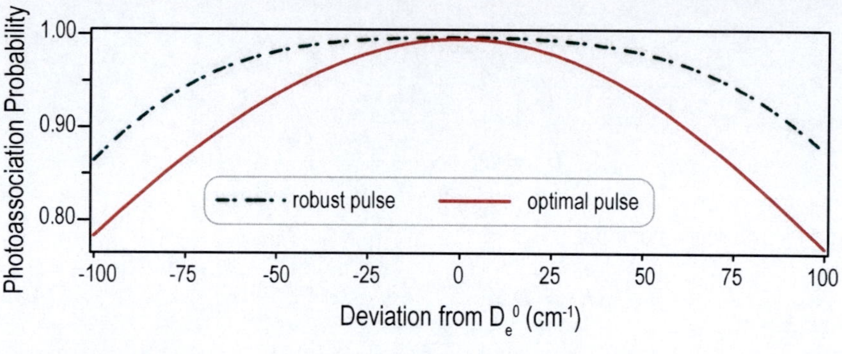

Fig. 4.7 Photoassociation probabilities versus deviation from D_e^0 under the action of the optimal laser pulse (black solid line) and the robust laser pulse (red dashed line)

The control field $u(t)$ can be efficiently optimized iteratively based on a monotonically convergent fast-kick-off TBQCP algorithm [36] with a zero-area constraint [55], and is updated using the following recurrence relation:

$$u^{[l]}(t) = u^{[l-1]}(t) + \eta S(t)\left[f_\mu^{[l]}(t) - \theta A_{l-1}\right], l = 1, 2, \ldots, \tag{4.17}$$

where $\eta > 0$ is the step size of updating the field, $\theta > 0$ penalizes the area of the field (we choose $\theta = 10^{-5}$ here), $u^{[l]}(t)$ is the field of the lth iteration, the index $l = 0$ corresponds to the trial field, $S(t)$ is the shape function which is usually chosen to be the form of a sine (cosine) square function to enforce a smooth switch on and off for the control pulse, and $A_l = \int_0^T u^{[l]}(t)dt$ is the area of $u^{[l]}(t)$.

In the calculations, the parameters of the potential curve and dipole moment are taken from [9], and we choose $R_0 = 13.4$ a.u. and $\sigma_0 = 5.12$ a.u. and $E_c = 0.025$ eV for the initial state. The target state is chosen to be the lowest vibrational state associated with each sample potential. The initial trial field $u^{[0]}(t)$ is chosen to have a cosine square envelope shape with chirped frequency

$$u^{[0]}(t) = u_0^{[0]} \cos^2\left[\frac{\pi(t - t_0)}{\tau}\right] \cos\left[\omega_0(t - t_0) + \gamma(t - t_0)^2\right], \tag{4.18}$$

where $u_0^{[0]} = 265.0$ MV/cm ($\approx 5.15 \times 10^{-2}$ a.u.) is the peak amplitude, $\tau = 1770$ fs is the duration of the pulse, $\omega_0 = 2543$ cm^{-1} ($\approx 1.16 \times 10^{-2}$ a.u.) is the carrier frequency, $t_0 = 900$ fs denotes the time corresponding to the peak amplitude, and $\gamma = 2.71 \times 10^{-7}$ a.u. is the chirped rate. The calculations are performed over the time interval [0, $T = 1800$] fs.

We first calculate the optimal laser pulses using only one sample with $D_e = D_e^0$. The step size η of optimal control algorithm is set to be 11.0. After 1126 iterations, an optimal laser pulse achieves a photoassociation probability of 0.995 from 0.015 based on the trial field. The photoassociation probabilities versus deviation Δ_{D_e} from

D_e^0 ($\Delta_{D_e} = D_e - D_e^0$) are shown in Fig. 4.7. The photoassociation probability drops as $|\Delta_{D_e}|$ increases, from 0.995 at $\Delta_{D_e} = 0$ cm^{-1} to less than 0.780 at $\Delta_{D_e} = \pm 100$ cm^{-1}.

Now we present numerical results for robust control with the uncertainties of the potential curve. We select $N = 11$ samples with $\mathcal{E} = 100$ cm^{-1} and $\eta = 1.0$ to maximize $K[T]$ in (4.16) with zero-area constraint. We choose the optimal laser pulse as the trial field and achieve $K[T] = 0.985$ after almost 50000 iterations. We calculate the photoassociation probabilities of the same 200 samples used for the optimal control in Fig. 4.7 to examine the robustness of the learned field. For comparison, the photoassociation probabilities versus Δ_{D_e} achieved by the robust laser pulse are also shown in Fig. 4.7, most of which are higher than those achieved by the optimal control field.

We may also consider possible uncertainties in the laser amplitudes, i.e., $\varepsilon u(t)$ with uncertainty bound $\mathcal{E}$ satisfying $\varepsilon \in [-\mathcal{E}, \mathcal{E}]$ $(0 < \mathcal{E} < 1)$. The Hamiltonian for each sample can be written as

$$H_j = -\frac{1}{2m}\frac{\partial^2}{\partial R^2} + V_0(R) - (1 + \varepsilon_j)\mu(R)u(t),\ j = 1, \ldots, N, \tag{4.19}$$

where

$$\varepsilon_j = -\mathcal{E} + \frac{2(j-1)\mathcal{E}}{N-1},\ j = 1, \ldots, N. \tag{4.20}$$

We may calculate the robust control pulse with the uncertainties in the laser amplitude taken into account by selecting $N = 11$ samples. With the trial field, the photoassociation probability goes down dramatically as ε increases or decreases from $\varepsilon = 0$. The value drops to only 0.03 at $\varepsilon = -0.1$ and 0.27 at $\varepsilon = 0.1$ with an average value of 0.53 over these 200 samples. However, when the robust control field is implemented, the average photoassociation probability goes up to 0.98, showing the clear advantage in robustness performance.

4.2.3 *Synchronizing Laser with Molecules for Charge Transfer*

Charge transfer processes play a fundamental role in the elementary processes of physics, chemistry, and biology [35]. For example, it is the main mechanism for recombination, ionization, and excitation in astrophysical and laboratory plasmas [8]. Laser fields have been used for modifying electron transfer processes in ion-atom collisions [7, 22, 23]. Symmetric collisions between protons and atomic hydrogen have been tested as a platform for the basic principles of laser-assisted electron transfer [5, 39]. In asymmetric ion-atom collisions, the charge exchange cross section may be strongly modified by a suitable laser field [46]. Even for a relatively weak laser intensity, a substantial increase of the charge transfer cross section has been observed [30, 43]. Optimal control theory can be applied to achieve a desired charge transfer outcome [70]. The solution of time-dependent Schrödinger equation depends on the initial condition. Therefore, a control field is effective only associated with a specific state at a specific time, and the efficiency of charge transfer may decrease significantly without the precise arrival of the control laser pulse. Although synchronizing the control pulse and collision plays a significant role, this feature cannot be directly controlled. Here, we present a method of searching for control pulses that are robust to synchronizing H^++D charge transfer processes. In particular, we consider constructing a robust field by an SLC method [3, 11, 12] in synchronizing the asymmetric H^++D(1s) collisional system [15, 16]. The samples are obtained by considering different pulse arrival times. Similar to Sect. 4.2.2, we adopt the fast-kick-off algorithm [36] based on the TBQCP method for simulation [24, 25, 37]. Furthermore, we impose a zero-area constraint on the control algorithm to ensure that the laser field can be readily implemented experimentally [55]. Our objective is to maximize the average probability over selected samples when the transient HD^+ complex dissociates into the designated outgoing channel at a large internuclear distance where the non-adiabatic coupling vanishes.

Assume that the scenario of colliding H^++D→ H+D^+ is described by a one-dimensional two-state model based on an adiabatic representation [15]. The zero point of energy is chosen as the asymptote of the $1s\sigma$ state. The input wave packet $\Phi_0^i(R)$ has a Gaussian form

$$\Phi_0^i(R) = \left(\frac{2}{\pi\sigma_i}\right)^{\frac{1}{4}} \exp\left[ik_i R - \left(\frac{R-R_i}{\sigma_i}\right)^2\right], \tag{4.21}$$

where R denotes the internuclear distance between the two atoms, σ_i the width, R_i the central position, and k_i the momentum. The collision energy E_i associated with this incoming Gaussian wave packet is

$$E_i = \frac{1}{2m}\left(k_i^2 + \frac{1}{\sigma_i^2}\right), \tag{4.22}$$

where m is the reduced mass of the collisional system.

We assume that the expected arrival time of the laser pulse is 0. The real time t^ε when the laser pulse actually arrives is earlier or later than its expected arrival time, i.e., $t^\varepsilon \in [-\mathcal{E}_t, \mathcal{E}_t]$ with $\mathcal{E}_t \geq 0$. Thus, t^ε can be regarded as the time deviation from the expected arrival time and $\mathcal{E}_t$ is the maximum time deviation. In this scenario, the state of the system $\phi(t = t^\varepsilon)$ depends on the time shift, and $\phi(t = t^\varepsilon)$ can be obtained by solving field-free Schrödinger equation forward ($t^\varepsilon > 0$) or backward ($t^\varepsilon < 0$) from $\phi(t = 0) = \Phi_0^i$. We employ an SLC design method to find a robust laser pulse against the time deviation. Here we evenly discretize t^ε into N samples, thus,

$$t_j^{\varepsilon} = -\mathcal{E}_t + \frac{2(j-1)\mathcal{E}_t}{N-1}$$

for $j = 1, \dots, N$ and $N > 1$. We let $t^{\varepsilon} = 0$ when $N = 1$. We use $\phi(t = t_j^{\varepsilon})$ as the initial states in the optimal control calculations to construct a robust field arriving at time 0, which can drive each sample to a desired target state with high probability.

The dynamics of an N-sample two-state model in the presence of the same laser field is governed by the time-dependent equation [70]

$$\mathrm{i}\frac{\partial}{\partial t}\psi(R, t) = H\psi(R, t),$$

where the Hamiltonian of the system $H = \mathrm{diag}(H_0, \dots, H_0)$ contains N blocks H_0,

$$\psi(R, t) = \begin{bmatrix} \psi_1(R, t) \\ \cdots \\ \psi_N(R, t) \end{bmatrix}$$

is the associated nuclear wave packet. Here

$$\psi_j(R, t) = \begin{bmatrix} \psi_j^g(R, t) \\ \psi_j^e(R, t) \end{bmatrix} \tag{4.23}$$

$(j = 1, \dots, N)$ denotes the wave packets of the jth sample, and the subscripts g and e correspond to the electronic states $1s\sigma$ and $2p\sigma$, respectively, and the Hamiltonian H_0 of each sample is identical and can be expressed as

$$\begin{aligned} H_0 = &\begin{bmatrix} V_g^a(R) + \Delta - \frac{1}{2m}[P_{ge}^2(R) + Q_{gg}(R)] & 0 \\ 0 & V_e^a(R) + \Delta - \frac{1}{2m}[P_{ge}^2(R) + Q_{ee}(R)] \end{bmatrix} \\ &- \frac{1}{2m}\left\{\begin{bmatrix} \frac{\partial}{\partial R} & 0 \\ 0 & \frac{\partial}{\partial R} \end{bmatrix} + \begin{bmatrix} 0 & P_{ge}(R) \\ P_{eg}(R) & 0 \end{bmatrix}\right\}^2 - u(t)\begin{bmatrix} \mu_{gg}(R) & \mu_{ge}(R) \\ \mu_{eg}(R) & \mu_{ee}(R) \end{bmatrix}. \end{aligned} \tag{4.24}$$

Here $V_g^a(R)$ and $V_e^a(R)$ are adiabatic potential curves, $\mu_{ge} = \mu_{eg}$ is the transition dipole moment, and μ_{gg} and μ_{ee} are the permanent dipole moments; see [68] for details. Moreover, $P_{ge} = -P_{eg}$, Q_{gg} and Q_{ee} account for the non-adiabatic effects resulting from the nuclear radial motion [68]. Here, we only consider the uncertainty in the laser arrival time, and the laser field $u(t)$ is assumed to be linearly polarized along the molecular axis. We use the second-order split-operator method [18, 19, 33] to solve (4.23) by an adiabatic-diabatic transformation [70] to obtain the wave packet $\psi(R, t)$ of the jth sample with its initial condition $\Phi^i = [\phi(t = t_1^{\varepsilon}), \dots, \phi(t = t_N^{\varepsilon})]$.

We consider the laser-assisted charge transfer H^++D→ H+D^+ of the jth sample with the associated initial state $\Phi_j^i(R) = \psi_j^g(R, 0)$ in the incoming channel $1s\sigma$ to a target state $\Phi_j^f(R) = \psi_j^e(R, T)$ in the outgoing channel $2p\sigma$ at some final time $T > 0$. Here $\Phi_j^i(R < R_{out}) = 0$ and $\Phi_j^f(R < R_{out}) = 0$, where R_{out} ≈26 a.u. and the non-adiabatic radial couplings and the transition dipole moment are negligible for

$R > R_{out}$. The average probability over the selected samples in the charge transfer state at the final time T may be defined as

$$K[T] = \frac{\sum_{j=1}^{N} \int_{R_{out}}^{\infty} |\Phi_j^f(R)|^2 dR}{N}. \tag{4.25}$$

Note that the same target time T is chosen for each member of the ensemble despite the varying arrival time of the field for each member. In simulation, we use an adaptive target scheme in which the target is updated according to the renormalized fragmentary yield in the exit channel at each iteration until an optimal control field is attained [67].

We employ the monotonically convergent fast-kick-off TBQCP algorithm [36] with zero-area constraint [55] to search for an optimal control $u(t)$ to maximize the probability $K[T]$. The wave packets are discretized into 5120 evenly spaced grid points extending to 300 a_0 in order to prevent the wave packet from encountering the boundary. The time step size is chosen as 0.01 fs to ensure good accuracy. The eigenvalues and eigenvectors of the bound states for the $1s\sigma$ potential are calculated by the FGH method [41]. The calculations are performed over the time interval $[-200, 200]$ fs. For the optimal control simulations, we choose $R_i = 35$ a.u. and $\sigma_i = 5.1$ a.u. for the initial state. We aim to design robust fields to achieve high probability charge transfer with a time arrival uncertainty of $t^{\varepsilon} \in [-20, 20]$ fs. Even though 20 fs is small physically, it is still a considerable portion of the reaction time. Numerical results show that the initial state significantly changes if the arrival time of the laser pulse is earlier or later than its expected arrival time zero [68].

The initial trial field $\varepsilon^{[0]}(t)$ is chosen to have a Gaussian envelope shape

$$u^{[0]}(t) = u_0^{[0]} \exp\left[-4\ln 2\left(\frac{t - t_0}{\tau}\right)^2\right] \cos[\omega_0(t - t_0)], \tag{4.26}$$

where $u_0^{[0]} = 1.94 \times 10^{-3}$ a.u. (≈ 10 MV/cm or the intensity $I_0 = 1.32 \times 10^{11}$ W/cm^2) is the peak amplitude, $\tau = 56.6$ fs is the full width at half maximum (FWHM), $\omega_0 = 2366.7$ cm^{-1} is the carrier frequency and the collision energy is E_i=0.22 eV. After 323 iterations, an optimal laser pulse is obtained, achieving a charge transfer probability of 0.995. The charge transfer probability $P(t^{\varepsilon})$ drops rapidly as $|t^{\varepsilon}|$ increases, from 0.995 at $t^{\varepsilon} = 0$ to 0.930 at $t^{\varepsilon} = -20$ fs and 0.936 at $t^{\varepsilon} = 20$ fs, respectively. Then we evenly select 200 samples in the interval [-20, 20] fs to test the robustness of the laser pulse. In contrast, we also calculate the charge transfer probabilities $P_{\text{robust}}(t^{\varepsilon})$ with these 200 samples by the robust laser field. The two boundary probabilities are 0.977 at $t^{\varepsilon} = -20$ fs and 0.982 at $t^{\varepsilon} = 20$ fs, and at most points $P_{\text{robust}}(t^{\varepsilon})$ are higher than their counterparts achieved using the optimal laser pulse, showing better robustness of the laser pulse learned using the SLC method. More details can be found in [68].

4.3 SLC for Robust Generation of Quantum Gates

An important task to implement quantum computation is the realization of quantum gates. It is well known that a suitable set of single-qubit and two-qubit quantum gates can accomplish universal quantum computation. A universal gate set may consist of a quantum phase gate (S gate), a Hadamard gate (H gate), a $\frac{\pi}{8}$ gate ($T_{\frac{\pi}{8}}$ gate), and a CNOT gate [44]. Realizing such a universal gate set is a fundamental objective in quantum computation.

In this section, we first apply the SLC method to guide the design of robust control fields for construction of universal quantum gates [14, 61] and then discuss the b-GRAPE algorithm [60] for robust quantum gate control. In the first task, we aim to generate the set of universal quantum gates $\{S, H, T_{\frac{\pi}{8}}, \text{CNOT}\}$. The results show that the designed control fields are insensitive to different uncertainties in the process of generating quantum gates. The quantum gates with the designed control fields can have improved robustness and reliability.

The evolution of a quantum gate $U(t)$ satisfies

$$\dot{U}(t) = -\mathrm{i}H(t)U(t), \quad U(0) = I. \tag{4.27}$$

Now the objective is to design the Hamiltonian $H(t)$ to robustly steer the unitary $U(t)$ from $U(0) = I$ to the desired target $U_F \in \{H, S, T_{\frac{\pi}{8}}, \text{CNOT}\}$, with high fidelity. The fidelity is defined as:

$$F(U_F, U(T)) = \frac{1}{2^k} |\langle U_F | U(T) \rangle|, \tag{4.28}$$

where k ($k = 1$ or 2 in this section) is the number of qubits involved in the quantum gate and $\langle U_F | U(T) \rangle \equiv \text{Tr}(U_F^\dagger U(T))$.

By applying a gradient-based optimization algorithm, we now consider the problem of generating a high fidelity quantum gate in a given time T. Assume that the Hamiltonian has the following form:

$$H(t) = H_0 + \sum_{m=1}^{M} u_m(t) H_m, \tag{4.29}$$

where H_0 is the free Hamiltonian, H_m ($m = 1, 2, \ldots M$) are related to the control Hamiltonian with the corresponding control pulses $u_m(t)$.

The performance function of the transfer process can be defined as

$$\|U_F - U(T)\|^2 = \|U_F\|^2 - 2\Re\langle U_F | U(T) \rangle + \|U(T)\|^2. \tag{4.30}$$

In practical applications, considering the possible existence of an arbitrary global phase factor $e^{\mathrm{i}\varphi}$, we minimize

$$\|U_F - e^{i\varphi}U(T)\|^2 = \|U_F\|^2 - 2\Re\langle U_F|e^{i\varphi}U(T)\rangle + \|e^{i\varphi}U(T)\|^2, \tag{4.31}$$

which is equivalent to maximize $\Re\langle U_F|e^{i\varphi}U(T)\rangle$. In order to eliminate the influence of the global phase factor, we maximize the performance function

$$J = |\langle U_F|e^{i\varphi}U(T)\rangle|^2. \tag{4.32}$$

Let U_j denote the unitary transformation when the jth pulse u^j is applied. We can decompose $U(T)$ as $U(T) = U_N \dots U_1$. Denote operators A_j and B_j as

$$A_j = U_j \dots U_1,$$

$$B_j = U_{j+1}^\dagger \dots U_N^\dagger U_F = A_j U(T)^\dagger U_F.$$

We have the following relationship:

$$J = \langle B_j|A_j\rangle\langle A_j|B_j\rangle. \tag{4.33}$$

The gradient $\frac{\partial J}{\partial u_m(j)}$ to the first order in Δt is given by

$$\begin{aligned}\frac{\partial J}{\partial u_m(j)} &= -\langle B_j|A_j\rangle\langle i\Delta t H_m A_j|B_j\rangle - \langle B_j|i\Delta t H_m A_j\rangle\langle A_j|B_j\rangle \\ &= -2\Re\{\langle B_j|i\Delta t H_m A_j\rangle\langle A_j|B_j\rangle\}.\end{aligned} \tag{4.34}$$

The optimal control field can be searched for by following the gradient.

The SLC method involves two steps of training and testing. In the training step, we select N samples to train the control fields. These samples are selected according to the distribution of the uncertain parameters (e.g., uniform distribution). For example, when

$$H(t) = (1+\varepsilon_0)H_0 + \sum_{m=1}^{M}(1+\varepsilon_m)u_m(t)H_m, \tag{4.35}$$

an augmented system can be constructed as follows:

$$\begin{bmatrix}\dot U_1(t)\\ \dot U_2(t)\\ \vdots \\ \dot U_N(t)\end{bmatrix} = -i\begin{bmatrix}((1+\varepsilon_{01})H_0 + \sum_{m=1}^{M}(1+\varepsilon_{m1})u_m(t)H_m)U_1(t)\\ ((1+\varepsilon_{02})H_0 + \sum_{m=1}^{M}(1+\varepsilon_{m2})u_m(t)H_m)U_2(t)\\ \vdots \\ ((1+\varepsilon_{0N})H_0 + \sum_{m=1}^{M}(1+\varepsilon_{mN})u_m(t)H_m)U_N(t)\end{bmatrix}, \tag{4.36}$$

where the Hamiltonian of the nth sample is

$$(1+\varepsilon_{0n})H_0 + \sum_{m=1}^{M}(1+\varepsilon_{mn})u_m(t)H_m,$$

with $n = 1, 2, \ldots, N$. The performance function of the augmented system is defined as $J_N(u)$

$$J_N(u) = \frac{1}{N}\sum_{n=1}^{N}\frac{1}{2^k}|\langle U_F|U_n(T)\rangle|. \tag{4.37}$$

The task of the training step is to find an optimal control u^* which maximizes the performance function above. The representative samples for these uncertain parameters can be selected according to the method in Chap. 3. In the testing step, we apply the control field u^* obtained in the training step to a large number of additional samples, which are randomly selected according to the uncertainty parameters. If the average fidelity of all the tested samples is satisfactory, we accept the designed control, which means the quantum gate we construct is robust. In this section, we use 2000 additional samples to test our designed control in this step.

Considering the existence of uncertainties for one-qubit quantum gates, we assume that the Hamiltonian can be described as

$$H(t) = (1 + \varepsilon_0)\omega_0\sigma_z + (1 + \varepsilon_1)\omega_x(t)\sigma_x, \tag{4.38}$$

and both uncertain parameters ε_0 and ε_1 have uniform distributions with the same bound on the uncertainties $\mathcal{E} = 0.2$ (i.e., 40% fluctuations, $\varepsilon_0 \in [-0.2, 0.2]$ and $\varepsilon_1 \in [-0.2, 0.2]$). Using the SLC method [3], an augmented system is constructed by selecting $N_0 = 5$ values for ε_0, and $N_1 = 5$ values for ε_1. The samples are selected as

$$(\varepsilon_0, \varepsilon_1) \in \{(-0.2 + 0.04(2m - 1), -0.2 + 0.04(2n - 1))|m, n = 1, 2, \ldots, 5\}.$$

Figure 4.8 shows the results for three classes of quantum gates: S, H, and $T_{\frac{\pi}{8}}$, respectively. After 100,000 iterations, the precision reaches 0.9976 for the S gate, 0.9979 for the H gate, and 0.9991 for the $T_{\frac{\pi}{8}}$ gate, respectively. The corresponding control fields are shown in Figs. 4.8b, d and f. Then the learned fields are applied to 2000 additional samples that are generated randomly by selecting values of the uncertainty parameters according to a uniform distribution. The average fidelity reaches 0.9973 for the S gate, 0.9976 for the H gate, and 0.9989 for the $T_{\frac{\pi}{8}}$ gate, respectively.

In the laboratory, it may be easier for some quantum systems to generate discrete control pulses with constant amplitudes. Here, we consider the performance using different numbers of control pulses to approximate the fields. For the S gate, the relationship of the number of pulses versus the average fidelity is shown in Fig. 4.9. From Fig. 4.9, it is clear that excellent performance can be achieved even if we only use $20 \sim 40$ control pulses to realize the approximation of the continuous control fields. Hence, we use 40 pulses to implement the control field in the following numerical calculations.

Now, we consider the problem of finding robust control pulses for generating quantum CNOT gates. In particular, we consider the example based on the two coupled superconducting phase qubits in [1], which has also been discussed for the

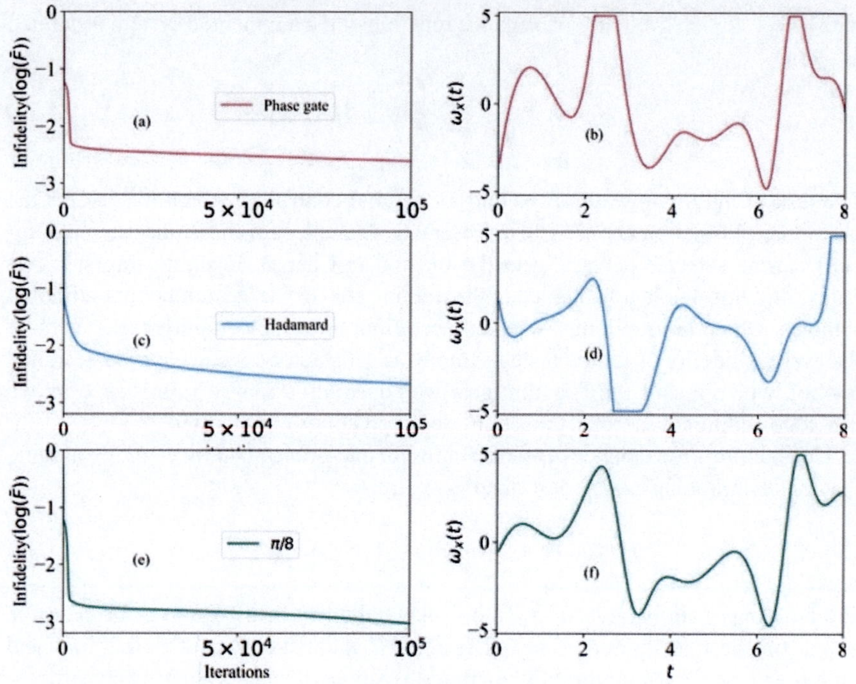

Fig. 4.8 Robust control performance for the S, H, $T_{\frac{\pi}{8}}$ gates with parameter uncertainties on ω_0 and ω_x. **a**, **c**, **e** The infidelity ($\log \bar{F}$) versus iterations; **b**, **d**, **f** Learned robust control fields

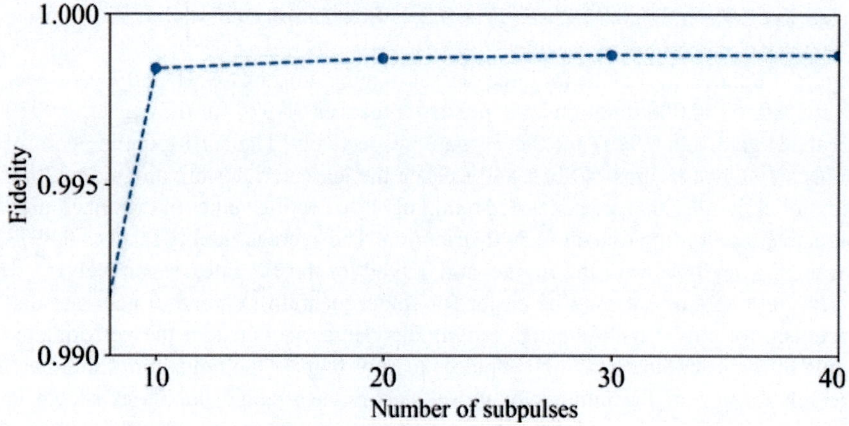

Fig. 4.9 Fidelity versus the number of sub-pulses for the S gate

robust control of quantum states in Sect. 4.2.1. Each phase qubit is a nonlinear resonator built from an Al/AlO$_x$/Al Josephson junction, and two qubits are coupled via a modular four-terminal device (for details, see Fig. 1 in [1]). This four-terminal device is constructed using two nontunable inductors, a fixed mutual inductance and a tunable inductance. The equivalent Hamiltonian can be described as

$$\begin{aligned} H = & \frac{\omega_1(t)}{2}\sigma_z^{(1)} + \frac{\omega_2(t)}{2}\sigma_z^{(2)} + \frac{\omega_3(t)}{2}\sigma_x^{(1)} + \frac{\omega_4(t)}{2}\sigma_x^{(2)} \\ & + \frac{\Omega_c(t)}{2}(\sigma_x^{(1)}\sigma_x^{(2)} + \frac{1}{6\sqrt{M_1 M_2}}\sigma_z^{(1)}\sigma_z^{(2)}), \end{aligned} \tag{4.39}$$

where M_1 and M_2 are the number of levels in the potentials of qubits 1 and 2. The typical values for M_1 and M_2 are $M_1 = M_2 = 5$. We assume that the Hamiltonian has the following form (due to possible uncertainties):

$$\begin{aligned} H = & \frac{(1+\varepsilon_1)\omega_1(t)}{2}\sigma_z^{(1)} + \frac{(1+\varepsilon_2)\omega_2(t)}{2}\sigma_z^{(2)} + \frac{\omega_3}{2}\sigma_x^{(1)} + \frac{\omega_4}{2}\sigma_x^{(2)} \\ & + \frac{(1+\varepsilon_3)\Omega_c(t)}{2}\left(\sigma_x^{(1)}\sigma_x^{(2)} + \frac{1}{30}\sigma_z^{(1)}\sigma_z^{(2)}\right) \end{aligned} \tag{4.40}$$

with $\varepsilon_j \in [-0.2, 0.2]$ $(j = 1, 2, 3)$. Here, we assume that the frequencies $\omega_1(t)$, $\omega_2(t) \in [-5, 5]$ GHz can be adjusted by changing the bias currents of the two phase qubits, and $\Omega_c(t) \in [-500, 500]$ MHz can be adjusted by changing the bias current in the coupler. Let $\omega_3 = \omega_4 = 2$ GHz, the operation time $T = 20$ ns is divided into 40 smaller time intervals, and the step size is 0.1. The initial control fields are $\omega_1(t) = \omega_2(t) = \sin t$ GHz, $\Omega_c(t) = 0.05 \sin t$ GHz. When the uncertainty bounds are 0.2, the results are shown in Fig. 4.10. In the training step, the precision of the CNOT gate can reach 0.9965. Then the average fidelity of 0.9961 can be reached for 2000 additional samples in the testing step.

Now we discuss the b-GRAPE algorithm which can be looked as a further development of the SLC method. Although the SLC method is effective to various tasks such as control and classification of inhomogeneous quantum ensembles, and control of quantum systems with uncertainties, in practice the applicability of the method is limited to low-dimensional quantum systems or uncertain quantum systems with few uncertainty parameters since the computational cost will quickly increase with the increase of uncertainty parameters. In order to solve this issue, Wu et al. [60] presented a batch-based GRAPE (b-GRAPE) algorithm to search robust control fields for quantum control problems inspired by deep neural networks. In this method, the search for robust quantum control fields is formulated as a supervised learning task where the time-ordered quantum evolution can correspond to a layer-ordered neural network, and the controlled quantum evolution can be thought of as training a deep neural network to accomplish this task. In that sense, the requirement of control robustness naturally corresponds to the training goal of achieving a highly generalizable neural network model [60]. Hence, some training techniques for improving the generalizability of deep neural networks can be used to search robust quantum control fields. In particular, the techniques of data augmentation and mini-batch optimization have been used in the development of b-GRAPE algorithm where an unlimited number of training samples is available for our quantum robust control problems,

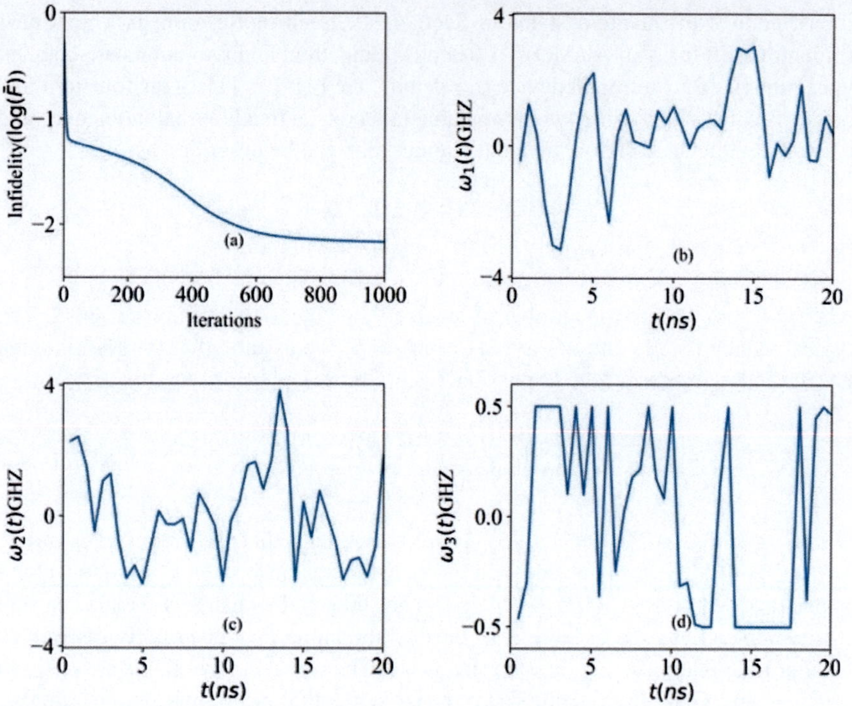

Fig. 4.10 Dong et al. [14] Robust control performance for CNOT gates, with parameter uncertainties on ω_1, ω_2 and Ω_c, $\mathcal{E} = 0.2$. **a** Infidelity ($\log \bar{F}$) versus iterations. **b–d** Learned robust control fields

and noisy and less stable training dynamics can effectively improve generalizability by pulling the search away from non-robust solutions.

Assume that the uncertain quantum system with uncertainty parameter ε is described by

$$\dot{U}(t, \varepsilon) = -\mathrm{i}H(u(t), \varepsilon)U(t), \tag{4.41}$$

where the uncertainty parameter has *a priori* probability distribution $P(\varepsilon)$. We may choose the following performance index function:

$$J = \frac{1}{B}\sum_{n=1}^{B} \|U(T, \varepsilon_n) - U_f\|^2, \tag{4.42}$$

where ε_n is a value of the uncertainty parameter sampled according to the probability distribution $P(\varepsilon)$. The b-GRAPE algorithm can be outlined as follows:

Step 1: Choose a proper batch size B, and sample the batches according to the probability distribution $P(\varepsilon)$, forming

$$S^{(j)} = \left\{\varepsilon_1^{(j)}, \ldots, \varepsilon_B^{(j)}\right\}$$

with the iteration index j.

Step 2: Calculate the gradient direction in each iteration:

$$g\left[u(t), S^{(j)}\right] = \frac{\delta J\left[u(t), S^{(j)}\right]}{\delta u(t)} = \frac{1}{B}\sum_{n=1}^{B} \frac{\delta \left\| U\left(T, \varepsilon_n^{(j)}\right) - U_f \right\|}{\delta u(t)}.$$

Step 3: Introduce a momentum term to stabilize the training dynamics to reduce the variance of the performance index function and update the control as follows:

$$u^{(j+1)}(t) = u^{(j)}(t) - \alpha_j \cdot \left\{\lambda g\left[u^{(j)}(t), S^{(j)}\right] + (1-\lambda) g\left[u^{(j-1)}(t), S^{(j-1)}\right]\right\}.$$

Compared with choosing fixed representative samples in the SLC method (also called as the s-GRAPE method in [60]), b-GRAPE randomly selecting samples for calculating the gradient direction, which has the potential to provide improved performance by exploring the richness and diversity of samples. This has also been demonstrated in the numerical results in [60]. Consider a three-qubit quantum system with parametric uncertainties in its Hamiltonian:

$$H(t) = \left(1 + \varepsilon^{(1)}\right)\sigma_{1z}\sigma_{2z} + \left(1 + \varepsilon^{(2)}\right)\sigma_{2z}\sigma_{3z} + \sum_{k=1}^{3}\left[u_{kx}(t)\sigma_{kx} + u_{ky}(t)\sigma_{ky}\right], \tag{4.43}$$

where $\sigma_{k\beta}$ with $k = 1, 2, 3$ and $\beta = x, y, z$ are the Pauli matrices for the kth qubit. The target is to prepare the Toffoli gate U_f on the three-qubit system.

For the following parameter setting: the time duration $T = 10$, each field being evenly discretized into $M = 100$ piecewise control segments, $|\varepsilon^{(1)}| \leq 0.2$ and $|\varepsilon^{(2)}| \leq 0.2$. The robustness performance was compared for GRAPE, b-GRAPE (B=1 and random mini-batch) and the SLC method (B=100 and fixed large batch). At the level set 0.001, the SLC approach enhances the robustness by about 4 times compared with GRAPE while b-GRAPE improves the robustness performance more than 10 times than the SLC approach [60]. Additional numerical results also show that b-GRAPE algorithm based on batch techniques can significantly enhance the robustness of quantum control problems while maintaining high fidelity. Moreover, the b-GRAPE algorithm has also been applied to enhance the control robustness against clock noise where clock noises may exist in clock signals in imperfect signal generators (e.g., arbitrary wave-form generators) and the robust controls were trained using randomly generated clock-noise samples [10].

4.4 SLC for Open Quantum Systems

For an open dissipative system, its dynamics can be described by a master equation. We will use an OPEN GRAPE algorithm to calculate the gradient (see [27, 28, 48, 50]). We assume that the state of the system is described by a master equation

$$\dot{\rho}(t) = -(\mathrm{i}\mathcal{H}(u(t)) + \Lambda(u(t)))\rho(t) \tag{4.44}$$

with the Hamiltonian superoperator $\mathcal{H}(u(t))(\cdot) = [H(u), \cdot]$, and the decoherence superoperator $\Lambda(u(t))$. The solution to the master equation is a linear map, according to $\rho(t) = \mathcal{G}(t)\rho(0)$. Hence, $\mathcal{G}(t)$ follows the operator equation

$$\dot{\mathcal{G}}(t) = -(\mathrm{i}\mathcal{H} + \Lambda)\mathcal{G}(t) \tag{4.45}$$

with $\mathcal{G}(0) = I$. The objective is to find a control field $u(t)$ to maximize the fidelity with a given final time T

$$F(U_F, \mathcal{G}(T)) = \frac{1}{2^q}|\mathrm{Tr}\{U_F^\dagger \mathcal{G}(T)\}|. \tag{4.46}$$

The gradient of $F(U_F, \mathcal{G}(T))$ can be calculated using the method in [48] and the control field can be updated using the gradient.

Now, we consider a flux qubit subject to decoherence to generate the S, H, and $\mathrm{T}_{\frac{\pi}{8}}$ quantum gates. We assume that the dynamics of the flux qubit can be described as

$$\dot{\rho}(t) = -\mathrm{i}[H(t), \rho(t)] + \Gamma_1 \mathcal{D}[\sigma_-]\rho(t) + \Gamma_\varphi \mathcal{D}[\sigma_z]\rho(t) \equiv \mathcal{L}\rho(t), \tag{4.47}$$

where

$$\mathcal{D}[c]\rho = c\rho c^\dagger - \frac{1}{2}c^\dagger c\rho - \frac{1}{2}\rho c^\dagger c.$$

Here, Γ_1 and Γ_φ are the relaxation rate and dephasing rate of the system, respectively. Considering the experiment [2], we choose $\Gamma_1 = 10^5\ \mathrm{s}^{-1}$ and $\Gamma_\varphi = 10^6\ \mathrm{s}^{-1}$. Let $T = 5$ ns and assume that the control Hamiltonian is described as

$$H(t) = u_x(t)\sigma_x + u_z(t)\sigma_z.$$

We assume that there exist uncertainties (with the uncertainty bound 0.2) in the relaxation rate and dephasing rate. Using the OPEN GRAPE algorithm [50], we can learn robust control fields for generating the three classes of quantum gates. The results are shown in Fig. 4.11. After 80 iterations, the fidelity of all three gates reaches 0.9948 using 40 control pulses for each class of quantum gates.

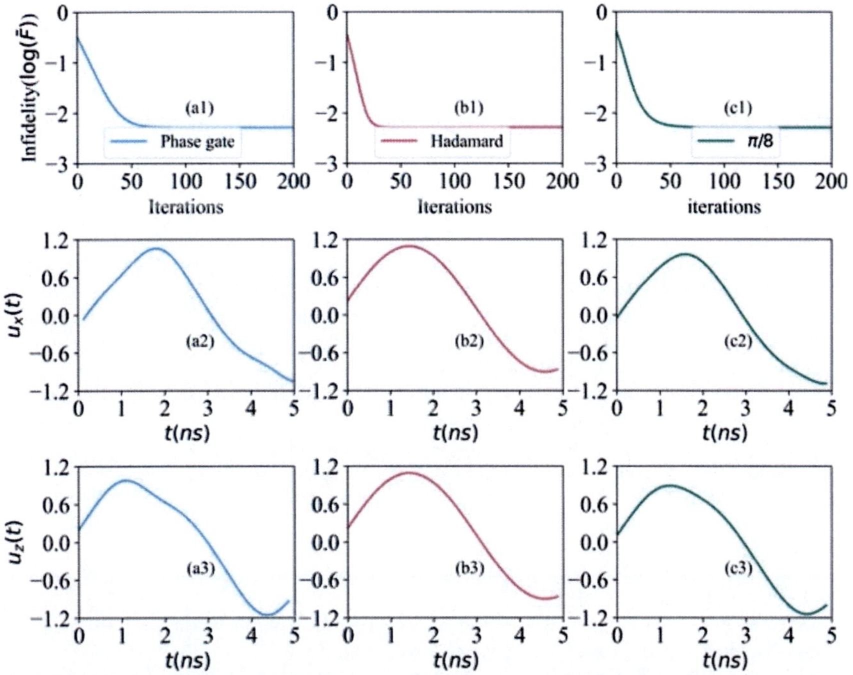

Fig. 4.11 Dong et al. [14] Robust control performance for the S, H, $T_{\frac{\pi}{8}}$ gates, with parameter uncertainties in the relaxation rate and dephasing rate for open quantum systems. **a1**, **b1**, **c1** Convergence for the H gate, the S gate, and the $T_{\frac{\pi}{8}}$ gate, respectively. The other sub-figures show the robust control pulses

4.5 Summary and Further Reading

In this chapter, we employed the SLC method to solve several classes of quantum robust control problems including state control of quantum superconducting circuits, robust photoassociation of O+H using an ultrafast laser, robustly synchronizing a laser with molecules for charge transfer, robust generation of quantum gates, and robust control of open quantum systems. We also briefly introduced the b-GRAPE algorithm where a similar idea of using samples to the SLC approach has been employed. b-GRAPE uses stochastic samples and takes advantages of several techniques in deep learning while the SLC approach employed deterministic samples. Future work includes further extending and improving these algorithms for more complex systems and more challenging quantum robust control problems.

Further reading may include [69] for a more comprehensive treatment of robust photoassociation of O+H, [68] for more results on synchronization of shaped laser pulses with molecules for charge transfer, and [61] for robust control of quantum unitary transformation. For stochastic optimization of robust controls, readers are referred to [60] for more results on the b-GRAPE algorithm, which has been extended to the a-GRAPE algorithm [20, 21], and [59] for applications to systems with random clock noises.

References

1. Bialczak RC, Ansmann M, Hofheinz M, Lenander M, Lucero E, Neeley M, O'Connell AD, Sank D, Wang H, Weides M, Wenner J, Yamamoto T, Cleland AN, Martinis JM (2011) Fast tunable coupler for superconducting qubits. Phys Rev Lett 106(6):060501
2. Bylander J, Gustavsson S, Yan F, Yoshihara F, Harrabi K, Fitch G, Cory DG, Nakamura Y, Tsai J-S, Oliver WD (2011) Noise spectroscopy through dynamical decoupling with a superconducting flux qubit. Nat Phys 7:565–570
3. Chen C, Dong D, Long R, Petersen IR, Rabitz H (2014) Sampling-based learning control of inhomogeneous quantum ensembles. Phys Rev A 89:023402
4. Chiorescu I, Bertet P, Semba K, Nakamura Y, Harmans CJPM, Mooij JE (2004) Coherent dynamics of a flux qubit coupled to a harmonic oscillator. Nature 431:159–162
5. Ciappina MF, Madsen LB (2008) Laser-assisted ion-atom collisions: plateau, cutoff, and multiphoton peaks. Phys Rev A 77:023412
6. Clarke J, Wilhelm FK (2008) Superconducting quantum bits. Nature 453:1031–1042
7. Copeland DA, Tang CL (1976) Photon-assisted nonresonant charge exchange: a simple molecular model. J Chem Phys 65:3161
8. Dalgarno A (1985) Charge transfer processes in astrophysical plasmas. Nuclear Instrum Methods Phys Res Sect B: Beam Interact Mater Atoms 9(4):655–657
9. de Lima EF, Ho T-S, Rabitz H (2011) Optimal laser control of molecular photoassociation along with vibrational stabilization. Chem Phys Lett 501(4–6):267–272
10. Ding H, Wu RB (2019) Robust quantum control against clock noises in multiqubit systems. Phys Rev A 100:022302
11. Dong D, Chen C, Qi B, Petersen IR, Nori F (2015) Robust manipulation of superconducting qubits in the presence of fluctuations. Sci Rep 5:7873 (Springer Nature)
12. Dong D, Mabrok MA, Petersen IR, Qi B, Chen C, Rabitz H (2015) Sampling-based learning control for quantum systems with uncertainties. IEEE Trans Control Syst Technol 23(6):2155–2166
13. Dong D, Petersen IR (2022) Quantum estimation, control and learning: opportunities and challenges. Ann Rev Control 54:243–251
14. Dong D, Wu C, Chen C, Qi B, Petersen IR, Nori F (2016) Learning robust pulses for generating universal quantum gates. Sci Rep 6:36090 (Springer Nature)
15. Esry BD, Sadeghpour HR (1999) Adiabatic formulation of heteronuclear hydrogen molecular ion. Phys Rev A 60(5):3604–3617
16. Esry BD, Sadeghpour HR, Wells E, Ben-Itzhak I (2000) Charge exchange in slow $H^+ + D(1s)$ collisions. J Phys B: At Mol Opt Phys 33(23):5329–5341
17. Falci G, D'Arrigo A, Mastellone A, Paladino E (2005) Initial decoherence in solid state qubits. Phys Rev Lett 94(16):167002
18. Feit MD, Fleck JA Jr (1983) Solution of the Schrödinger equation by a spectral method II: Vibrational energy levels of triatomic molecules. J Chem Phys 78(1):301
19. Feit MD, Fleck JA Jr (1984) Wave packet dynamics and chaos in the Hénon-Heiles system. J Chem Phys 80(6):2578
20. Ge X, Ding H, Rabitz H, Wu RB (2020) Robust quantum control in games: an adversarial learning approach. Phys Rev A 101:052317
21. Ge X, Wu RB (2021) Risk-sensitive optimization for robust quantum controls. Phys Rev A 104:012422
22. Hall FHJ, Aymar M, Bouloufa-Maafa N, Dulieu O, Willitsch S (2011) Light-assisted ion-Neutral reactive processes in the cold regime: radiative molecule formation versus charge exchange. Phys Rev Lett 107(24):243202

23. Ho T-S, Laughlin C, Chu S-I (1985) Laser-assisted charge-transfer reactions (Li^{3+}+H): coupled dressed-quasimolecular-state approach. Phys Rev A 32(1):122–133
24. Ho T-S, Rabitz H (2010) Accelerated monotonic convergence of optimal control over quantum dynamics. Phys Rev E 82(2):026703
25. Ho T-S, Rabitz H, Chu S-I (2011) A general formulation of monotonically convergent algorithms in the control of quantum dynamics beyond the linear dipole interaction. Comput Phys Commun 182(1):14–17
26. Hofheinz M, Wang H, Ansmann M, Bialczak RC, Lucero E, Neeley M, O'Connell AD, Sank D, Wenner J, Martinis JM, Cleland AN (2009) Synthesizing arbitrary quantum states in a superconducting resonator. Nature 459:546–549
27. Huang S-Y, Goan H-S (2014) Optimal control for fast and high-fidelity quantum gates in coupled superconducting flux qubits. Phys Rev A 90(1):012318
28. Hwang B, Goan H-S (2012) Optimal control for non-Marknovian open quantum systems. Phys Rev A 85:032321
29. Jones KM, Tiesinga E, Lett PD, Julienne PS (2006) Ultracold photoassociation spectroscopy: long-range molecules and atomic scattering. Rev Mod Phys 78:483–535
30. Kirchner T (2004) Laser-field-induced modifications of electron-transfer processes in ion-atom collisions. Phys Rev A 69(6):063412
31. Koch CP, Luc-Koenig E, Masnou-Seeuws F (2006) Making ultracold molecules in a two-color pump-dump photoassociation scheme using chirped pulses. Phys Rev A 73(3):033408
32. Koch CP, Kosloff R, Masnou-Seeuws F (2006) Short-pulse photoassociation in rubidium below the D_1 line. Phys Rev A 73(4):043409
33. Kosloff R (1988) Time-dependent quantum-mechanical methods for molecular dynamics. J Phys Chem 92:2087–2100
34. Kosut RL, Grace MD, Brif C (2013) Robust control of quantum gates via sequential convex programming. Phys Rev A 88(5):052326
35. Kuznetsov AM (1995) Charge transfer in physics, chemistry and biology: physical mechanisms of elementary processes and an introduction to the theory. Taylor & Francis, London
36. Liao S-L, Ho T-S, Chu S-I, Rabitz H (2011) Fast-kick-off monotonically convergent algorithm for searching optimal control fields. Phys Rev A 84:031401
37. Liao S-L, Ho T-S, Rabitz H, Chu S-I (2013) Maximum attainable field-free molecular orientation of a thermal ensemble with near-single-cycle THz pulses. Phys Rev A 87:013429
38. Liu YX, You JQ, Wei LF, Sun CP, Nori F (2005) Optical selection rules and phase-dependent adiabatic state control in a superconducting quantum circuit. Phys Rev Lett 95:087001
39. Madsen LB, Hansen JP, Kocbach L (2002) Excitation in ion-atom collisions inside subfemtosecond laser pulses. Phys Rev Lett 89(9):093202
40. Makhlin Y, Schön G, Shnirman A (2001) Quantum-state engineering with Josephson-junction devices. Rev Mod Phys 73:357–400
41. Marston CC, Balint-Kurti GG (1989) The Fourier grid Hamiltonian method for bound state eigenvalues and eigenfunctions. J Chem Phys 91:3571
42. Montangero S, Calarco T, Fazio R (2007) Robust optimal quantum gates for Josephson charge qubits. Phys Rev Lett 99(17):170501
43. Niederhausen T, Thumm U (2006) Capture and ionization in laser-assisted proton-hydrogen collisions. Phys Rev A 73:041404
44. Nielsen MA, Chuang IL (2010) Quantum computation and quantum information. Cambridge University Press, Cambridge
45. Pashkin YuA, Yamamoto T, Astafiev O, Nakamura Y, Averin DV, Tsai JS (2003) Quantum oscillations in two coupled charge qubits. Nature 421:823–826
46. Pindzola MS, Minami T, Schultz DR (2003) Laser-modified charge-transfer processes in proton collisions with lithium atoms. Phys Rev A 68:013404
47. Pravia MA, Boulant N, Emerson J, Fortunato EM, Havel TF, Cory DG, Farid A (2003) Robust control of quantum information. J Chem Phys 119:9993–10001
48. Rebentrost P, Serban I, Schulte-Herbrüggen T, Wilhelm FK (2009) Optimal control of a qubit coupled to a non-Markovian environment. Phys Rev Lett 102:090401

49. Schoelkopf RJ, Girvin SM (2008) Wiring up quantum systems. Nature 451:664–669
50. Schulte-Herbrüggen T, Spörl A, Khaneja N, Glaser SJ (2011) Optimal control for generating quantum gates in open dissipative systems. J Phys B: At Mol Opt Phys 44(15):154013
51. Shapiro EA, Shapiro M, Pe'er A, Ye J (2007) Photoassociation adiabatic passage of ultracold Rb atoms to form ultracold Rb_2 molecules. Phys Rev A 75(1):013405
52. Sillanpää MA, Park JI, Simmonds RW (2007) Coherent quantum state storage and transfer between two phase qubits via a resonant cavity. Nature 449:438–442
53. Slichter DH, Vijay R, Weber SJ, Boutin S, Boissonneault M, Gambetta JM, Blais A, Siddiqi I (2012) Measurement-induced qubit state mixing in circuit QED from up-converted dephasing noise. Phys Rev Lett 109(15):153601
54. Steffen M, Ansmann M, Bialczak RC, Katz N, Lucero E, McDermott R, Neeley M, Weig EM, Cleland AN, Martinis JM (2006) Measurement of the entanglement of two superconducting qubits via state tomography. Science 313:1423–1425
55. Sugny D, Vranckx S, Ndong M, Vaeck N, Atabek O (2014) Control of molecular dynamics with zero-area fields: application to molecular orientation and photofragmentation. Phys Rev A 90:053404
56. Vijay R, Macklin C, Slichter DH, Weber SJ, Murch KW, Naik R, Korotkov AN, Siddiqi I (2012) Stabilizing Rabi oscillations in a superconducting qubit using quantum feedback. Nature 490:77–80
57. Wendin G, Shumeiko VS (2006) Handbook of theoretical and computational nanotechnology. American Scientific Publishers, Karlsruhe (Chap. 12)
58. Wu RB, Chu B, Owens DH, Rabitz H (2018) Data-driven gradient algorithm for high-precision quantum control. Phys Rev A 97:042122
59. Wu RB, Ding HJ (2019) Robust quantum control against clock noises in multiqubit systems. Phys Rev A 100:022302
60. Wu RB, Ding H, Dong D, Wang X (2019) Learning robust and high-precision quantum controls. Phys Rev A 99:042327
61. Wu C, Qi B, Chen C, Dong D (2017) Robust learning control design for quantum unitary transformations. IEEE Trans Cybern 47(12):4405–4417
62. Xiang Z-L, Ashhab S, You JQ, Nori F (2013) Hybrid quantum circuits: superconducting circuits interacting with other quantum systems. Rev Mod Phys 85:623–653
63. You JQ, Nori F (2005) Superconducting circuits and quantum information. Phys Today 58:42–47
64. You JQ, Nori F (2011) Atomic physics and quantum optics using superconducting circuits. Nature 474:589–597
65. You JQ, Tsai JS, Nori F (2003) Controllable manipulation and entanglement of macroscopic quantum states in coupled charge qubits. Phys Rev B 68(2):024510
66. Zare RN (1998) Laser control of chemical reactions. Science 279:1875–1879
67. Zhang W, Dong D, Petersen IR (2018) Adaptive target scheme for learning control of quantum systems. IEEE Trans Control Syst Technol 26:1259–1271
68. Zhang W, Dong D, Petersen IR, Rabitz H (2016) Sampling-based robust control in synchronizing collision with shaped laser pulses: an application in charge transfer for $H^+ + D \rightarrow H + D^+$. RSC Adv 6:92962–92969
69. Zhang W, Dong D, Petersen IR, Rabitz H (2017) Robust control of photoassociation of slow O + H collision. Chem Phys 483–484:149–155
70. Zhang W, Shu C-C, Ho T-S, Rabitz H, Cong S-L (2014) Optimal control of charge transfer for slow H^++D collisions with shaped laser pulses. J Chem Phys 140(9):093404

Chapter 5
Machine Learning for Quantum Control

Abstract This chapter presents results on learning control of quantum systems. In Sect. 5.2, two differential evolution (DE) algorithms are proposed and numerical results on quantum control via DE are presented for inhomogeneous open quantum ensembles and synchronization of quantum networks.[a] In Sect. 5.3 closed-loop learning control using DE is applied to ultrafast quantum engineering, and experimental results on optimal and robust control of molecules using femtosecond laser pulses are presented.[b] Numerical and experimental results on learning control design of quantum autoencoders are presented in Sect. 5.4.[c] In Sect. 5.5, some progress on quantum control using reinforcement learning is introduced.[d]

5.1 Introduction

Gradient algorithms (e.g., in Chaps. 3 and 4) have shown powerful capability for numerically finding optimal or robust controls due to their excellent performance [41]. In many practical applications, it may be difficult to obtain the gradient information or there exist local optima in complex quantum control problems. For these situations, a natural idea is to employ stochastic search algorithms to seek good controls. Evolutionary computation including genetic algorithm (GA) and DE has been widely used in the area of quantum control. In these evolutionary computation methods,

[a] Sections 5.2 and 5.3.3 contain materials reprinted, with permission, from IEEE Transactions on Cybernetics [26] © 2020 IEEE.
[b] Section 5.3.2 contains materials reprinted, with permission, from IEEE Transactions on Control Systems Technology [25] © 2020 IEEE.
[c] Section 5.4 contains materials reprinted, with permission, from Physical Review A [32] ©2020 American Physical Society.
[d] Section 5.5 contains materials reprinted, with permission, from IEEE Transactions on Neural Networks and Learning Systems [13] © 2014 IEEE.

D. Dong and I. R. Petersen, *Learning and Robust Control in Quantum Technology*, Communications and Control Engineering,
https://doi.org/10.1007/978-3-031-20245-2_5

crossover, mutation and selection operations are iteratively implemented to search for good solutions (optimal controls) in a parameter space [24]. For example, a subspace-selective self-adaptive differential evolution algorithm has been proposed to achieve a high-fidelity single-shot Toffoli gate and single-shot 3-qubit gates [107, 108]. GA has been widely used in the area of quantum control and has achieved success in learning control of molecular systems [74]. Several promising evolution algorithms have been comparatively investigated in [109], and it was found that DE usually outperformed GA and particle swarm optimization for hard quantum control problems. In the chapter, we first present two improved DE algorithms including ***MS*_DE algorithm** (DE algorithm with mixed strategies) and ***msMS*_DE algorithm** (multiple samples and mixed-strategy DE algorithm). We present numerical results on the application of the *msMS*_DE algorithm to control of inhomogeneous open quantum ensembles and robust synchronization of quantum networks with uncertainties.

The gradient-based learning results in Chaps. 3 and 4, and the above-mentioned numerical results mainly involve open-loop control strategies. Evolutionary computation has demonstrated powerful capability when it is integrated into closed-loop control design. Closed-loop learning control, where each cycle of the closed-loop is executed on a new sample, has achieved great successes in laser control of laboratory chemical reactions [40, 74]. A closed-loop learning control procedure generally involves three components [23, 74]: (i) a trial laser control input, (ii) the laboratory generation of the control that is applied to the sample and subsequently observed for its impact, and (iii) a learning algorithm to suggest the form of the next control input by considering the prior experiments. A feature of a good closed-loop learning control design is its insensitivity to the initial trials. A key task is to develop a good learning algorithm for ensuring that the learning process converges to achieve a predetermined objective. GA, DE and several rapid convergence algorithms have been developed for this task [8, 40]. The optimal control problem is usually formulated as solving an optimization problem by maximizing a functional which is related to some variables such as the control inputs, quantum states and control time but may have no analytical form. Closed-loop learning control is especially effective for fragmentation control of molecules using femtosecond laser pulses. Femtosecond (fs) (1 fs $= 10^{-15}$ s) lasers [102] have been widely applied in the quantum control of molecular dynamics. Here, we employ the *MS*_DE algorithm to achieve optimal control of $Pr(hfac)_3$ molecules using fs laser pulses. Also, we employ the *msMS*_DE algorithm to another experimental quantum control problem, where the goal is to identify a robust solution (shaped fs laser pulse) that can maximize the CH_2Br^+/CH_2I^+ product ratio from the fragmentation of the CH_2BrI molecule. Moreover, we develop a closed-loop learning control scheme for assisting in control design of optimal quantum autoencoders, compare the performance of several machine learning algorithms including stochastic gradient algorithms, GA, DE and evolution strategy (ES) algorithms, and present experimental results of quantum autoencoders on quantum optical systems.

Reinforcement learning (RL) [88] is another important machine learning approach, and it addresses the problem of how an active agent can learn to approximate an optimal strategy while interacting with its environment. It is a model-free feedback-based approach and works well even when the system model is unknown or with uncertainties. RL first was used to quantum control problems in [22], and then a fidelity-based probabilistic Q-learning approach was presented to naturally solve the balance problem between exploration and exploitation and was applied to learning control of quantum systems in [13]. By utilizing two-stage learning with teacher and student networks and a reward quantifying the capability to recover the quantum information stored in a quantum system, the work [28] showed how a network-based "agent" in RL can discover good quantum-error-correction strategies to protect qubits against noise. In [64], deep reinforcement learning was employed to simultaneously optimize the speed and fidelity of quantum computation against both leakage and stochastic control errors. In this chapter, we also give a brief introduction to the progress of quantum control via reinforcement learning [9].

5.2 Differential Evolution for Quantum Control: Numerical Results

5.2.1 *Differential Evolution*

Differential evolution (DE) has been widely used in the continuous search domain and has achieved considerable success in science and engineering applications [16, 31, 86]. In DE, a population is composed of a group of individual trial solutions or parameter vectors, usually represented in a real-valued vector $X = [x_1, x_2, \ldots, x_D]^T$. In the process of searching, an objective function regarding a target vector X is defined as $J(X)$. Then, the learning process works by generating variations of the individuals within the given parameter space and selecting better ones to be carried into the next generation, until a "best" individual is obtained. Consider that DE searches for a global optimum point in a D-dimensional real parameter space $\mathbb{R}^D$. We can summarize its main steps as follows:

(a) **Initialization.** The population (i.e., target vector) at the current generation is denoted as

$$X_{i,G} = \begin{bmatrix} x^1_{i,G} \\ \cdots \\ x^D_{i,G} \end{bmatrix}, \quad i = 1, \ldots, NP$$

and let $x^j_{i,G} \in [x^j_{\min}, x^j_{\max}]$, $(j = 1, 2, \ldots, D)$. Usually, the population (at $G = 0$) is initialized in a uniform way [57]:

$$x^j_{i,0} = x^j_{\min} + \text{rand}(0, 1) \cdot \left(x^j_{\max} - x^j_{\min}\right), \quad j = 1, 2, \ldots, D, \tag{5.1}$$

where rand(0, 1) is a uniformly distributed random number, which helps guarantee that the vectors cover the range of the parameter space.

(b) **Mutation.** The core idea of the "mutation" operation is to generate mutant vectors from the existing target vectors. For example, by choosing three other distinct parameter vectors from the current generation (say, X_{r_1}, X_{r_2}, X_{r_3}), we could formulate a donor vector $V_{i,G+1}$ as

$$V_{i,G} = X_{r_1,G} + F \cdot (X_{r_2,G} - X_{r_3,G}), \tag{5.2}$$

where the indices $r_1, r_2, r_3 \in \{1, \ldots, NP\}$ are mutually exclusive integers from $[1, NP]$ and $r_1, r_2, r_3 \neq i$. Besides, the scaling factor F is normally set between 0.4 and 1.

(c) **Crossover.** In DE, a mutation phase is usually followed by a crossover operation, as it generates trial vectors from mutant vector $V_{i,G}$ and target vector $X_{i,G}$. There are two typical crossover operations including exponential crossover and binomial crossover. They are functionally equivalent to each other, and the binomial one could be expressed as:

$$y_{i,G}^j = \begin{cases} v_{i,G}^j, & \text{if } \mathrm{rand}(j) \leq CR \text{ or } j = \mathrm{rand}(1, D), \\ x_{i,G}^j, & \text{if } \mathrm{rand}(j) > CR \text{ and } j \neq \mathrm{rand}(1, D), \end{cases} \tag{5.3}$$

where the predefined parameter CR controls the potential diversity of the evolving population. The condition $j = \mathrm{rand}(1, D)$ is introduced to ensure that the trial vector is different from its corresponding target vector by at least one parameter.

(d) **Selection.** After the crossover, a selection process is adopted to determine the individuals of the next generation from the target vectors and the trial vectors by comparing their fitness functions. For a maximization problem, if the new trial vector yields an equal or higher value of the objective function, it survives into the next generation; otherwise the target vector retains. This can be described as:

$$X_{i,G+1} = \begin{cases} Y_{i,G}, & \text{if } J(Y_{i,G}) \geq J(X_{i,G}), \\ X_{i,G}, & \text{otherwise.} \end{cases} \tag{5.4}$$

Except from the initialization step, the other three steps are repeated generation after generation until some specific termination criteria are satisfied. As for control parameters of DE, we sample F by a normal distribution with mean value 0.5 and standard deviation 0.3, denoted by $N(0.5, 0.3)$. Similarly, the value of CR is sampled by a normal distribution denoted as $N(0.5, 0.1)$. Considering that CR has the practical meaning of probability, those values falling out $[0, 1]$ should be abandoned and new values should be regenerated. Consequently, a set of F and CR values are assigned to each target vector for performing mutation, crossover and selection.

According to previous studies [73, 86], mutation operation aims at generating variations for the population, and therefore the adopted mutation strategy can have key impact on its searching performance. Existing results have shown that different mutation strategies exhibit various optimization effects on different searching problems, and they are suitable for solving different specific optimization problems [3, 16, 62]. In [53], a DE algorithm with mixed strategies has been demonstrated to be a promising candidate for quantum control problems. We denote the mutation strategy using the notation DE/*a*/*b*, where *a* represents a string denoting the base vector to be perturbed, *b* is the number of difference vectors considered for perturbation of *a*. To guarantee the optimization effects, we investigate several strategies and finally decide to employ the following four effective yet diverse strategy candidates [3, 53]:

Strategy 1: DE/rand/1

$$V_i = X_{r_1} + F \cdot (X_{r_2} - X_{r_3}). \tag{5.5}$$

Strategy 2: DE/rand to best/2

$$V_i = X_i + F \cdot (X_{\text{best}} - X_i) + F \cdot (X_{r_1} - X_{r_2}) + F \cdot (X_{r_3} - X_{r_4}). \tag{5.6}$$

Strategy 3: DE/rand/2

$$V_i = X_{r_1} + F \cdot (X_{r_2} - X_{r_3}) + F \cdot (X_{r_4} - X_{r_5}). \tag{5.7}$$

Strategy 4: DE/current-to-rand/1

$$V_i = X_i + K \cdot (X_{r_1} - X_i) + F \cdot (X_{r_2} - X_{r_3}). \tag{5.8}$$

The indices r_1, r_2, r_3, r_4 and r_5 are mutually exclusive integers randomly chosen from the range $[1, NP]$, and all of them are different from the index i. X_{best} is the best individual vector with the best fitness (i.e., the highest objective function value for a maximization problem) in the population. To eliminate one additional parameter, the control parameter K in (5.8) could be set as $K = 0.5$.

In this chapter, we present two improved DE algorithms for solving quantum control problems. The first is called *MS*_DE algorithm which is described in Algorithm 5.1, where Normrnd means F and CR values are randomly sampled from a normal distribution. In the *MS*_DE algorithm, a mutation scheme from a candidate pool is selected and then crossover operation is determined. These algorithms are also compared with several other DE algorithms such as basic DE algorithm and *ms*_DE algorithm (DE algorithm with multiple samples). In our implementation, the first three mutation schemes are combined with a binomial crossover operation, while the fourth scheme directly generates trial vectors without crossover.

Another DE algorithm we will adopt is *msMS*_DE algorithm. In order to design appropriate control laws to achieve good robustness performance, we integrate the idea of SLC [14] into the *msMS*_DE algorithm. First, we prepare N samples $\Theta_k = (\theta_0, \theta_1, \ldots, \theta_M)$ $(k = 1, 2, \ldots, N)$ with different values of the uncertain parameters. We compute the fitness values of these sample vectors. Then, we evaluate the average fitness value $\bar{J}$ for these samples, and $\bar{J}$ is defined as follows:

$$\bar{J}(Y_{i,G}) = \frac{1}{N} \sum_{k=1}^{N} J(Y_{i,G}, \Theta_k),$$

Algorithm 5.1 Algorithmic description of *MS*_DE

Set the generation number $G = 0$
for $i = 1$ to NP **do**
  initialize $X_{i,0}$ and obtain fitness function $J(X_{i,0})$
end for
Set the vector with the maximum fitness as $X_{\text{best},0}$
repeat (for each generation $G = 0, 1, \ldots, G_{\max}$)
  repeat (for each vector X_i, $i = 1, 2, \ldots, NP$)
    Set parameters $F_{i,G} = \text{Normrnd}(0.5, 0.3)$ and $CR_{i,G} = \text{Normrnd}(0.5, 0.1)$
    while $CR_{i,G} < 0$ or $CR_{i,G} > 1$ **do**
      $CR_{i,G} = \text{Normrnd}(0.5, 0.1)$
    end while
    randomly choose a **strategy** from candidate pool and obtain mutant vectors $V_{i,G}$ according to (5.5)–(5.8)
    if strategy $\in \{1, 2, 3\}$ **then**
      obtain $Y_{i,G}$ according to Eq. (5.3)
    end if
    if strategy $\in \{4\}$ **then** $Y_{i,G} = V_{i,G}$
    end if
    if $J(Y_{i,G}) \geq J(X_{i,G})$ **then**
      $X_{i,G+1} \leftarrow Y_{i,G}, \;\; J(X_{i,G+1}) \leftarrow J(Y_{i,G})$.
    end if
    Update the best vector $X_{\text{best},G}$ and $i \leftarrow i + 1$
  until $i = NP$
  $G \leftarrow G + 1$
until $G = G_{\max}$

$$\bar{J}(X_{i,G}) = \frac{1}{N}\sum_{k=1}^{N} J(X_{i,G}, \Theta_k).$$

In the *msMS*_DE algorithm, similarly in the *MS*_DE algorithm, we use a normal distribution $N(0.5, 0.3)$ to approximate the parameter F and let CR obey a normal distribution denoted by $N(0.5, 0.1)$ [73]. Consequently, a set of F and CR values are randomly sampled from a normal distribution and applied to each target vector in the current population. We may obtain some extraordinary values far from $[-0.4, 1.4]$ for the scale factor F, and we usually accept them to increase diversity. The *msMS*_DE algorithm is outlined in Algorithm 5.2.

In simulations, after we obtain the nominal value of an individual, we may generate other samples by perturbing the nominal value and then calculate the average fitness function. In experiments, we need to measure the fitness of each sample and then calculate the average fitness. Usually, a larger number of samples N may lead to better robustness performance [14]. However, the computational or experimental time will significantly increase with an increase in the size N. In this chapter, we use three samples for each uncertain parameter to reduce computational and experimental time. In the proposed *msMS*_DE algorithm, we preset a maximum generation $G_{\max}$ as the termination criterion. During the implementation of the algorithm, the population evolves until the learning process reaches $G = G_{\max}$.

Algorithm 5.2 Algorithmic description of *msMS*_DE

Set the generation number $G = 0$
for $i = 1$ to NP **do**
  for $j = 1$ to D **do**
    $x_{i,G}^{j} = x_{\min}^{j} + \text{rand}(0, 1) \cdot (x_{\max}^{j} - x_{\min}^{j})$
  end for
end for
Initialize fitness $J(X_{i,G})$ and evaluate vector by f and mark the best vector with maximum J as $X_{\text{best},G}$
repeat (for each generation $G = 1, 2, \ldots, G_{\max}$)
  repeat (for each vector X_i, $i = 1, 2, \ldots, NP$)
    Set parameters $F_{i,G} = \text{Normrnd}(0.5, 0.3)$ and $CR_{i,G} = \text{Normrnd}(0.5, 0.1)$
    while $CR_{i,G} < 0$ or $CR_{i,G} > 1$ **do**
      $CR_{i,G} = \text{Normrnd}(0.5, 0.1)$
    end while
    randomly choose a **strategy** from candidate pool and obtain mutant vectors $V_{i,G}$ according to (5.5)–(5.8)
    if strategy $\in \{1, 2, 3\}$ **then**
      obtain $Y_{i,G}$ according to Eq. (5.3)
    end if
    if strategy $\in \{4\}$ **then** $Y_{i,G} = V_{i,G}$
    end if
    for each sample Θ_k, $(k = 1, 2, \ldots, N)$ **do**
      evaluate the fitness function $f(Y_{i,G}, \Theta_k)$
    end for
    Compute $\bar{J}(Y_{i,G}) = \frac{1}{N}\sum_{k=1}^{N} J(U_{i,G}, \Theta_k)$
    if $\bar{J}(Y_{i,G}) \geq \bar{J}(X_{i,G})$ **then**
      $X_{i,G+1} \leftarrow U_{i,G}, \quad \bar{J}(X_{i,G+1}) \leftarrow \bar{J}(Y_{i,G})$.
    end if
    Renew the best vector $X_{\text{best},G}$ and $i \leftarrow i + 1$
  until $i = NP$
  $G \leftarrow G + 1$
until $G = G_{\max}$

5.2.2 *DE for Control of Open Quantum Ensembles*

In Chap. 3, we presented an SLC method to achieve high-fidelity control of inhomogeneous quantum ensembles. In those results, decoherence and dissipation were usually not considered [65–67]. The existence of decoherence and dissipation may irreversibly lead a quantum ensemble to becoming an open system [7], and the manipulation of inhomogeneous open quantum ensembles becomes more challenging than that without considering decoherence. Here, we employ the *msMS*_DE algorithm in Sect. 5.2.1 to search for robust control fields for inhomogeneous open quantum ensembles aimed at achieving enhanced control performance. We assume that each member in the inhomogeneous ensemble can be described by a Markovian master equation in the Lindblad form as in [48, 99]

$$\dot{\rho}(t) = -\mathrm{i}[H(t), \rho(t)] + \sum_k \gamma_k \mathcal{D}[L_k]\rho(t), \tag{5.9}$$

where the non-negative coefficients γ_k specify the relevant relaxation rates, L_k are appropriate Lindblad operators and

$$\mathcal{D}[L_k]\rho = \left(L_k \rho L_k^\dagger - \frac{1}{2} L_k^\dagger L_k \rho - \frac{1}{2} \rho L_k^\dagger L_k \right).$$

For a member in the inhomogeneous open quantum ensemble, the Hamiltonian can be described in the form

$$H_\varepsilon(t) = g_0(\varepsilon_0) H_0 + \sum_{j=1}^{M} g_j(\varepsilon_j) u_j(t) H_j, \tag{5.10}$$

where we assume that M control Hamiltonians are used. Let $\varepsilon = (\varepsilon_0, \varepsilon_1, \ldots, \varepsilon_M)$ and the functions $g_j(\varepsilon_j)$ $(j = 0, 1, \ldots, M)$ characterize possible inhomogeneities. For example, $g_0(\varepsilon_0)$ corresponds to inhomogeneity in the free Hamiltonian (e.g., due to chemical shift in NMR). $g_j(\varepsilon_j)$ $(j = 1, \ldots, M)$ can characterize imprecise parameters in the dipole approximation or possible multiplicative noises in the control fields. We assume that $g_j(\varepsilon_j)$ $(j = 0, 1, \ldots, M)$ are continuous functions of ε_j, and the parameters ε_j could be time-dependent and $\varepsilon_j \in [-\mathcal{E}_j, \mathcal{E}_j]$. For simplicity, we assume that $g_j(\varepsilon_j) = 1 + \varepsilon_j$ in this class of quantum control problems, and $\mathcal{E}_0 = \ldots = \mathcal{E}_j = \ldots = \mathcal{E}_M = \mathcal{E}$.

For an open quantum system in (5.9), we may define a coherent vector as

$$\mathbf{z} := (\mathrm{Tr}(O_l \rho), \mathrm{Tr}(O_2 \rho), \ldots, \mathrm{Tr}(O_m \rho))^T,$$

where $\mathrm{i}O_1, \mathrm{i}O_2, \ldots \mathrm{i}O_m$ $(m = n^2 - 1)$ are orthogonal generators of the special unitary group $\mathbf{su}(n)$ with degree n. Its density operator can be written as:

$$\rho = \frac{I}{n} + \frac{1}{2} \sum_{l=1}^{m} z_l O_l. \tag{5.11}$$

Substituting (5.11) into (5.9), the evolution of the coherent vector $\mathbf{y}$ can be described as:

$$\dot{\mathbf{y}} = (\mathcal{L}_{H_0} + \mathcal{L}_D)\mathbf{z} + \sum_{j=1}^{M} u_j \mathcal{L}_{H_j} \mathbf{z} + l_0, \tag{5.12}$$

where the superoperators $\mathcal{L}_{H_0}$, $\mathcal{L}_D$, $\mathcal{L}_{H_j}$ $(j = 1, 2, \ldots, M)$ and the term l_0 are explained in detail in [43, 105]. We choose the objective function $J(u)$ to be maximized as follows [105]:

$$J(u) = 1 - \frac{n}{8(n-1)} \|\mathbf{z}_f - \mathbf{z}(T)\|^2, \tag{5.13}$$

where $\|\mathbf{z}\|^2 = \mathbf{z}^T \mathbf{z}$ is a vector norm and it is clear that $J(u) \in [0, 1]$. Also, $\mathbf{z}_f$ and $\mathbf{z}(T)$ are the target state and the final state of the quantum system in terms of coherent vectors, respectively.

The control of an inhomogeneous open quantum ensemble can be formulated as:

$$\max_u \ J(u) := \max_u \mathbb{E}[J_\varepsilon(u)]$$

s.t.

$$\begin{cases} \dot{\mathbf{z}}_\varepsilon(t) = ((1+\varepsilon_0)\mathcal{L}_{H_0} + \mathcal{L}_D + (1+\varepsilon_j)\sum_{j=1}^{M} u_j(t)\mathcal{L}_{H_j})\mathbf{z}_\varepsilon(t) + l_0, \\ \mathbf{z}_\varepsilon(0) = \mathbf{z}_0, \quad t \in [0, T] \\ \varepsilon_j \in [-\mathcal{E}, \mathcal{E}], \quad j = 0, 1, \ldots M \end{cases} \tag{5.14}$$

where $J_\varepsilon(u)$ is the objective function for given ε and $\mathbb{E}[J_\varepsilon(u)]$ denotes the average performance of $J(u)$ over the parameter inhomogeneities ε.

We consider an open two-level ensemble with inhomogeneous parameter bound $\mathcal{E} = 0.2$. The members of this ensemble are governed by the following Hamiltonian:

$$H(t) = (1+\varepsilon_0)\frac{1}{2}\sigma_z + (1+\varepsilon_1)u(t)(\sigma_x \cos\varphi + \sigma_y \sin\varphi), \tag{5.15}$$

where $\varphi = 0.8897$, $u(t) \in [-10, 10]$. For simplicity, we let the decoherence coefficients $\gamma_k = 1$ and the Lindblad operators are given by [37]

$$L_1 = \begin{bmatrix} 0 & 0 \\ 0.1 & 0 \end{bmatrix}, \ L_2 = \begin{bmatrix} 0 & 0.2 \\ 0 & 0 \end{bmatrix}, \ L_3 = \begin{bmatrix} 0.2 & 0 \\ 0 & 0 \end{bmatrix}, \tag{5.16}$$

where the items L_1 and L_2 correspond to relaxation and L_3 item characterizes the dephasing process. For a two-level quantum ensemble, O_1, O_2 and O_3 in (5.11) can be chosen as σ_x, σ_y and σ_z. The coherent vector for the density matrix is the Bloch vector

$$\mathbf{z} = \begin{bmatrix} z_1 \\ z_2 \\ z_3 \end{bmatrix} = \begin{bmatrix} \mathrm{Tr}(\sigma_x \rho) \\ \mathrm{Tr}(\sigma_y \rho) \\ \mathrm{Tr}(\sigma_z \rho) \end{bmatrix}.$$

The dynamical equation for $\mathbf{z}$ can be written as

$$\begin{aligned} \dot{\mathbf{z}}(t) = & \begin{bmatrix} -0.045 & -\theta_0 & 0 \\ \theta_0 & -0.045 & 0 \\ 0 & 0 & -0.05 \end{bmatrix} \mathbf{z}(t) + \begin{bmatrix} 0 \\ 0 \\ 0.03 \end{bmatrix} \\ & + \theta_1 u(t) \begin{bmatrix} 0 & 0 & -2\sin\varphi \\ 0 & 0 & 2\cos\varphi \\ 2\sin\varphi & -2\cos\varphi & 0 \end{bmatrix} \mathbf{z}(t). \end{aligned} \tag{5.17}$$

The average fitness function is given as

$$E[J_\varepsilon(u)] = \frac{1}{N}\sum_{N_0}\sum_{N_1}\left[1 - \frac{1}{4}\left\|\mathbf{z}_\mathrm{f} - \mathbf{z}_{\varepsilon_0,\varepsilon_1}(T)\right\|^2\right], \tag{5.18}$$

where N_0 is the number of samples for ε_0, N_1 is the number of samples for ε_1, and N is the total number of the chosen samples. An upper bound of the fitness function is 1 although we do not a priori know the maximum that can be achieved. In the *msMS*_DE algorithm, we choose three samples for each parameter, and here we have $N = 9$. During learning control of the inhomogeneous quantum ensemble, we employ DE algorithms to seek the optimal control $u^*(t)$. Then, we apply the optimal control field to additional samples with inhomogeneous parameters $(\varepsilon_0, \varepsilon_1)$ following uniform distributions within $[-0.2, 0.2]$ to test its performance. We assume that the initial state ρ_0 and the target state ρ_f are, respectively,

$$\rho_0 = \begin{bmatrix} 1 & 0 \\ 0 & 0 \end{bmatrix}, \quad \rho_f = \begin{bmatrix} 0 & 0 \\ 0 & 1 \end{bmatrix}. \tag{5.19}$$

The target time $T = 10$ and the time interval $[0,\ T]$ is equally divided into $D = 200$ time steps, and $\Delta t = 0.05$. The population size is set as $NP = 50$ for all the algorithms in this example. The simulation is implemented on a MATLAB platform (version 8.3.0.532). The hardware environment for simulation is Intel(R)-Core(TM) i7-6700K CPU, dominant frequency @4.00 GHz, and 16G(ARM).

To demonstrate the performance of the proposed *msMS*_DE algorithm for the control problem of inhomogeneous quantum ensembles, we make a performance comparison between it and *ms*_DE (i.e., using the average fitness function of multiple samples) with various parameters. We first present the results for the traditional DE (i.e., "DE/rand/1/bin") using multiple samples with three typical sets of control parameters. Three cases with different control parameters are labeled as "*ms*_DE1" ($F = 0.9, CR = 0.1$), "*ms*_DE2" ($F = 0.9, CR = 0.9$) and "*ms*_DE3" ($F = 0.5$, $CR = 0.3$), and the training performance is presented in Fig. 5.1a. It is clear that *ms*_DE1 and *ms*_DE3 have better performance than *ms*_DE2 for the quantum control problem. *ms*_DE1 can achieve the highest fitness of 0.9566 among these three cases. We then compare the training performance of *ms*_DE1, GA and *msMS*_DE, and the results are illustrated in Fig. 5.1b. The *msMS*_DE algorithm achieves the highest fitness of $J_{\mathrm{max}} = 0.9798$, while *ms*_DE1 and GA converge to a maximum value of 0.9566 and 0.9667, respectively. A comparison of testing performance for 2000 additional samples (i.e., 2000 quantum systems generated according to the inhomogeneous ensemble) and training times between DE1 (using one sample), *ms*_DE1, *ms*_DE2, *ms*_DE3, GA (with crossover probability $P_\mathrm{c} = 0.8$ and mutation probability $P_\mathrm{m} = 0.05$) and *msMS*_DE in Table 5.1 shows that *msMS*_DE is superior to *ms*_DE and DE1 in terms of fidelity. More numerical results also show that *msMS*_DE can usually find the control field with the best robustness among these algorithms because *msMS*_DE employs mixed mutation strategies as well as

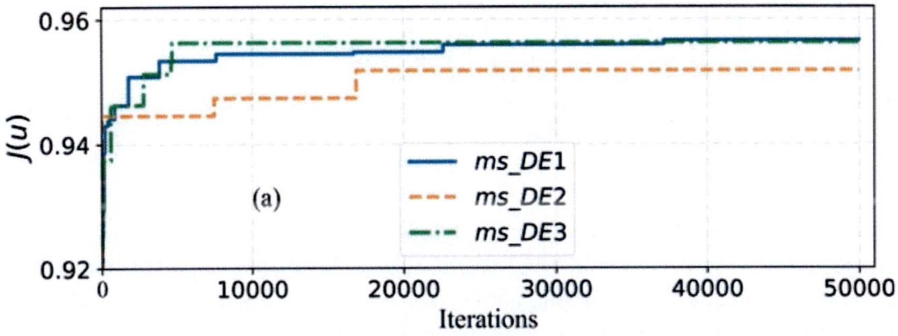

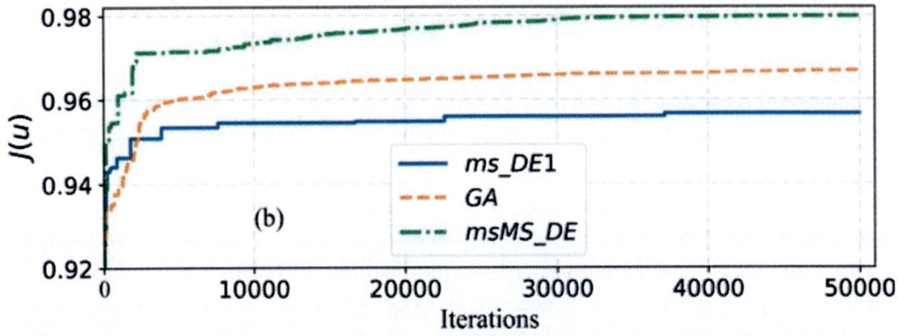

Fig. 5.1 **a** The training performance of the two-level open quantum ensemble via *ms*_DE1 ($F = 0.9, CR = 0.1$), *ms*_DE2 ($F = 0.9, CR = 0.9$) and *ms*_DE3 ($F = 0.5, CR = 0.3$). **b** The training performance of *ms*_DE1, GA and *msMS*_DE

Table 5.1 Performance comparison of different algorithms

Algorithm	Parameters	N	$\bar{\mathbf{J}}(u)$
DE1	$CR = 0.1, F = 0.9$	1	0.9408
*ms*_DE1	$CR = 0.1, F = 0.9$	9	0.9610
*ms*_DE2	$CR = 0.9, F = 0.9$	9	0.9537
*ms*_DE3	$CR = 0.3, F = 0.5$	9	0.9601
GA	$P_c = 0.8, P_m = 0.05$	9	0.9691
*msMS*_DE	$F = N(0.5, 0.3)$, $CR = N(0.5, 0.1)$	9	0.9803

average performance using multiple samples. *ms*_DE1, *ms*_DE2, *ms*_DE3, GA and *msMS*_DE also take a similar time to find an optimal solution for the ensemble control problem. For example, *msMS*_DE takes 9 h 20 min and GA takes 10 h 18 min 14 s.

5.2.3 DE for Synchronization of a Quantum Network

In this section, we consider the problem to drive a quantum network into a consensus state in the presence of uncertainties. Achieving consensus is one of primary objectives in distributed coordination and control of classical (non-quantum) networked systems [61]. Consensus usually means that all of the nodes in a network hold the same state. Recent developments in quantum technology have made it significant and feasible to analyze quantum networks where each node (agent) represents a quantum system such as a photon, an electron, a spin system or a superconducting qubit [58, 82]. Consensus of quantum networks may have potential applications in promising quantum communication networks, distributed quantum computation and one-way quantum computation [58, 82]. Since the nodes in a quantum network are quantum systems, some unique characteristics such as quantum entanglement and measurement back action, different from classical multi-agent systems [49], should be carefully considered, and the analysis and control of quantum networks raise new challenges. Some results have been presented for the consensus problem of quantum networks. For example, consensus algorithms for classical systems have been generalized to non-commutative spaces and the asymptotic convergence to the consensus state of a fully mixed state has been analyzed in [80]. The work in [58, 59, 93] presented a series of results on consensus of quantum networks including several different definitions for quantum consensus, quantum gossip algorithms and quantum consensus results within a group-theoretic framework. A systematic investigation on consensus of quantum networks with continuous-time dynamics within the framework of graph theory has been presented in [82]. Here, we consider the basic problem of finding a robust control law to steer a quantum network to a reduced state consensus (as defined in [58]) and do not consider the distributed solutions to achieving quantum consensus. In particular, we employ the proposed *msMS*_DE for driving a superconducting qubit network with uncertainties into a reduced state consensus.

In order to present the definition of reduced state consensus, we need to use the concept of partial trace mentioned in Chap. 2 which is defined as follows:

Definition 5.1 (*Partial trace*) [63] Let $\mathcal{H}_A$ and $\mathcal{H}_B$ be the state spaces of two quantum systems A and B, respectively. Their composite system is described as a density operator ρ^{AB}. The partial trace over system B denoted as $\mathrm{Tr}_{\mathcal{H}_B}$ is given in the following form:

$$\mathrm{Tr}_{\mathcal{H}_B}(|a_1\rangle\langle a_2| \otimes |b_1\rangle\langle b_2|) = |a_1\rangle\langle a_2|\mathrm{Tr}(|b_1\rangle\langle b_2|), \tag{5.20}$$

where the vectors $|a_1\rangle, |a_2\rangle \in \mathcal{H}_A$, and the vectors $|b_1\rangle, |b_2\rangle \in \mathcal{H}_B$. When the composite system is in the state ρ^{AB}, the reduced density operator for system A is defined as $\rho_A = \mathrm{Tr}_{\mathcal{H}_B}(\rho^{AB})$ and the reduced density operator for system B is defined as $\rho_B = \mathrm{Tr}_{\mathcal{H}_A}(\rho^{AB})$.

The reduced state consensus for a quantum network can be defined as follows:

Definition 5.2 (*Reduced state consensus*) [58] A quantum network consisting of m nodes with the state $\bar{\rho}$ is in a reduced state consensus if

$$\bar{\rho}_1 = \bar{\rho}_2 = \ldots = \bar{\rho}_m,$$

where $\bar{\rho}_j = \mathrm{Tr}_{\otimes_{k \neq j} \mathcal{H}_k}(\bar{\rho})$ $(j = 1, 2, \ldots, m)$ is defined as the reduced density operator for node j and can be calculated according to Definition 5.1.

We aim to steer a quantum network into a consensus state in Definition 5.2. In practical applications, the existence of intrinsic and extrinsic noises, inaccuracies (e.g., variation in the coupling between nodes) and uncertainties (e.g., fluctuations in control fields) in quantum networks is unavoidable. We assume that the Hamiltonian with uncertainties can be written as

$$H_\varepsilon(t) = (1 + \varepsilon_0) H_0 + \sum_{j=1}^{M} (1 + \varepsilon_j) u_j(t) H_j. \tag{5.21}$$

The problem can be formulated as follows:

$$\max_u \; J(u) := \max_u \mathbb{E}[J_\varepsilon(u, \bar{\rho})]$$

s.t.

$$\begin{cases} \dot{\rho}(t) = -\mathrm{i}\left[(1 + \varepsilon_0) H_0 + \sum_{j=1}^{M} (1 + \varepsilon_j) u_j(t) H_j, \rho(t)\right], \\ \rho(0) = \rho^0, \quad t \in [0, T] \\ \varepsilon_j \in [-\mathcal{E}, \mathcal{E}], \quad j = 0, 1, 2, \ldots M \end{cases} \tag{5.22}$$

where $J_\varepsilon(u, \bar{\rho})$ is the objective function for given ε and $\bar{\rho}$, $\mathbb{E}[J_\varepsilon(u, \bar{\rho})]$ denotes the average performance function with respect to the parameter variations ε, the target consensus state is $\bar{\rho}$, and $\mathcal{E} \in [0, 1]$ is the bound of the parameter uncertainties.

Here, we consider a quantum network that consists of superconducting qubits as its nodes. Assume that a quantum network consists of three superconducting qubits with control fields acting on all qubits. We denote

$$\sigma_x^{(12)} = \sigma_x \otimes \sigma_x \otimes I,$$

$$\sigma_x^{(23)} = I \otimes \sigma_x \otimes \sigma_x,$$

$$\sigma_x^{(13)} = \sigma_x \otimes I \otimes \sigma_x.$$

Its free Hamiltonian can be described as

$$H_0 = \omega_{12} \sigma_x^{(12)} + \omega_{23} \sigma_x^{(23)} + \omega_{13} \sigma_x^{(13)}. \tag{5.23}$$

Let

$$\sigma_x^{(1)} = \sigma_x \otimes I \otimes I,$$

$$\sigma_x^{(2)} = I \otimes \sigma_x \otimes I,$$

$$\sigma_x^{(3)} = I \otimes I \otimes \sigma_x,$$

and

$$\sigma_z^{(1)} = \sigma_z \otimes I \otimes I,$$

$$\sigma_z^{(2)} = I \otimes \sigma_z \otimes I,$$

$$\sigma_z^{(3)} = I \otimes I \otimes \sigma_z.$$

We have the control Hamiltonian in the following form:

$$H_u(t) = u_1^x \sigma_x^{(1)} + u_1^z \sigma_z^{(1)} + u_2^x \sigma_x^{(2)} + u_2^z \sigma_z^{(2)} + u_3^x \sigma_x^{(3)} + u_3^z \sigma_z^{(3)}. \quad (5.24)$$

The goal is to drive the quantum network from an arbitrary initial state (usually three qubits having different reduced states) to a consensus state. Furthermore, if we withdraw the external control fields, the quantum network will remain in the consensus state under the free Hamiltonian. Let $\mathbf{1}_n$ be the n-dimensional matrix with all of its elements being 1. Assume that the target state is $\bar{\rho} = \frac{1}{8}\mathbf{1}_8$. We have the following result whose proof can be found in [26]:

Proposition 5.1 *The state $\bar{\rho} = \frac{1}{8}\mathbf{1}_8$ is a consensus state for the 3-qubit network. Also, $\bar{\rho}$ is invariant under the action of free Hamiltonian*

$$H_0 = \omega_{12}\sigma_x^{(12)} + \omega_{23}\sigma_x^{(23)} + \omega_{13}\sigma_x^{(13)}.$$

The initial state is set as $\bar{\rho}^0 = \rho_1^0 \otimes \rho_2^0 \otimes \rho_3^0$ where

$$\rho_1^0 = \begin{bmatrix} 1 & 0 \\ 0 & 0 \end{bmatrix}, \ \rho_2^0 = \begin{bmatrix} \frac{1}{2} & -\frac{1}{2} \\ -\frac{1}{2} & \frac{1}{2} \end{bmatrix}, \ \rho_3^0 = \begin{bmatrix} 0 & 0 \\ 0 & 1 \end{bmatrix}.$$

In practical applications, there may exist variations in magnetic fields and electric fields in superconducting qubits. The practical control Hamiltonian is assumed to be

$$\begin{aligned} H_u(t) = &(1+\varepsilon_x)u_1^x \sigma_x^{(1)} + (1+\varepsilon_z)u_1^z \sigma_z^{(1)} + (1+\varepsilon_x)u_2^x \sigma_x^{(2)} + (1+\varepsilon_z)u_2^z \sigma_z^{(2)} \\ &+ (1+\varepsilon_x)u_3^x \sigma_x^{(3)} + (1+\varepsilon_z)u_3^z \sigma_z^{(3)}. \end{aligned} \quad (5.25)$$

We apply the *msMS_DE* algorithm to search for a robust control field to reach a consensus state in the above quantum network. The stimulation parameters are set as: the population size $NP = 100$, the time interval $[0, 20]$ ns is equally divided into 100 smaller time steps (i.e., $D = 100$), the control field components are u_1^x, u_1^z, u_2^x, u_2^z, u_3^x, $u_3^z \in [0, 1]$ GHz. Considering that we can manipulate a superconducting

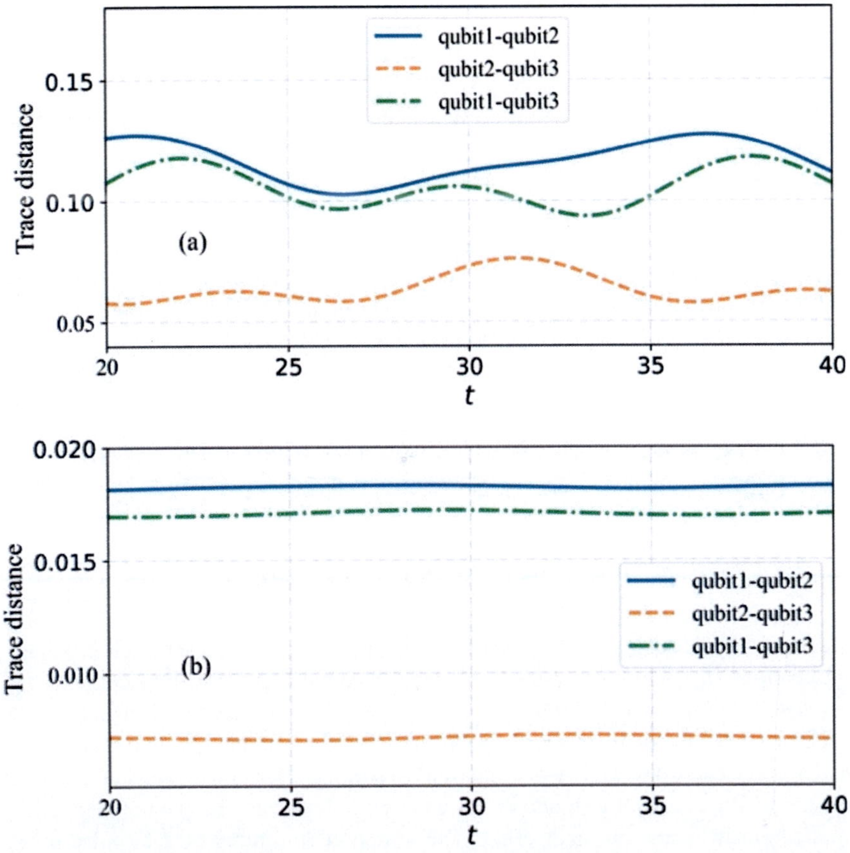

Fig. 5.2 Trace distances between different qubits in the qubit network with free Hamiltonian. **a** Average evolution curves of trace distances between the three qubits for 2000 samples (DE1). **b** Average evolution curves of trace distances between the three qubits for 2000 samples (*msMS*_DE)

circuit at a nanosecond scale and the coupling between two superconducting qubits can be at the scale of 100 MHz [21, 106], let $\omega_{12} = \omega_{23} = \omega_{13} = 0.1$ GHz. We assume that $\varepsilon_x \in [-0.02, 0.02]$ and $\varepsilon_z \in [-0.02, 0.02]$ (i.e., $\mathcal{E} = 0.02$). For each uncertain parameter, we choose three samples and have $N = 9$ samples for training. We employ DE1 (one sample) for comparison. *msMS*_DE achieves a rather high fitness $J_{\max} = 0.9988$ (where 1 is an upper bound of J), while DE1 achieves the fitness of $J_{\max} = 0.9561$. Based on the control fields from DE1 and *msMS*_ED, we test 2000 additional samples using the trace distance defined as (for $i = 1, 2, 3$)

$$\left\| \rho_i - \frac{1}{2}\mathbf{1}_2 \right\|_{\mathrm{Tr}} = \frac{1}{2}\mathrm{Tr}\sqrt{\left(\rho_i - \frac{1}{2}\mathbf{1}_2\right)^{\dagger}\left(\rho_i - \frac{1}{2}\mathbf{1}_2\right)} = \frac{1}{2}\sum_{j=1}^{2}|\lambda_j|,$$

where λ_j are eigenvalues of $\rho_i - \frac{1}{2}\mathbf{1}_2$. Since the maximum trace distance between two quantum states may be 1, we define the relative error between two quantum states ρ_i and ρ_j as $\|\rho_i - \rho_j\|_{\mathrm{Tr}} \times 100\%$. The relative error between each qubit and its target state always remains below 1.2% for the case using *msMS*_DE while the relative error between each qubit and its target state may exceed 10.0% for the case using DE1. We further show the trace distances between the quantum states of different qubits after the control Hamiltonian is withdrawn in Fig. 5.2. It is clear that the relative errors between the reduced states are always below 2.0% for the case of *msMS*_DE while the relative errors may exceed 12.0% for the case of DE1. The results demonstrate that the approximate consensus state achieved using *msMS*_DE has much better stability than that obtained using DE1.

5.3 DE-Based Control Applications in Ultrafast Quantum Engineering

5.3.1 *Ultrafast Quantum Control Engineering*

Modifying quantum mechanical systems with applied ultrafast laser fields has many applications in quantum physics, molecular chemistry and quantum engineering. Femtosecond (fs) lasers [102] have found wide applications in controlling molecular dynamics because of their short pulse duration, which is comparable to the time scales of the electronic and nuclear motions of a molecule. Femtosecond laser pulses are promising as a tool to modify molecules and create new products for practical application in quantum engineering. The temporal structures of a fs pulse could be manipulated by pulse shaping techniques [102], which is typically achieved by modulating the phase and/or amplitude of the laser frequency components with a computer programmable spatial light modulator (SLM) before recombination into a "shaped pulse". In quantum control experiments using ultrafast laser pulses, a practical approach is to use closed-loop learning control [74] to find an optimal field that can steer the quantum system toward a desired outcome. In closed-loop learning control, the learning process can be conceptually expressed as follows: First, one applies trial input pulses to the molecules subject to control and observes the results. Second, a learning algorithm suggests better control inputs based on the prior experiments. Third, one applies "better" control inputs to new molecules [74]. In this chapter, the experimental system of fs laser quantum control in the Department of Chemistry at Princeton University that we used can be conceptually illustrated in Fig. 5.3. The system consists of three key components: (1) a fs laser system, (2) a pulse shaper, and (3) a time-of-flight mass spectrometry (TOF-MS). In particular, the fs laser system (KMlab, Dragon) consists of a Ti:sapphire oscillator and an amplifier, which produces 1 mJ, 25 fs pulses centered at 790 nm. The laser pulses from the fs laser system are introduced into a pulse shaper that is equipped a programmable dual-mask liquid crystal spatial light modulator. The interaction between the spatial light

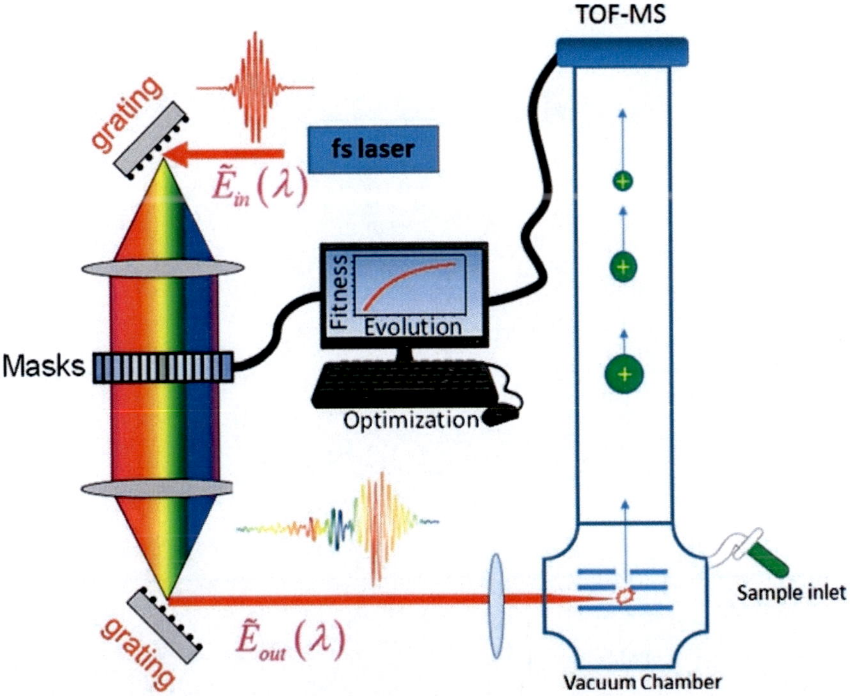

Fig. 5.3 Illustration of experimental setup of the fs laser system using closed-loop learning control

modulator and the learning algorithm is accomplished by LabVIEW software. The spatial light modulator has 640 pixels with 0.2 nm/pixel resolution and can modulate amplitude and phase independently [90, 91]. Every eight adjacent pixels are bundled together to form an array of 80 "grouped pixels". Each array of 80 "grouped pixels" corresponds to a control variable, which can be used to adjust the amplitude and phase values. In these experiments, we fix all the amplitude values at 1 (i.e., fixed energy) and only adjust the phase values to generate different control pulses (also known as a phase-only control strategy).

In this closed-loop learning control strategy, an evolutionary algorithm (e.g., GA) is often employed to assist the search for an optimal pulse. In this chapter, we employ the *MS*_DE and *msMS*_DE algorithms to investigate two experimental quantum control problems. In order to achieve good performance, we first need to identify a reference phase mask on the SLM that gives the shortest transform limited (TL) pulse, which can be obtained from optimizing the signal of two photon absorption (TPA). Optimal pulses with reference to the TL pulse are not subject to the variations of the reference pulse and are more meaningful to undergo further analysis and comparison. We also use the two DE algorithms for optimizing the TPA signal to identify the shortest pulse that removes the residual high-order dispersion in the amplifier output.

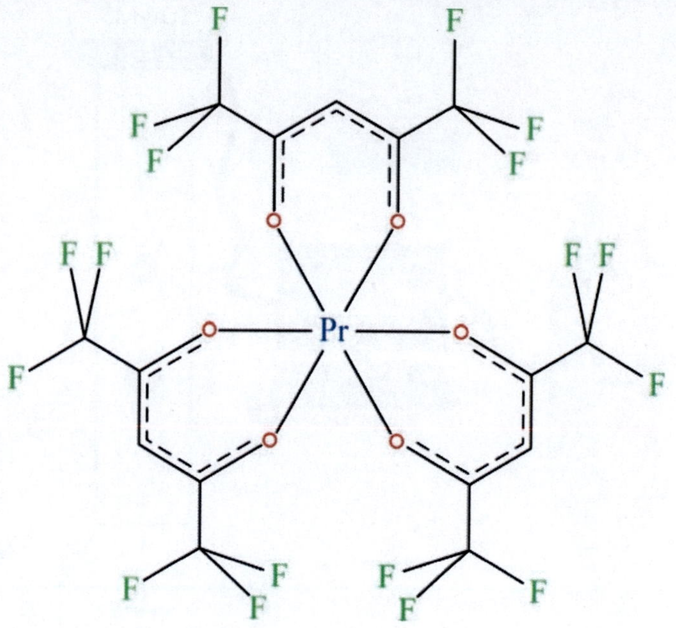

Fig. 5.4 Structure of a Pr(hfac)$_3$ molecule

5.3.2 Fragmentation Control of Pr(hfac)$_3$ Using fs Laser

Fluorinated praseodymium complexes Pr(hfac)$_3$ (hfac = hexafluoroacetylacetonate) molecules are a common precursor for making thin films of praseodymium materials with metal-organic chemical vapor deposition, because of their high thermal stability and volatility [76, 84] and superior transport properties [56, 72]. The molecular structure of Pr(hfac)$_3$ is shown in Fig. 5.4. Even though Pr(hfac)$_3$ is an oxygen-coordinated complex, the praseodymium oxides are not easy to observe using Pr(hfac)$_3$ as a precursor in prior laser-dissociation experiments. Very small amounts of oxide fragments from Pr(hfac)$_3$ were previously reported with continuous-wave and nanosecond lasers [60]. However, Pr(hfac)$_3$ is still an excellent candidate for deposition of praseodymium fluorides [55, 56]. The formation of fluorides was explained in [89], where a unimolecular reaction was initiated by rotation of the $C_\alpha - C(O)$ bond bringing the CF_3 group into proximity to the metal.

Using intense and ultrashort fs laser pulses, it is possible to observe a strong PrO^+ peak with the precursor Pr(hfac)$_3$. The shaped laser pulses on the fs timescale greatly restrict the $C_\alpha - C(O)$ bond rotation and enhance PrO^+ generation. The results explain why PrO^+ was rarely observed under continuous-wave and nanosecond laser beams in previous studies. The purity of the thin praseodymium oxides film and the efficiency to generate oxides are two interesting and valuable problems. Finding the best shaped pulses to optimize the PrO^+/PrF^+ ratio is a challenging task. We employ

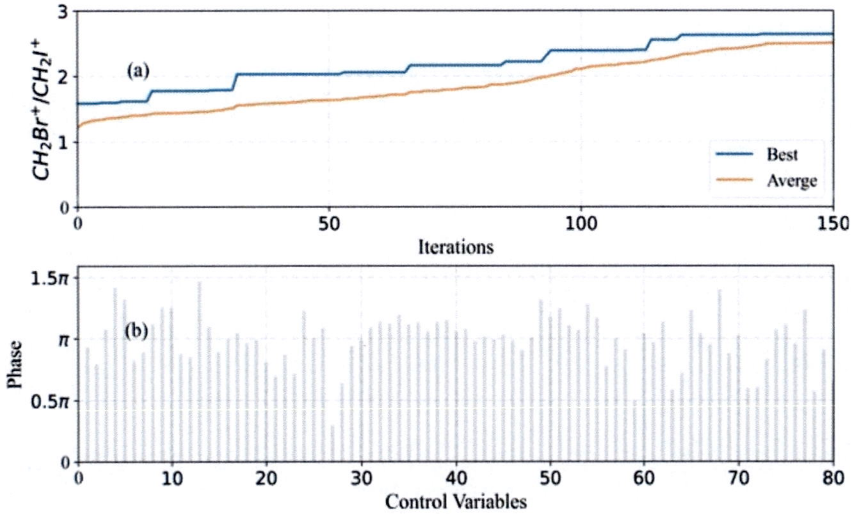

Fig. 5.5 Experimental result on the fs laser control system for optimizing the ratio between the products PrO^+ and PrF^+ using *MS*_DE without constraint on the phase [25]. **a** Ratio PrO^+/PrF^+ versus iterations, where "Best" represents the maximum fitness and "Average" represents the average fitness of all individuals during each iteration. **b** Optimized phases of 80 control variables for the optimal result that corresponds to the maximum fitness

the *MS*_DE to find an optimal field to control the PrO^+/PrF^+ fragmentation ratio in $Pr(hfac)_3$ molecules.

In the experiment, the solid $Pr(hfac)_3$ molecule samples are heated and vaporized into the gas phase in a vacuum chamber with the pressure 1.3×10^{-7} Torr. The shaped laser pulses out of the shaper are focused into the vacuum chamber, where photoionization and photofragmentation occur for the gas-phase $Pr(hfac)_3$ molecules. The fragment ions from these gas-phase $Pr(hfac)_3$ molecules are separated with a set of ion lens and pass through a TOF tube before being collected with a microchannel plate detector. The mass spectrometry signals are recorded with a fast oscilloscope, which accumulates 3000 laser shots in one second before the average signal is sent to a personal computer for further analysis. A small fraction of the beam ($< 5\%$) is separated from the main beam and focused into a DET25K Thorlab photodiode. The photodiode collects signals arising from two photon absorption for optimizing a given photofragment ratio of $Pr(hfac)_3$ molecules.

Before implementing the experiments, we first employ the *MS*_DE algorithm to optimize the TPA signal. Then we consider the fragmentation control of $Pr(hfac)_3$ molecules, where the fitness is defined as the photofragment ratio of PrO^+/PrF^+, i.e., $J = PrO^+/PrF^+$. We aim to maximize the objective function J. The control variables are the phases of fs laser pulses, and the *MS*_DE algorithm is employed to optimize the phases of 80 control variables. In the learning algorithm, the parameters are set as follows: $D = 80$ and $NP = 30$.

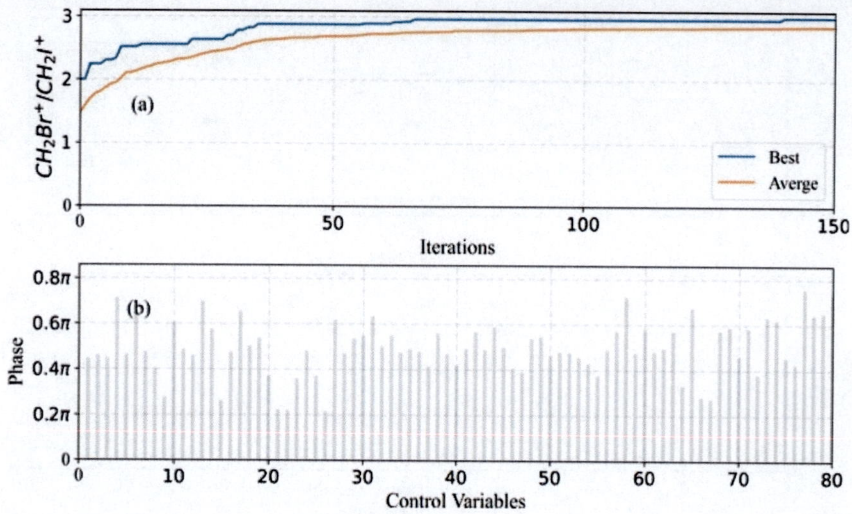

Fig. 5.6 Experimental result on the fs laser control system for optimizing the ratio between the products PrO^+ and PrF^+ using *MS*_DE when the phase is constrained in $[0, \pi]$. **a** Ratio PrO^+/PrF^+ versus iterations, where "Best" represents the maximum fitness and "Average" represents the average fitness of all individuals during each iteration. **b** Optimized phases of 80 control variables for the optimal result that corresponds to the maximum fitness

In the first experiment, we assume that there are no constraints on the phase values; that is, the phase values may take on arbitrary values between 0 and 2π. An experimentally acceptable termination condition of 1000 generations (iterations) is used. For 1000 iterations, it approximately takes twelve hours to run the experiment. For each generation, a total of 30,000 signal measurements are made. Figure 5.5 shows the experimental results using the *MS*_DE algorithm, where the ratio PrO^+/PrF^+ as the fitness function is shown in Fig. 5.5a and the 80 optimized phases for the final optimal result are presented in Fig. 5.5b. In Fig. 5.5a, "Best" represents the maximum fitness and "Average" represents the average fitness of all individuals during each iteration. With 553 iterations, *MS*_DE can find an optimized pulse to make PrO^+/PrF^+ achieve 3.067. After 553 iterations, the maximum ratio remains unchanged.

In the second experiment, we assume that the phase values can only vary between 0 and π. A termination condition of 200 generations (iterations) has been used to save the experiment time. Figure 5.6 shows the results from the *MS*_DE algorithm, where the average ratio PrO^+/PrF^+ as the fitness function is shown in Fig. 5.6a and the 80 optimized phases for the final optimal result are presented in Fig. 5.6b. *MS*_DE can find an optimized pulse to make PrO^+/PrF^+ achieve 3.037. Even though the constraint of phase values lying only between 0 and π, the ratio PrO^+/PrF^+ can still reach 99% of the ratio in the case without phase constraint at 186 iterations.

In two additional experiments, we assume that the phase values can only vary between 0 and $\frac{\pi}{2}$, and between 0 and $\frac{\pi}{4}$, respectively. The termination conditions of 200 generations (iterations) have been used in the two experiments. The results

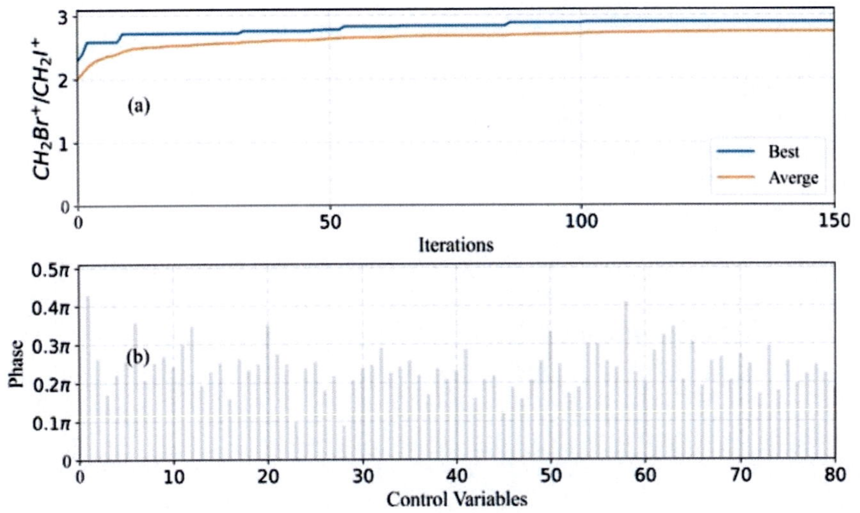

Fig. 5.7 Experimental result on the fs laser control system for optimizing the ratio between the products PrO^+ and PrF^+ using *MS*_DE when the phase is constrained in $[0, \frac{\pi}{2}]$. **a** Ratio PrO^+/PrF^+ versus iterations, where "Best" represents the maximum fitness and "Average" represents the average fitness of all individuals during each iteration. **b** Optimized phases of 80 control variables for the optimal result that corresponds to the maximum fitness

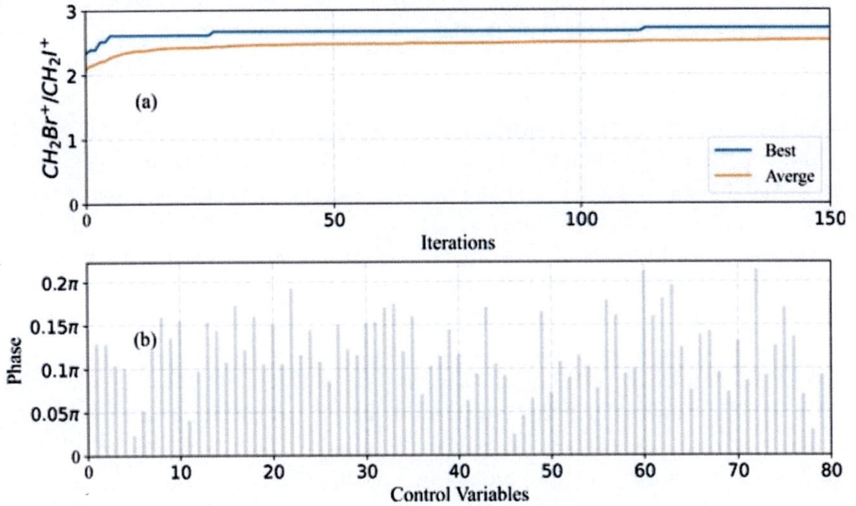

Fig. 5.8 Experimental result on the fs laser control system for optimizing the ratio between the products PrO^+ and PrF^+ using *MS*_DE when the phase is constrained in $[0, \frac{\pi}{4}]$. **a** Ratio PrO^+/PrF^+ versus iterations, where "Best" represents the maximum fitness and "Average" represents the average fitness of all individuals during each iteration. **b** Optimized phases of 80 control variables for the optimal result that corresponds to the maximum fitness

are shown in Figs. 5.7 and 5.8. From Fig. 5.7a, the *MS*_DE algorithm can find an optimized pulse to make PrO^+/PrF^+ achieve 2.898 when the phase values are constrained between 0 and $\frac{\pi}{2}$. The ratio PrO^+/PrF^+ can achieve 2.715 if the phase values are constrained between 0 and $\frac{\pi}{4}$ as shown in Fig. 5.8a. From these results, it is clear that the *MS*_DE algorithm can assist in finding good fs laser pulses to optimize the product ratio PrO^+/PrF^+ even when different constraints are placed on the amplitude and phase values of the fs laser pulses.

5.3.3 Robust Control of Photofragmentation Using fs Laser

There are few quantum control experiments using fs laser pulses that investigated robustness to variations in the control. In this section, we employ the *msMS*_DE algorithm to experimentally identify a robust solution (shaped fs laser pulse) that can maximize the CH_2Br^+/CH_2I^+ product ratio from the fragmentation of the CH_2BrI molecule. The consideration of robustness would also ensure good transferability of the experimental results or photonic reagents [92] to another laboratory. That is, an optimal pulse identified from one laser system would also perform well (if not optimal) when transferred to another system despite the minor differences or uncertainties in the control parameters (i.e., the spectral phases on the SLM) and even in the second laser system setup. Here, CH_2BrI is chosen as the target molecule. As a family member of halomethane molecules, whose dissociative products play a central role in ozone depletion, CH_2BrI has attracted wide attention because of its importance in environmental chemistry. In addition, it is one of the simplest prototype molecules containing different bonds, a stronger C–Br bond and a weaker C–I bond, which is ideal for the study of controlling selective bond-breaking. Under strong fs laser pulses, CH_2BrI molecules will undergo ionization and dissociation, and their charged products can be separated and detected with a TOF-MS. In particular, we choose to optimize the photoproduct ratio of CH_2Br^+/CH_2I^+ as our control objective, which corresponds to breaking the weak C–I bond versus the strong C–Br bond. We apply closed-loop learning control, using the proposed *msMS*_DE algorithm, to search for a robust ultrafast laser pulse that maximizes this ratio.

We first optimize the TPA signal using the parameter setting: $D = 80$, $NP = 30$ and $N = 3$. The fitness function corresponds to the TPA signal, and an average fitness of three samples is used. Three samples for each individual are selected as follows: The first sample comes from the current individual, denoted as

$$X_i^1 = \begin{bmatrix} x_i^1 \\ x_i^2 \\ \dots \\ x_i^{80} \end{bmatrix},$$

the second sample is generated by adding a random fluctuation between 0 and 0.1π to each component of the current individual, i.e.,

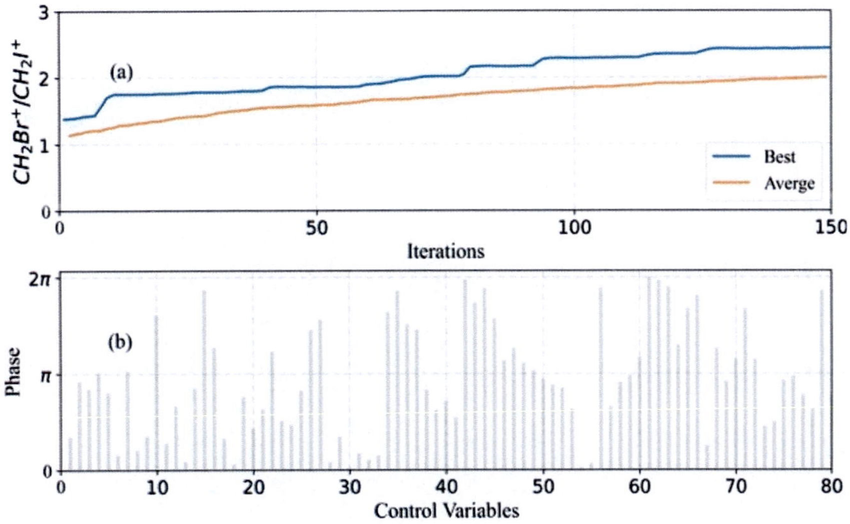

Fig. 5.9 Experimental result for optimizing the ratio between the products CH_2Br^+ and CH_2I^+ using DE1 [26]. **a** Ratio CH_2Br^+/CH_2I^+ versus iterations, where "Best" represents the maximum fitness and "Average" represents the average fitness of all individuals during each iteration. **b** Optimized phases of 80 control variables for the final optimal result corresponding to the maximum fitness

$$X_i^2 = \begin{bmatrix} x_i^1 + 0.05\text{rand}(0,1) \times 2\pi \\ x_i^2 + 0.05\text{rand}(0,1) \times 2\pi \\ \cdots \\ x_i^{80} + 0.05\text{rand}(0,1) \times 2\pi \end{bmatrix},$$

and the third sample is selected as

$$X_i^3 = \begin{bmatrix} x_i^1 - 0.05\text{rand}(0,1) \times 2\pi \\ x_i^2 - 0.05\text{rand}(0,1) \times 2\pi \\ \cdots \\ x_i^{80} - 0.05\text{rand}(0,1) \times 2\pi \end{bmatrix}.$$

This means that each control variable is permitted to have up to 5% (of the maximum phase) additive noise. An experimentally reasonable termination condition of 150 generations (iterations) is used. For 150 iterations, it approximately takes five and a half hours to run the experiment. After 150 generations, the best average TPA signal for three samples can reach 1.35.

After optimizing the TPA signal, we consider the fragmentation control of CH_2BrI, where the fitness is defined as maximization of the photofragment ratio of CH_2Br^+/CH_2I^+, while the control variables are the phases and the parameter setting is $D = 80$ and $NP = 30$. DE algorithms are employed to optimize the phases of 80 control variables. Here, we apply both DE1 and *msMS_DE* for comparison.

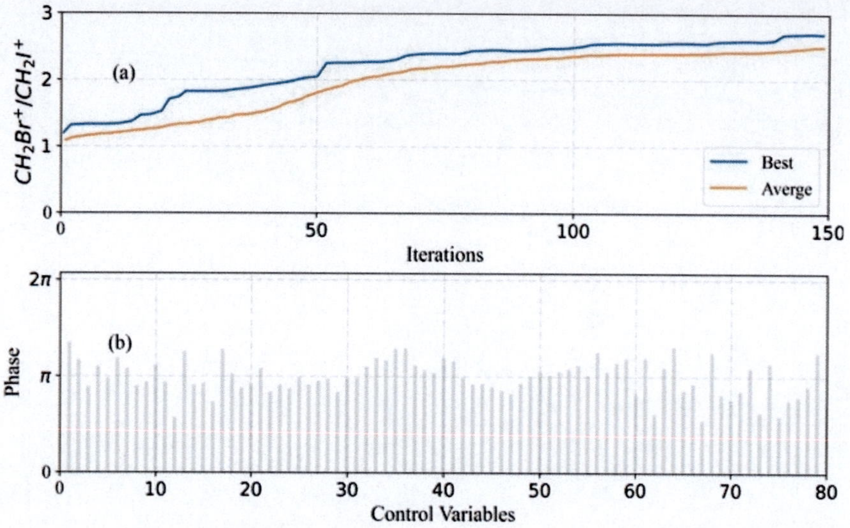

Fig. 5.10 Experimental result for optimizing the ratio between the products CH_2Br^+ and CH_2I^+ using *msMS*_DE [26]. **a** Ratio CH_2Br^+/CH_2I^+ versus iterations, where "Best" represents the maximum fitness and "Average" represents the average fitness of all individuals during each iteration. **b** Optimized phases of 80 control variables for the final optimal result corresponding to the maximum fitness

Figure 5.9 shows the experimental results using the DE1 algorithm, where the ratio CH_2Br^+/CH_2I^+ as the fitness function is presented in Fig. 5.9a and the optimized phases of 80 control variables for the final optimal result is given in Fig. 5.9b. In Fig. 5.9a, "Best" represents the maximum fitness and "Average" represents the average fitness of all individuals during each iteration. With 150 iterations, DE1 can find an optimized pulse to make CH_2Br^+/CH_2I^+ to achieve 2.41.

Figure 5.10 shows the results from the *msMS*_DE algorithm, in which three samples are measured in each experiment. Three samples for each individual are selected using the same method as that in the experiments of optimizing TPA signals. With 150 iterations, *msMS*_DE can find an optimal pulse for making the average CH_2Br^+/CH_2I^+ of three samples to achieve 2.67. The experimental results are shown in Fig. 5.10, where the average ratio CH_2Br^+/CH_2I^+ of three samples as the fitness function is presented in Fig. 5.10a and the optimized phases of 80 control variables for the final optimal result are given in Fig. 5.10b.

After we obtain the optimal fs control pulses using DE1 and *msMS*_DE, we can test the performance of the optimal pulses. We consider 100 testing samples with random noises between -7.5 and $+7.5\%$ (with respect to the maximum phase 2π) for the optimized fs pulse. If we denote the best individual as

$$X_{\text{best}} = \begin{bmatrix} x_b^1 \\ x_b^2 \\ \dots \\ x_b^{80} \end{bmatrix},$$

these 100 testing samples can be written as (for $k = 1, \dots, 100$)

$$X_s^k = \begin{bmatrix} x_b^1 + 0.075(2\text{rand}(0, 1) - 1) \times 2\pi \\ x_b^2 + 0.075(2\text{rand}(0, 1) - 1) \times 2\pi \\ \dots \\ x_b^{80} + 0.075(2\text{rand}(0, 1) - 1) \times 2\pi \end{bmatrix}.$$

The average CH_2Br^+/CH_2I^+ of the 100 testing samples can achieve 2.61 for the pulse from *msMS_DE* while the average CH_2Br^+/CH_2I^+ is only 2.12 for the pulse from DE1. It is clearly evident that *msMS_DE* outperforms DE1 in terms of reaching a better objective fitness value in the presence of phase noise (e.g., shot-to-shot variations in reproducing laser source).

5.4 Learning Control Design of Quantum Autoencoders

5.4.1 Quantum Autoencoders and Compression Rate

Information compression is one of the fundamental tasks in classical information theory [33, 35, 69, 81, 100, 113], and various compression methods have found wide applications such as text coding [33, 100] and image compression [35, 69]. In the quantum domain, the compression of quantum data has aroused widespread attention [38, 39] because it is highly valuable for effective utilization of precious quantum resources and efficient reduction of quantum memory in quantum communication networks, distributed quantum computation and quantum simulation [70]. Many methods of quantum information compression have been proposed [4, 50, 71, 78, 103, 104] among which quantum autoencoders [6, 18, 45, 70, 77, 85, 95] are a promising approach due to their capability of learning the data structure.

A traditional autoencoder aims at compressing data into a lower-dimensional space. As shown in Fig. 5.11a, the input information can be compressed into a lower-dimensional space with the encoder E, and the decoder D can reconstruct the input data at the output. Autoencoders form one of the core issues in machine learning and have wide applications [29, 46]. In recent years, quantum machine learning, which combines both quantum physics and machine learning, shows powerful capability in various applications [5, 20, 47, 68, 75]. Autoencoders for quantum data can be looked as a subfield of quantum machine learning and aim at reallocating quantum information for efficient utilization of quantum resources. During this process, the input state is divided into two parts, where the latent state contains the important

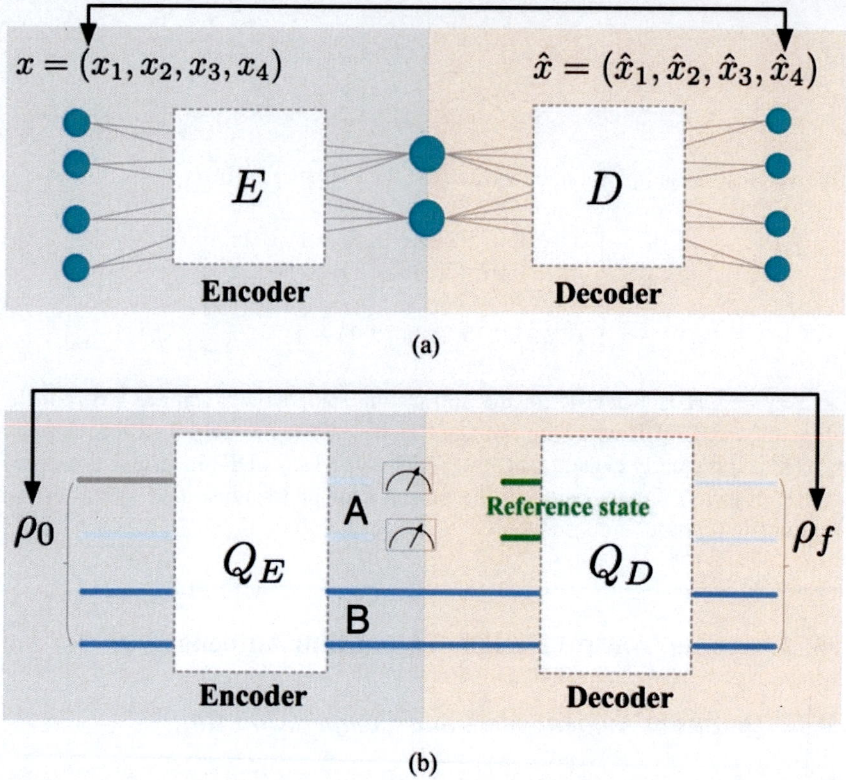

Fig. 5.11 **a** Neural network representation of a classical 4–2–4 autoencoder. **b** Quantum circuit representation of a quantum 4–2–4 autoencoder

information and the trash state represents the redundant information. Autoencoders for quantum data have received much attention in the field of quantum information in recent years [18, 45, 70, 77, 85, 95]. For example, the work [95] introduced a feedforward quantum neural network for quantum data compression. A simple autoencoder framework was formulated using a programmable circuit for compressing the ground states of the Hubbard model and molecular Hamiltonians in [77]. For dimension reduction of qudits, a quantum autoencoder with low-level errors has been experimentally realized in [70] where the occupation probability of "junk" mode was utilized as the cost function. The work [85] designed a quantum optical neural network and implemented simulations for quantum optical state compression. The connection between a quantum autoencoder and an approximate quantum adder has been analyzed [45], and an experimental result on quantum autoencoders based on the Rigetti cloud quantum computer has been presented in [18]. A novel quantum autoencoder was developed to denoise Greenberger–Horne–Zeilinger (GHZ) states subject to spin-flip errors and random unitary noise [6].

For a quantum device to realize an autoencoder, as illustrated in Fig. 5.11b, a quantum encoder $Q_{\rm E}$ (corresponding to a unitary operator) is trained where measurement results are considered and an optimization algorithm is employed to iteratively optimize $Q_{\rm E}$. Among the investigation of quantum autoencoders, the compression rate is a significant indicator on the efficiency of quantum autoencoders. A given set of states may only admit a certain compression rate, which is closely related with the inner pattern or the structure of input states [77]. It is highly desirable to determine general criteria for the compression rate of a quantum autoencoder. For example, the compression rate can be defined as the fidelity between the input state ρ_0 and the output state $\rho_{\rm f}$. We analyze the relationship between the compression rate and the inner structure of the input states and present numerical and experimental implementation to achieve quantum information compression using a trained quantum autoencoder.

A quantum autoencoder aims at encoding n-qubit state ρ_0 into d-qubit state ρ_B in the latent space (B space) where $n > d$, and recovering to n-qubit state $\rho_{\rm f}$ using a decoder $Q_{\rm D}$. As an illustrative example, a graphical representation for a 4–2–4 quantum autoencoder is described in Fig. 5.11b, where $n = 4$ and $d = 2$. The network includes two parts: (1) Encoder $Q_{\rm E}$ (generally taken as a unitary transformation) reorganizes the 4-qubit input state ρ_0 onto the inner layer of the latent qubits, followed by discarding superfluous information contained in some of the input nodes. For example, this can be realized by tracing out the qubits representing these nodes. (2) Decoder $Q_{\rm D}$ (another unitary transformation) reconstructs the 4-qubit state $\rho_{\rm f}$ by using the combinations of the latent state and ancillary fresh qubits (initialized to the reference state). The goal of the quantum autoencoder is to maximize the overlap between the recovered state and the original state. In this work, we mainly consider input pure states and unitary transformation maps for $Q_{\rm E}$ and $Q_{\rm D}$.

Different from a classical autoencoder, where both the encoding and decoding operations are usually performed to optimize the parameters for E and D, the training of a quantum autoencoder can be reduced to optimizing the encoding transformation $Q_{\rm E}$. After encoding quantum states from the data set using a trained unitary $Q_{\rm E}$, quantum states can be naturally decoded by acting with $Q_{\rm D} = Q_{\rm E}^{\dagger}$ [70]. To train the quantum autoencoder, a practical way is to feed a set of input states into the network. As such, the task of a quantum autoencoder is to design an encoding map $Q_{\rm E}$ to compress a given set of quantum states to optimize the compression rate.

In order to find an optimal quantum encoder, it is natural to use the fidelity between the recovered state $\rho_{\rm f}$ and the initial state ρ_0 as an objective function (compression rate) for optimization. For simplifying the optimization problem, we here choose the overlap between the trash state in A space (representing the superfluous information) and the reference state $\rho_{\rm ref}$ to quantify the efficiency of a quantum autoencoder. Denote the reference state as $\rho_{\rm ref} = |\Psi_{\rm ref}\rangle\langle\Psi_{\rm ref}|$, where $|\Psi_{\rm ref}\rangle$ is an arbitrary fixed pure state. The objective function may be defined as

$$J = F(\rho_{\rm ref}, \rho_A) = \langle\Psi_{\rm ref}|{\rm Tr}_B(U\rho_0 U^{\dagger})|\Psi_{\rm ref}\rangle. \tag{5.26}$$

Now, we consider the compression of a set of input states using the same unitary operator U (corresponding to the quantum encoder $Q_{\rm E}$). Let $\{p_i, |\psi_0^i\rangle\}_{i=1}^{K}$ be an ensemble of input pure states, and we may write the corresponding density operator as follows:

$$\rho = \sum_{i=1}^{K} p_i |\psi_0^i\rangle\langle\psi_0^i|. \tag{5.27}$$

For each input state $|\psi_0^i\rangle$, the performance is evaluated by

$$J(U, |\psi_0^i\rangle) = \langle\Psi_{\text{ref}}|\text{Tr}_B(U|\psi_0^i\rangle\langle\psi_0^i|U^\dagger)|\Psi_{\text{ref}}\rangle. \tag{5.28}$$

The overall objective function can be written as

$$\begin{aligned} J(U) &= \sum_{i=1}^{K} p_i \langle\Psi_{\text{ref}}|\text{Tr}_B(U|\psi_0^i\rangle\langle\psi_0^i|U^\dagger])|\Psi_{\text{ref}}\rangle \\ &= \langle\Psi_{\text{ref}}|\text{Tr}_B\left(\sum_{i=1}^{K} p_i U|\psi_0^i\rangle\langle\psi_0^i|U^\dagger\right)|\Psi_{\text{ref}}\rangle \\ &= \langle\Psi_{\text{ref}}|\text{Tr}_B(U\rho U^\dagger)|\Psi_{\text{ref}}\rangle. \end{aligned} \tag{5.29}$$

In this chapter, the fidelity between $|\Psi_{\text{ref}}\rangle$ and the trash state $\text{Tr}_B(U\rho U^\dagger)$ is defined as the compression rate. Theoretically, $|\Psi_{\text{ref}}\rangle$ can be any pure state, but it is usually set as a pure state that is easy to generate in physical implementation, such as $|00\ldots0\rangle_A$.

By means of eigen-decomposition, the following conclusion has been proved in [32]: A perfect quantum autoencoder can be achieved if the number of maximum linearly independent vectors from the input states is no more than the dimension of the latent space. Further an upper bound on the compression rate for a given quantum autoencoder has been established in [54], which is determined by the eigenvalues of the density matrix representation of the input states (for more details, please see [54]).

Now, we present a learning control method for training the autoencoder to approach or achieve the maximal compression rate. A quantum autoencoder aims at compressing a set of input states $\{p_k, |\psi_0^k\rangle\}_{k=1}^K$ using the same control strategy $\mathbf{u} := \{u_j(t)\}_{j=1}^M$, with an average loss function regarding different input states. The problem of finding the unitary operator $\Phi_{\text{U}}(\mathbf{u})$ for a quantum autoencoder can be formulated as

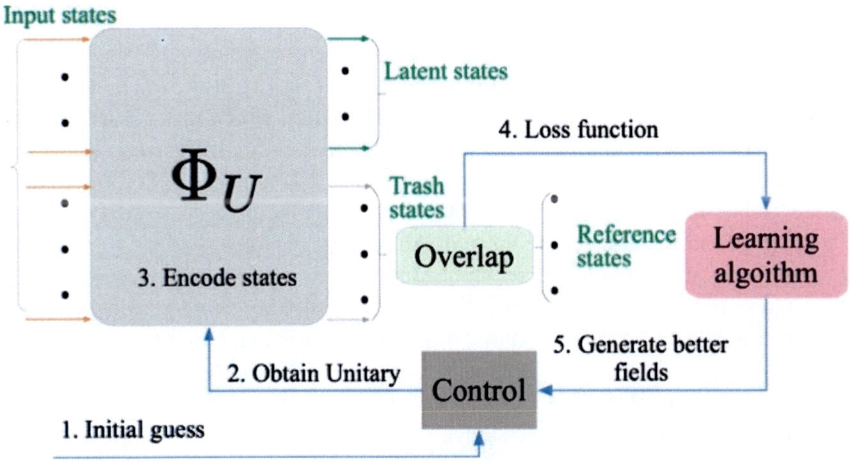

Fig. 5.12 Procedure for the closed-loop learning control approach to a quantum autoencoder

$$\max_{\mathbf{u}} J(\mathbf{u}) := \max_{\mathbf{u}} \mathbb{E}[J(\mathbf{u}, |\psi_0^k\rangle)] = \max_{\mathbf{u}} \sum_{k=1}^{K} p_i J(\mathbf{u}, |\psi_0^k\rangle)$$

s.t.

$$\begin{cases} \dfrac{\mathrm{d}}{\mathrm{d}t}\Phi_{\mathrm{U}}(\mathbf{u}(t)) = -\mathrm{i}\left(H_0 + \displaystyle\sum_{j=1}^{M} u_j(t)H_j\right)\Phi_{\mathrm{U}}(\mathbf{u}(t)), \quad t \in [0, T], \\ |\psi_T^k\rangle = \Phi_{\mathrm{U}}(\mathbf{u}(T))|\psi_0^k\rangle, \quad k = 1, 2, \ldots, Q, \\ J(\mathbf{u}, |\psi_0^k\rangle) = \langle\Psi_{\mathrm{ref}}|\mathrm{Tr}_B(|\psi_T^k\rangle\langle\psi_T^k|)\Psi_{\mathrm{ref}}\rangle. \end{cases} \tag{5.30}$$

To find a numerical optimal control solution for problem (5.30), we adopt a closed-loop learning control approach as shown in Fig. 5.12 to train the quantum autoencoder. The approach starts from an initial guess, and employs learning algorithms to suggest a better control strategy, based on the learning performance of the prior trial. For each trial, it is an open-loop control process, while the control performance is sent back to the learning algorithm to guide the optimization for the control strategy [15]. The general procedure is summarized as follows:

Step 1: Generate an initial guess of feasible control field $\mathbf{u}$.
Step 2: Obtain the unitary operator $\Phi_{\mathrm{U}}(\mathbf{u})$.
Step 3: Perform the unitary transformation $\Phi_{\mathrm{U}}(\mathbf{u})$ for all the input states and obtain their trash states.
Step 4: Compute the average objective function $J(\mathbf{u})$.
Step 5: Suggest a better control field $\mathbf{u}$ using machine learning algorithms and go to Step 2.

A crucial step of the general procedure lies in Step 5 that suggests better control fields. For example, gradient (usually Gradient Descent) methods (GD) have gained wide popularity due to their computational efficiency [14, 41]. However, they usually rely on accurate system models for gradient evaluation, which may result in local traps. Stochastic searching methods such as evolutionary algorithms can step over local maxima and have been widely used in complex quantum control problems due to their global searching abilities [2, 109, 111]. Among them, GA has achieved great success in closed-loop learning control of laboratory quantum systems [40, 94]. DE has emerged as another powerful technique in real optimization problems, especially for quantum robust control problems [26, 51]. Evolutionary strategy (ES) methods have been applied in quantum control experiments [83] and exhibit an advantage in exploring unknown environments in games [79]. In this chapter, we compare the performance of employing GD, DE, GA, ES to optimize a quantum autoencoder. For DE algorithm, we use DE/rand/1 strategy. The other algorithms used are outlined in the Appendix of this chapter. The termination criterion for training the quantum autoencoder can be set as the maximum number of iterations, or the gap of $J(\mathbf{u})$ between a number of successive iterations below a predefined small threshold.

5.4.2 Numerical Results

In this section, numerical results on 2-qubit and 3-qubit systems are presented. The input states are generated randomly and independently several times to form a set of pure states with equal probabilities. For each compression task, numerical simulation is implemented for 20 runs independently, and each run deals with different input states.

For a 2-qubit system, we assume that its Hamiltonian is described as

$$H(t) = H_0 + \sum_{j=1}^{4} u_j(t) H_j,$$

where the free Hamiltonian is given as $H_0 = \sigma_z \otimes \sigma_z$. The control Hamiltonian operators are

$$H_1 = \sigma_x \otimes I,$$

$$H_2 = I \otimes \sigma_x,$$

$$H_3 = \sigma_y \otimes I,$$

$$H_4 = I \otimes \sigma_y.$$

The control time duration $T_{\mathrm{f}} = 1.1$ a.u. is equally divided into 20 subintervals. The control fields are constrained as $u_1, u_2, u_3, u_4 \in [-4, 4]$.

For 3-qubit system, we assume that its Hamiltonian is

$$H(t) = H_0 + \sum_{j=1}^{6} u_j(t) H_j.$$

Denote

$$\sigma_x^{(12)} = \sigma_x \otimes \sigma_x \otimes I,$$

$$\sigma_x^{(23)} = I \otimes \sigma_x \otimes \sigma_x,$$

$$\sigma_x^{(13)} = \sigma_x \otimes I \otimes \sigma_x.$$

The free Hamiltonian is

$$H_0 = 0.1\sigma_x^{(12)} + 0.1\sigma_x^{(23)} + 0.1\sigma_x^{(13)}.$$

Let

$$\sigma_x^{(1)} = \sigma_x \otimes I \otimes I,$$

$$\sigma_x^{(2)} = I \otimes \sigma_x \otimes I,$$

$$\sigma_x^{(3)} = I \otimes I \otimes \sigma_x,$$

and

$$\sigma_z^{(1)} = \sigma_z \otimes I \otimes I,$$

$$\sigma_z^{(2)} = I \otimes \sigma_z \otimes I,$$

$$\sigma_z^{(3)} = I \otimes I \otimes \sigma_z.$$

The control Hamiltonian is

$$H_c(t) = u_1^x \sigma_x^{(1)} + u_1^z \sigma_z^{(1)} + u_2^x \sigma_x^{(2)} + u_2^z \sigma_z^{(2)} + u_3^x \sigma_x^{(3)} + u_3^z \sigma_z^{(3)}.$$

The control time duration $T_f = 20$ a.u. is equally divided into 100 subintervals. The control field is constrained as $u_1^x, u_1^z, u_2^x, u_2^z, u_3^x, u_3^z \in [0, 1]$.

Here, we compare the performance of four learning algorithms on 2-qubit and 3-qubit systems. Firstly, the training performance for compressing 2-qubit into 1-qubit and 3-qubit into 2-qubit states for 20 runs are shown in Fig. 5.13. According to the results in Fig. 5.13a, b, the four algorithms achieve similar performance with almost the same fidelity and convergence rate on 2-qubit systems. To be specific, DE is the fastest, while ES is the slowest. However, the performance of four algorithms on 3-qubit systems is quite different, where ES and GD exhibit comparative performance, while GA and DE fall far behind in Fig. 5.13c, d. In addition, we measure the average

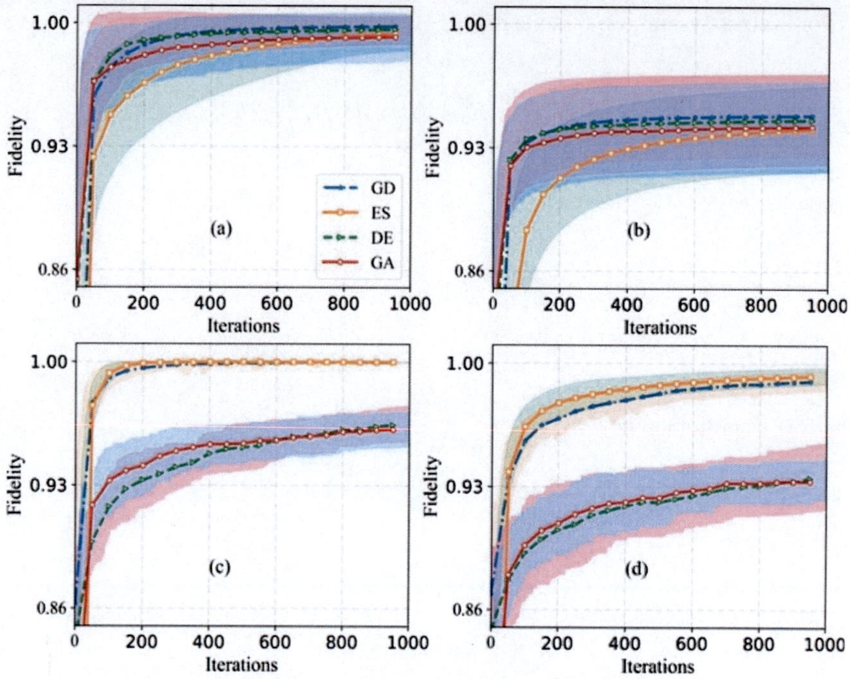

Fig. 5.13 Learning performance by four learning algorithms. **a** 2-qubit → 1-qubit ($K = 2$). **b** 2-qubit → 1-qubit ($K = 4$). **c** 3-qubit → 2-qubit ($K = 2$). **d** 3-qubit → 2-qubit ($K = 4$)

Table 5.2 Statistical results of best fidelities for 3-qubit systems [54]

	$K = 2$		$K = 4$	
Method	Mean	Std	Mean	Std
GD	0.999999	1.641E−12	0.999934	1.640E−04
ES	0.999999	2.881E−10	0.999955	4.716E−05
DE	0.993053	5.150E−03	0.948686	1.914E−02
GA	0.974116	7.659E−03	0.927634	1.232E−02

recovered fidelities between ρ_0 and ρ_f using the optimal control strategies searched for by the four learning algorithms for 3-qubit case. From Table 5.2, the performance of GD regarding both mean and variance ranks first, with a nearly perfect value for 3-qubit input states ($K = 2, 4$), closely followed by the results for ES. The high variances of DE and GA for 3-qubit systems reveal that they are not as robust as GD and ES, which achieve small variance values. The results show that GD and ES are more powerful and robust in solving the optimization problem of the quantum autoencoder. For more details of parameter settings and numerical results, please refer to [54].

5.4.3 Experimental Results on Quantum Optical Systems

In this section, we focus on the experimental realization of a quantum autoencoder to compress 2-qubit states into 1-qubit states. It is well known that any binary quantum alternative of a photon can serve as a qubit [32, 97]. Thus, the polarization and path degrees of freedom can serve as two qubits. Let $|R\rangle/|L\rangle$ be two eigenstates for the path qubit and $|H\rangle/|V\rangle$ be two eigenstates for the polarization qubit. Photons can be physical carriers of 2-qubit states, such as $|RH\rangle$ and $|LV\rangle$.

The experimental setup is illustrated in Fig. 5.14. It is divided into four modules.

(1) **State preparation** (Fig. 5.14a, b)**:** A Sagnac interferometer is used to generate phase-stable 2-qubit states. Photon pairs with wavelength $\lambda = 808$ nm are created using type-I spontaneous parametric down-conversion (SPDC) in a nonlinear crystal (BBO), which is pumped by a 40-mW beam at 404 nm. The two photons pass through two interference filters whose full width at half maximum is 3 nm. One is sent to a single photon counting module (SPCM) to act as a trigger. The other is prepared in the state of highly pure horizonal polarization state $|H\rangle$ through a polarizer beam splitter (PBS). Then a half-wave plate (HWP) along with a PBS controls the path qubit of the photon. In each path, an HWP and a quarter-wave plate (QWP) are used to control the polarization of the photon.
(2) **Parameterized unitary gate** (Fig. 5.14c)**:** Another interferometer is used to generate a 2-qubit unitary gate. In particular, four unitary polarization operators V_1, V_2, V_{R} and V_{L} are used, and each of them is composed of two QWPs, an HWP, and a phase shifter (PS) consisting of a pair of wedge-shaped plates, which are controlled by a computer. A special beam splitter cube that is half PBS-coated and half coated by a non-polarizer beam splitter (NBS) is applied in the junction of two Sagnac interferometers. By combining path unitary gates with polarization gates, 2-qubit universal parameterized unitary gates can be realized, and then these unitary gates can be characterized by process tomography using various algorithms such as maximum-likelihood estimation [36] and linear regression estimation [97]. Here, we estimate the process matrix using the maximum-likelihood method [36]. Please refer to [27] for detailed information.
(3) **Measurements** (Fig. 5.14d, e)**:** Local measurements on polarization can be achieved with the combination of a QWP, an HWP and a PBS. The typical count rate is set as 3000 photons per second.
(4) **Optimization routine** (Fig. 5.14f)**:** A computer collects the coincidence and employs machine learning algorithms to optimize the rotations of the wave plates of the 2-qubit unitary gate.

Now we turn to the core task of encoding quantum information into a lower dimension. The goal is to find a 2-qubit unitary operator U which can encode two 2-qubit states $|\varphi_1\rangle$, $|\varphi_2\rangle$ into two 1-qubit states $|\varphi_1'\rangle$, $|\varphi_2'\rangle$. We may encode two 2-qubit states $|RH\rangle$, $|LV\rangle$ into states $|R\rangle|\varphi_1'\rangle$, $|R\rangle|\varphi_2'\rangle$. Here $|R\rangle/|L\rangle$ stands for the path qubit and $|H\rangle/|V\rangle$ stands for the polarization qubit. Thus, we can trash the path qubit and obtain the compressed states $|\varphi_1'\rangle$, $|\varphi_2'\rangle$ which maintain the original quantum

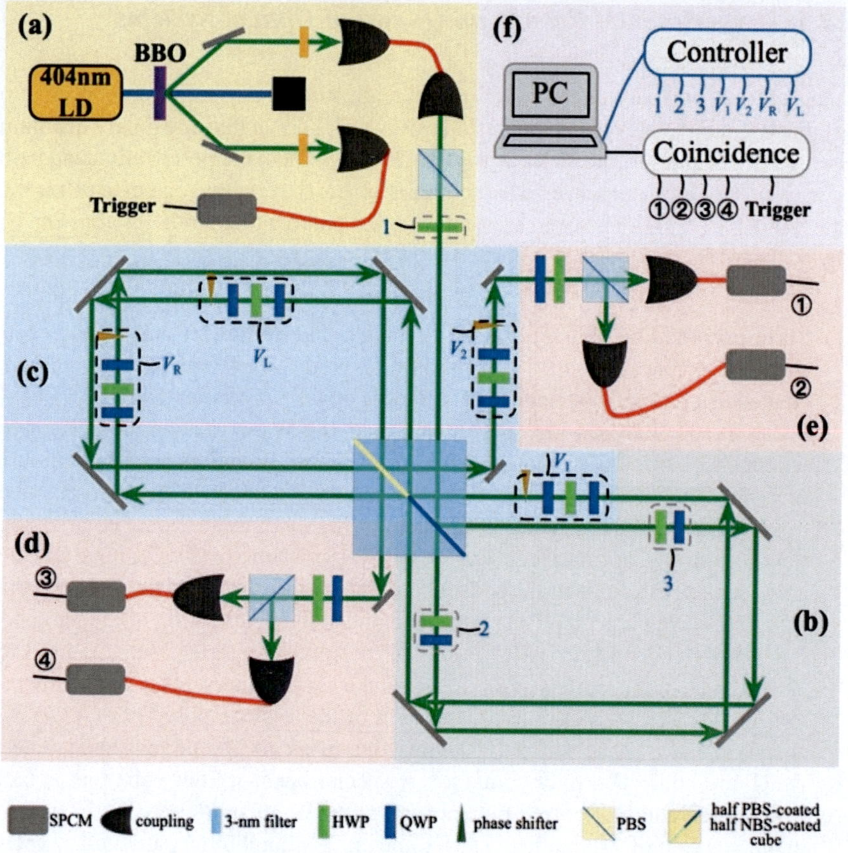

Fig. 5.14 Schematic of the experimental setup for a quantum autoencoder. The setup is divided into four modules. **a**, **b** State preparation, **c** parameterized unitary gate, **d**, **e** measurements, **f** optimizing routine

information in the polarization qubit. Similarly, encoding the information into a path qubit is also feasible. Using the stochastic gradient algorithm in the Appendix of this chapter, we efficiently train the parametrized unitary operator U to achieve the goal. Figure 5.15a shows the results of encoding

$$\left\{\frac{1}{\sqrt{2}}(|RH\rangle + |LH\rangle), \frac{1}{\sqrt{2}}(|RV\rangle + |LV\rangle)\right\}$$

into the polarization qubit. Here, infidelity is the cost function in the algorithm and iterations indicate the training process. The results of encoding another set of states

$$\left\{\frac{1}{4}(|RH\rangle - \mathrm{i}|RV\rangle - |LH\rangle + \mathrm{i}|LV\rangle), \frac{1}{4}(|RH\rangle + \mathrm{i}|RV\rangle + |LH\rangle + \mathrm{i}|LV\rangle)\right\}$$

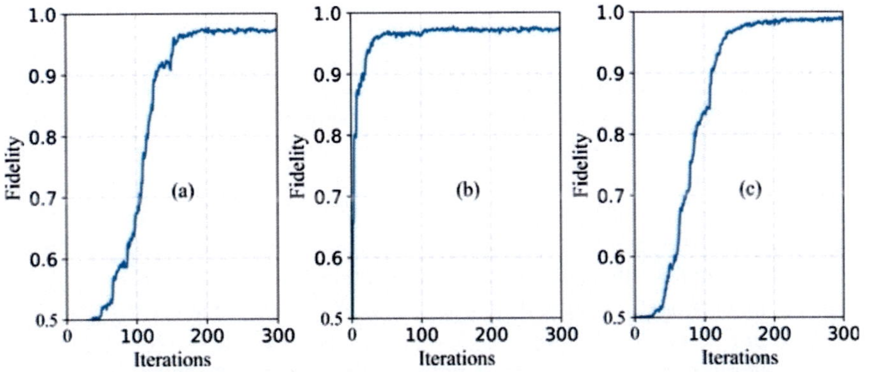

Fig. 5.15 Experimental results of encoding two 2-qubit states into two 1-qubit polarization states for different initial states. **a** Input states $\{\frac{1}{\sqrt{2}}(|RH\rangle + |LH\rangle), \frac{1}{\sqrt{2}}(|RV\rangle + |LV\rangle)\}$. **b** Input states $\{\frac{1}{4}(|RH\rangle - \mathrm{i}|RV\rangle - |LH\rangle + \mathrm{i}|LV\rangle), \frac{1}{4}(|RH\rangle + \mathrm{i}|RV\rangle + |LH\rangle + \mathrm{i}|LV\rangle)\}$. **c** Input states $\{\frac{1}{4}(|RH\rangle - \mathrm{i}|RV\rangle + |LH\rangle + \mathrm{i}|LV\rangle), \frac{1}{4}(|RH\rangle + \mathrm{i}|RV\rangle + |LH\rangle + \mathrm{i}|LV\rangle)\}$

into the polarization qubit are shown in Fig. 5.15b. Figure 5.15c shows the results of encoding

$$\left\{\frac{1}{4}(|RH\rangle - \mathrm{i}|RV\rangle + |LH\rangle + \mathrm{i}|LV\rangle), \frac{1}{4}(|RH\rangle + \mathrm{i}|RV\rangle + |LH\rangle + \mathrm{i}|LV\rangle)\right\}$$

into the polarization qubit. The input states in the experiments are generally linearly independent. Hence, the maximum number of linearly independent vectors among the input states equals the dimension of the latent space, which means that a perfect quantum autoencoder can be theoretically achieved. Here, the performance of the quantum autoencoder in this paper is related to the experimental conditions such as imperfect NBS-coated surface, unbalanced coupling efficiency, and uneven wave plates. Even under these imperfect conditions, the fidelities in Fig. 5.15 can still approach 1 after 300 iterations.

5.5 Reinforcement Learning for Quantum Control

5.5.1 Q-Learning for Quantum Control

Reinforcement learning (RL) [88] is an important approach to machine learning, control engineering, operations research, etc. RL algorithms, such as the temporal difference (TD) algorithms [87] and Q-learning algorithms [98], have been extensively studied in various aspects and widely used in intelligent control and industrial applications [19, 44, 96]. However, there exist several difficulties in developing practical applications from RL methods. These difficult issues include the tradeoff between exploration and exploitation, function approximation methods and speedup of the learning process. Hence, new ideas are necessary to improve reinforcement learning performance. In [20], two features (i.e., quantum parallelism and probabilistic phenomena) from the superposition of probability amplitudes in quantum computation [63] were bought to improve TD learning algorithms [12, 20, 22]. Inspired by [20], here we focus only on the probabilistic essence of decision-making in Q-learning with fidelity-directed exploration strategy, and propose a fidelity-based probabilistic Q-learning method for the control design of quantum systems.

Q-learning can acquire optimal control policies from delayed rewards, even when the agent has no prior knowledge of the environment. For the discrete case, a Q-learning algorithm assumes that the state set S and action set A can be divided into discrete values. At a certain step t, the *agent* observes the state s_t and then chooses an action a_t. After executing the action, the agent receives a reward r_{t+1}, which reflects how good that action is (in a short-term sense). The state will change into the next state s_{t+1} under action a_t. Then the agent will choose the next action a_{t+1} according to the best known knowledge. The goal of Q-learning is to learn a policy

$$\pi : S \times \cup_{i \in S} A_{(i)} \rightarrow [0, 1],$$

so that the expected sum of discounted rewards for each state will be maximized:

$$Q^{\pi}_{(s,a)} = r^a_s + \gamma \sum_{s'} p^a_{ss'} \sum_{a'} p^{\pi}(s', a') Q^{\pi}_{(s',a')}, \tag{5.31}$$

where $\gamma \in [0, 1)$ is a discount factor,

$$p^a_{ss'} = \Pr\{s_{t+1} = s' | s_t = s, a_t = a\}$$

is the probability for state transition from s to s' with action a, $p^{\pi}(s', a')$ is the probability of selecting action a' for state s' under policy π and

$$r^a_s = E\{r_{t+1} | s_t = s, a_t = a\}$$

is an expected one-step reward. $Q_{(s,a)}$ is called the value function of the state-action pair (s, a). Let α_t be the learning rate. The one-step updating rule of Q-learning may be described as:

$$Q(s_t, a_t) \leftarrow (1 - \alpha_t) Q(s_t, a_t) + \alpha_t (r_{t+1} + \gamma \max_{a'} Q(s_{t+1}, a')). \tag{5.32}$$

The optimal value function $Q^*_{(s,a)}$ satisfies the Bellman equation [88]:

$$Q^*_{(s,a)} = \max_{\pi} Q_{(s,a)} = r^a_s + \gamma \sum_{s'} p^a_{ss'} \max_{a'} Q^{\pi}_{(s',a')}. \tag{5.33}$$

More details about Q-learning can be found in [88, 98].

To efficiently approach the optimal policy

$$\pi^* = \arg\max_{\pi} Q^{\pi}_{(s,a)} (\forall s \in S),$$

where π^* is the optimal policy when $Q^{\pi}_{(s,a)}$ is maximized, Q-learning always needs a certain exploration strategy (i.e., the action selection method). One widely used action selection method is the ε-greedy ($\varepsilon \in [0, 1)$) approach [88], where the optimal action is selected with probability $(1 - \varepsilon)$ and a random action is selected with probability ε. The exploration probability ε can also be reduced over time, which moves the learning from exploration to exploitation. The ε-greedy method is simple and effective, but it has the drawback that when the learning system explores, it chooses equally among all actions. This means that the learning system makes no difference between the worst action and the next-to-best action. Another problem is that it is difficult to choose a proper value for the parameter ε to achieve the optimal balance between exploration and exploitation.

Another kind of action selection methods involves randomized strategies, such as the Softmax method [88] and the simulated annealing method [30]. Such methods use a positive parameter τ called the temperature and choose an action a with the probability proportional to $\mathrm{e}^{Q(s,a)/\tau}$. Compared with the ε-greedy method, the "best" action is still given the highest selection probability, but all the others are ranked and weighted according to their estimated Q-values. It can also move from exploration to exploitation by adjusting the "temperature" parameter τ. It is natural to sample actions according to this distribution, but it is very difficult to set and adjust a good parameter τ and may converge slowly. Another shortcoming is that it does not work well when the Q-values of the actions are close and the best action cannot be separated from the others. Moreover, when the parameter τ is reduced over time to acquire more exploitation, there is no effective mechanism to guarantee reexploration when necessary.

We focus on exploration strategies (i.e., action selection methods), to better balancing between exploration and exploitation [10, 30, 34, 42]. In particular, we present a novel fidelity-based probabilistic Q-learning (FPQL) algorithm where a probabilistic action selection method is used as a more effective exploration strategy to improve the performance of Q-learning for complex learning control problems. As demonstrated in [13], the ε-greedy method uses a prefixed exploration policy and the action a_1 with the maximum of Q-value ($Q(s, a_1)$) is selected with the probability of $(1 - \varepsilon)$ and all the other actions ($a_2 \sim a_m$) are selected with the probability of $\frac{\varepsilon}{m-1}$, respectively. In the Softmax method, the action a_i, $i = 1, 2, \ldots, m$, is selected with the probability of

$$\frac{\mathrm{e}^{Q(s,a_i)/\tau}}{\sum_{j=1}^{m} \mathrm{e}^{Q(s,a_j)/\tau}}.$$

In our FPQL, the action selection probability distribution is dynamically updated along with the learning process instead of being computed from the estimated Q-

values and a temperature parameter, and the fidelity is used to direct the probability distribution and to strengthen the learning effects with regulation on the updating process.

Assume that the policy to be learned $\pi : S \times A \rightarrow [0, 1]$ is represented using the probability distribution of the state-action space

$$\pi : P^{\pi} = (p^{\pi}(s, a))_{n \times m}, \tag{5.34}$$

where $s \in S$, $a \in A$ and for a certain state s, the probability distribution on the action set A is

$$p_s^{\pi} = \{p^{\pi}(s, a_i)\}, i = 1, 2, \ldots, m.$$

The lookup table for the Q-values and the probability distribution are of the form

$$\begin{pmatrix} & \mathbf{a_1} & \mathbf{a_2} & \cdots & \mathbf{a_m} \\ \mathbf{s_1} & \begin{bmatrix} Q_{(s_1,a_1)} \\ p^{\pi}_{(s_1,a_1)} \end{bmatrix} & \begin{bmatrix} Q_{(s_1,a_2)} \\ p^{\pi}_{(s_1,a_2)} \end{bmatrix} & \cdots & \begin{bmatrix} Q_{(s_1,a_m)} \\ p^{\pi}_{(s_1,a_m)} \end{bmatrix} \\ \mathbf{s_2} & \begin{bmatrix} Q_{(s_2,a_1)} \\ p^{\pi}_{(s_2,a_1)} \end{bmatrix} & \ddots & & \vdots \\ \vdots & \vdots & & \ddots & \vdots \\ \mathbf{s_n} & \begin{bmatrix} Q_{(s_n,a_1)} \\ p^{\pi}_{(s_n,a_1)} \end{bmatrix} & \cdots & \cdots & \begin{bmatrix} Q_{(s_n,a_m)} \\ p^{\pi}_{(s_n,a_m)} \end{bmatrix} \end{pmatrix}. \tag{5.35}$$

In the probabilistic action selection method, one selects an action a (under policy π) at a certain state s with the probability according to the probability distribution on the action set A, i.e.,

$$a_s^{\pi} = f^{\pi}(s) = \begin{cases} a_1 & \text{with probability } p^{\pi}(s, a_1), \\ a_2 & \text{with probability } p^{\pi}(s, a_2), \\ \vdots & \\ a_m & \text{with probability } p^{\pi}(s, a_m). \end{cases} \tag{5.36}$$

Such a probabilistic action selection method leads to a natural probabilistic exploration strategy for Q-learning.

The goal of FPQL is to learn a mapping from states to actions. The one-step updating rule of FPQL for $Q_{(s,a)}$ is the same as that of Q-learning. Besides the updating of $Q_{(s,a)}$, the probability distribution is also updated for each learning step. After the execution of the action a_t for state $s = s_t$, the corresponding probability $p(s_t, a_t)$ is updated according to the immediate reward r_{t+1}, the estimated value of $Q_{(s_{t+1},a')}$ for next state $s' = s_{t+1}$ and the fidelity $F(s_{t+1}, s_{\text{target}})$ between the state s_{t+1} and the target state s_{target}. That is

$$p(s_t, a_t) \leftarrow p(s_t, a_t) + k \left(r_{t+1} + \max_{a'} Q(s_{t+1}, a') + F(s_{t+1}, s_{\text{target}}) \right), \tag{5.37}$$

where k ($k \geq 0$) is an updating step size and the probability distribution of actions at state $s = s_t$

$$\{p(s, a_1), p(s, a_2), \ldots, p(s, a_m)\}$$

is normalized after each updating. The parameter setting of k is accomplished by experience and generally can be set as the same as the learning rate α_t. The variation of k in a relatively large range will only slightly affect the learning process because the probability distribution $\{p(s, a_1), p(s, a_2), \ldots, p(s, a_m)\}$ is normalized after each updating step. For example, k may be set as 0.01 [13]. The specification of the fidelity $F(s_{t+1}, s_{\text{target}})$ is defined regarding the objective of the learning control task. Here, the fidelity of quantum pure states is adopted for the learning control of quantum systems. The procedure of FPQL algorithm is shown as Algorithm 5.3 [13].

Algorithm 5.3 Fidelity-based probabilistic Q-learning

Initialize $Q(s, a)$ arbitrarily
Initialize the policy $\pi : P^{\pi} = (p^{\pi}(s, a))_{n \times m}$ to be evaluated
repeat (for each episode)
  Initialize $t = 1$, s_t
  repeat (for each step of episode)
    $a_t \leftarrow$ action a_i with probability $p(s_t, a_i)$ for s_t
    Take action a_t, observe reward r_{t+1}, and next state s_{t+1}
    $Q(s_t, a_t) \leftarrow Q(s_t, a_t) + \alpha_t \delta_{t+1}^{Q}$
    where $\delta_{t+1}^{Q} = r_{t+1} + \gamma \max_{a'} Q(s_{t+1}, a') - Q(s_t, a_t)$
    $p(s_t, a_t) \leftarrow p(s_t, a_t) + k\delta_{t+1}^{p}$
    where $\delta_{t+1}^{p} = r_{t+1} + \max_{a'} Q(s_{t+1}, a') + F(s_{t+1}, s_{\text{target}})$
    Normalize $\{p(s_t, a_i)|_{i=1,2,\ldots,m}\}$
    $t \leftarrow t + 1$
  until s_{t+1} is terminal
until the learning process ends

Compared with existing exploration strategies, such as ε-greedy, Softmax and simulated annealing methods, the probabilistic exploration strategy has the following merits: (i) The learning algorithm possesses a more reasonable credit assignment approach using a probabilistic method, and the action selection method is more natural without too much difficulty for parameter setting. The only parameter to be set is the step size k. The parameter k will not substantially affect the algorithm performance, because the action selection probabilities for a certain state are relative and will be normalized after each updating step. (ii) The method provides a natural reexploring mechanism; i.e., when the environment changes, the policy also changes along with the online learning process. Such a reexploring mechanism is difficult to implement for the existing exploration strategies (e.g., ε-greedy) [13]. Compared with basic Q-learning algorithms, the main feature of the FPQL algorithm is the straightforward probabilistic exploration strategy and the reinforcement strategy is also applied to dynamically update the probability distribution of action selection. The agent selects actions based on the variable probability distribution over an admissible action set at a certain state. Such an action selection method keeps a proper chance of exploration instead of obeying the policy learned so far and makes a good tradeoff between exploration and exploitation using probability.

In [13], the learning control problems of a spin-$\frac{1}{2}$ system and a Λ-type atomic system have been investigated using the FPQL algorithm. In these results, the control objective is to control the system from a given initial state $s_{\text{initial}} = |\psi_{\text{initial}}\rangle$ to a desired target state $s_{\text{target}} = |\psi_{\text{target}}\rangle$. The state set consists of some quantum states $S = \{s_i = |\psi_{\text{i}}\rangle\}$, $i = 1, 2, \ldots, n$, and the action set consists of some constant control pulses $A = \{a_j = u_j\}$, $j = 1, 2, \ldots, m$. The reward is set as $r = -1$ for each control step until it reaches the target state and a reward $r = 1000$ is obtained. Numerical results show that FPQL outperforms standard Q-learning (QL). FPQL quickly finds the optimal control sequence after less than 50 episodes, while QL needs more than 200 episodes for spin systems. For each episode in the learning process, QL also needs much more steps to find the target state. For the quantum control problem of a three-level Λ-type atomic system, the learning process converges after about 300 episodes using FPQL, while QL needs about 2000 episodes. The performance is very sensitive to the exploration behavior in the state-action space for QL, while the FPQL method shows an almost monotonically improved learning behavior. The numerical results in [13] demonstrate that the FPQL method contributes to a more effective tradeoff between exploration and exploitation than standard Q-learning. It can quickly converge to the optimal policy and significantly outperforms standard Q-learning. The results also demonstrate that FPQL is an alternative effective approach for quantum control design.

Besides the above results, several other works have also explored the use of Q-learning in solving quantum control problems. For example, the work [9] showed that the performance of Q-learning is comparable to optimal control approaches in the task of finding short and high-fidelity control protocols from an initial state to a given target state in nonintegrable many-body quantum systems of interacting qubits. In the implementation, the state space consists of all tuples $(t, u(t))$ of time t and the corresponding magnetic field $u(t)$, and the action space consists of all jumps δu in $u(t)$. The reward is set as $r = 0$ for each step until the end of each episode, the agent obtains a reward defined by the fidelity of the final state with the target state. Single qubits and many-coupled qubits were considered and the results showed that Q-learning can also help identify variational protocols with nearly optimal fidelity even in the glassy phase. In [112], Q-learning was compared with stochastic gradient descent algorithm, Krotov algorithm and policy gradient RL for quantum state preparation of single qubits and multiple qubits. In the work, the state space consists of some quantum states and all allowed values of the control field form the action space. The reward is calculated using the fidelity F between the current state and the target state according to $r = 10$ when $0.5 < F \leq 0.9$, $r = 100$ when $0.9 < F \leq 0.999$ and $r = 5000$ when $0.999 < F \leq 1$. The results showed that, compared with stochastic gradient descent and Krotov algorithms, Q-learning and policy gradient RL can adaptively reduce the number of pieces required during the optimization process and naturally work with restricted sets of actions.

5.5.2 Deep Reinforcement Learning for Quantum Control

The traditional approaches to solve RL problems cannot work well in the high-dimensional real world. Ever since deep learning has yielded state-of-the-art achievements in a variety of fields, it provides an alternative way to obtain effective and flexible representations for the learning agent through integrating deep learning into RL. Various DRL approaches have been developed where neural networks are used to approximate the value function [5]. DRL has also been applied to the area of quantum control [1, 28, 52, 64]. For example, in [64], DRL was employed to simultaneously optimize the speed and fidelity of quantum computation against both leakage and stochastic control errors. A universal quantum control framework was presented to improve control robustness by adding control noise into training environments for RL agents trained with trusted-region-policy-optimization. In the work [64], the cost function was defined as [64]

$$J(\alpha, \beta, \gamma, \kappa) = \alpha[1 - F[(U(T))]] + \beta L_{\text{tot}} + \gamma \int_0^T u^2(t)\mathrm{d}t + \kappa T,$$

where $\alpha, \beta, \gamma, \kappa$ represent the weight coefficients penalizing the gate infidelity, leakage errors, the control constraints and total runtime, respectively. $F[(U(T))]$ denotes the fidelity of quantum gates, and L_{tot} characterizes the accumulated leakage errors. In the algorithm implementation, the control trajectory is encoded by one neural network and the control cost function is encoded by a second neural network. Through training both neural networks under a stochastic environment mimicking noisy control actuation, a robust control can be obtained. The work [1] employed dueling double deep Q-learning to find an optimal control for Hadamard gates and CNOT gates. In [112], deep Q-learning was compared with basic Q-learning, the stochastic gradient descent algorithm and Krotov algorithm for quantum state control problems, and the results showed that, in general, deep Q-learning has the best performance among these algorithms. In [52], a deep reinforcement learning approach was proposed for achieving a fast and precise quantum control by constructing a curriculum consisting of a set of intermediate tasks, where the tasks among a curriculum can be statically determined before the learning process or dynamically generated during the learning process. By transferring knowledge between two successive tasks and sequencing tasks according to their difficulties, the proposed curriculum-based DRL method enables the agent to focus on easy tasks in the early stage, then move onto difficult tasks, and eventually approaches the final task. Numerical comparison with the traditional methods including GD, GA and several other DRL methods demonstrated that the curriculum-based DRL exhibits improved quantum control performance and also provides an efficient way to identify optimal strategies with few control pulses [52].

5.6 Summary and Further Reading

In this chapter, we discussed control of quantum systems using various machine learning algorithms including DE, GA, ES and RL. We presented two improved DE algorithms and applied them to different quantum control tasks. Numerical results including DE for open quantum ensemble control and quantum network synchronization, and RL for quantum state control were presented. The performance of various machine learning algorithms was compared for control design of quantum autoencoders. Experimental results on fragmentation control of molecules using fs laser pulses and closed-loop learning control of quantum autoencoders were presented. Future work includes further exploring the applications of state-of-the-art machine learning algorithms to challenging quantum control tasks.

Further reading may include [26] for a more comprehensive treatment of learning control using DE for open quantum ensembles, quantum network synchronization and robust fragmentation control experiments, [13] for more results on quantum state control using FPQL, [54] for more detailed numerical results on learning control of quantum autoencoders, [32] for more experimental results on learning control of quantum autoencoders, [17, 101] for other online learning algorithms including c-GRAPE and d-GRAPE algorithms and [11] for the application of online learning algorithms to calibrate pulse distortions in superconducting quantum systems.

Appendix

Here, we provide the pseudocode for several machine learning algorithms that are used for quantum autoencoders. Denote the control fields as a column vector x and let $J(x)$ be the loss function to be optimized. The goal of the algorithms is to find the optimal vector x^* such that $J(x^*)$ achieves the optimal value. Considering the physical restriction of control fields, the parameters are usually initialized as

$$x = \mathbf{u}_{\min} + \text{rand}(0, 1)(\mathbf{u}_{\max} - \mathbf{u}_{\min}). \tag{5.38}$$

Given the learning rate α, the main procedure of GD is as follows:

Step 1: Randomly initialize the parameters based on (5.38).
Step 2: Compute the gradient information $\frac{\partial J(x)}{\partial x}$.
Step 3: Update the parameter $x \leftarrow x + \alpha \frac{\partial J(x)}{\partial x}$.

To approximately calculate the gradients, we may generate a random unit vector $v^j = [0, \ldots, 0, 1, 0, \ldots, 0]^T$ and then obtain two new vectors $x_+^j = x + \beta v^j$ and $x_-^j = x - \beta v^j$, respectively. Then gradients are obtained according to

$$\frac{\partial J(x)}{\partial x^j} \cong \frac{J(x_+^j) - J(x_-^j)}{2\beta},$$

where β is a perturbation factor.

Given the population size NP, the crossing-over rate P_c, the mutation rate P_m, the main procedure of GA is as follows:

Step 1: Randomly generate NP individuals $\{x_i\}$ based on (5.38) and constitute $S = \{x_1, \ldots, x_{NP}\}$.
Step 2: Rank $\{x_i \in S\}$ according to $\{J(x_i)\}$ (descending).
Step 3: Select top $\lceil NP(1 - P_c) \rceil$ vectors to constitute S_1.
Step 4: Sample $NP - \lceil NP(1 - P_c) \rceil$ vectors from S with probability $P(x_i) = \frac{J(x_i)}{\sum_{j=1}^{NP} J(x_j)}$, to constitute S_2.
Step 5: Randomly pair vectors among S_2 and perform $\lceil \frac{N_2}{2} \rceil$ times of crossover to renew the vectors in S_2.
Step 6: Mutate vectors in S_2 with probability P_m.
Step 7: Obtain new generation $S \leftarrow \{S_1, S_2\}$.
Step 8: If convergent, go to Step 9, otherwise go to Step 2.
Step 9: Optimal control parameters $x^* = \arg\max_{x_i} (J(x_i))$.

Given the population size NP, the learning rate α, the momentum coefficient β, and the permutation factor δ, the main procedure of ES is as follows:

Step 1: Initialize gradient $\mathrm{d}J = 0$ and momentum $\mathrm{d}v = 0$.
Step 2: Initialize the mean vector $\bar{x}$ according to (5.38).
Step 3: Repeat for each individual $i = 1, \ldots, NP$.
Step 3.1: Sample variation $\varepsilon_i \sim N(0, I)$.
Step 3.2: Set mutation variant as $X_i \leftarrow \bar{x} + \delta\varepsilon_i$.
Step 4: Obtain gradient $\mathrm{d}J \leftarrow \frac{1}{NP\delta} \sum_{j=1}^{NP} J(X_i)\varepsilon_j$.
Step 5: Obtain momentum $\mathrm{d}v \leftarrow \beta \mathrm{d}v + (1 - \beta)\mathrm{d}J$.
Step 6: Update the new mean vector $\bar{x} \leftarrow \bar{x} + \alpha \mathrm{d}v$.
Step 7: If convergent, go to Step 8; otherwise go to Step 3.
Step 8: Optimal control parameters $x^* = \bar{x}$.

References

1. An Z, Zhou DL (2019) Deep reinforcement learning for quantum gate control. Europhys Lett 126(6):60002

2. Atabek O, Dion CM, Yedder ABH (2003) Evolutionary algorithms for the optimal laser control of molecular orientation. J Phys B At Mol Opt Phys 36(23):4667
3. Becerra RL, Coello Coello CA (2006) Cultured differential evolution for constrained optimization. Comput Methods Appl Mech Eng 195(33–36):4303–4322
4. Bennett CH, Harrow AW, Lloyd S (2006) Universal quantum data compression via nondestructive tomography. Phys Rev A 73(3):032336
5. Biamonte J, Wittek P, Pancotti N, Rebentrost P, Wiebe N, Lloyd S (2017) Quantum machine learning. Nature 549:195–202
6. Bondarenko D, Feldmann P (2020) Quantum autoencoders to denoise quantum data. Phys Rev Lett 124(13):130502
7. Breuer H-P, Petruccione F (2002) The theory of open quantum systems, 1st edn. Oxford University Press
8. Brif C, Chakrabarti R, Rabitz H (2010) Control of quantum phenomena: past, present and future. New J Phys 12:075008
9. Bukov M, Day AGR, Sels D, Weinberg P, Polkovnikov A, Mehta P (2018) Reinforcement learning in different phases of quantum control. Phys Rev X 8:031086
10. Cai C, Liao X, Carin L (2009) Learn to explore and exploit in POMDPs. In: Proceedings of the 22nd conference on neural information processing systems, Vancouver, BC, Canada, Dec 2009, pp 198–206
11. Cao X, Chu B, Peng Z, Liu Y-X, Wu RB (2022) Learning to calibrate quantum control pulses. IEEE Trans Control Syst Technol 30:193–201
12. Chen C, Dong D, Chen Z (2006) Quantum computation for action selection using reinforcement learning. Int J Quantum Inf 4(6):1071–1083
13. Chen C, Dong D, Li HX, Chu J, Tarn TJ (2014) Fidelity-based probabilistic Q-learning for control of quantum systems. IEEE Trans Neural Netw Learn Syst 25:920–933
14. Chen C, Dong D, Long R, Petersen IR, Rabitz H (2014) Sampling-based learning control of inhomogeneous quantum ensembles. Phys Rev A 89:023402
15. Chen C, Wang L, Wang Y (2013) Closed-loop and robust control of quantum systems. Sci World J 869285
16. Das S, Suganthan PN (2011) Differential evolution: a survey of the state-of-the-art. IEEE Trans Evol Comput 15(1):4–31
17. Ding H, Chu B, Qi B, Wu RB (2021) Collaborative learning of high-precision quantum control and tomography. Phys Rev Appl 16:014056
18. Ding Y, Lamata L, Sanz M, Chen X, Solano E (2019) Experimental implementation of a quantum autoencoder via quantum adders. Adv Quantum Technol 2:1800065
19. Dong D, Chen C, Chu J, Tarn TJ (2012) Robust quantum-inspired reinforcement learning for robot navigation. IEEE/ASME Trans Mechatron 17:86–97
20. Dong D, Chen C, Li H, Tarn TJ (2008) Quantum reinforcement learning. IEEE Trans Syst Man Cybern Part B Cybern 38(5):1207–1220
21. Dong D, Chen C, Qi B, Petersen IR, Nori F (2015) Robust manipulation of superconducting qubits in the presence of fluctuations. Sci Rep 5:7873 (Springer Nature)
22. Dong D, Chen C, Tarn TJ, Pechen A, Rabitz H (2008) Incoherent control of quantum systems with wavefunction controllable subspaces via quantum reinforcement learning. IEEE Trans Syst Man Cybern Part B Cybern 38:957–962
23. Dong D, Petersen IR (2010) Quantum control theory and applications: a survey. IET Control Theory Appl 4:2651–2671
24. Dong D, Petersen IR (2022) Quantum estimation, control and learning: opportunities and challenges. Annu Rev Control 54:243–251
25. Dong D, Shu C-C, Chen J, Xing X, Ma H, Guo Y, Rabitz H (2021) Learning control of quantum systems using frequency-domain optimization algorithms. IEEE Trans Control Syst Technol 29(4):1791–1798
26. Dong D, Xing X, Ma H, Chen C, Liu Z, Rabitz H (2020) Learning-based quantum robust control: algorithm, applications, and experiments. IEEE Trans Cybern 50:3581–3593

27. Englert B, Kurtsiefer C, Weinfurter H (2001) Universal unitary gate for single-photon two-qubit states. Phys Rev A 63:032303
28. Fösel T, Tighineanu P, Weiss T, Marquardt F (2018) Reinforcement learning with neural networks for quantum feedback. Phys Rev X 8:031084
29. Goodfellow I, Bengio Y, Courville A (2016) Deep learning. MIT Press, Cambridge
30. Guo MZ, Liu Y, Malec J (2004) A new Q-learning algorithm based on the metropolis criterion. IEEE Trans Syst Man Cybern B 34:2140–2143
31. Hamza NM, Essam DL, Sarker RA (2016) Constraint consensus mutation-based differential evolution for constrained optimization. IEEE Trans Evol Comput 20(3):447–459
32. Huang C, Ma H, Yin Q, Tang J, Dong D, Chen C, Xiang G-Y, Li CF, Guo GC (2020) Realization of a quantum autoencoder for lossless compression of quantum data. Phys Rev A 102:032412
33. Huffman DA (1952) A method for the construction of minimum redundancy codes. Proc IRE 40:1098–1101
34. Iwata K, Ikeda K, Sakai H (2004) A new criterion using information gain for action selection strategy in reinforcement learning. IEEE Trans Neural Netw 15:792–799
35. Jain AK (1981) Image data compression: a review. Proc IEEE 69(3):349–389
36. Ježek M, Fiurášek J, Hradil Z (2003) Quantum inference of states and processes. Phys Rev A 68:012305
37. Jirari H, Pötz W (2005) Optimal coherent control of dissipative N-level systems. Phys Rev A 72:013409
38. Jozsa R, Horodecki M, Horodecki P, Horodecki R (1998) Universal quantum information compression. Phys Rev Lett 81:1714
39. Jozsa R, Schumacher B (1994) A new proof of the quantum noiseless coding theorem. J Mod Opt 41:2343–2349
40. Judson RS, Rabitz H (1992) Teaching lasers to control molecules. Phys Rev Lett 68:1500–1503
41. Khaneja N, Reiss T, Kehlet C, Schulte-Herbrüggen T, Glaser SJ (2005) Optimal control of coupled spin dynamics: design of NMR pulse sequences by gradient ascent algorithms. J Magn Reson 172(2):296–305
42. Kim D, Kim KE, Poupart P (2012) Cost-sensitive exploration in Bayesian reinforcement learning. In: Proceedings of neural information processing systems, Lake Tahoe, Nevada, 3–8 Dec 2012
43. Kimura G (2003) The Bloch vector for N-level systems. Phys Lett A 314:339–349
44. Kondo T, Ito K (2004) A reinforcement learning with evolutionary state recruitment strategy for autonomous mobile robots control. Robot Auton Syst 46:111–124
45. Lamata L, Alvarez-Rodriguez U, Martín-Guerrero JD, Sanz M, Solano E (2018) Quantum autoencoders via quantum adders with genetic algorithms. Quantum Sci Technol 4:014007
46. LeCun Y, Bengio Y, Hinton G (2015) Deep learning. Nature 521:436–444
47. Li J-A, Dong D, Wei Z, Liu Y, Pan Y, Nori F, Zhang X (2020) Quantum reinforcement learning during human decision-making. Nat Hum Behav 4:294
48. Lindblad G (1976) On the generators of quantum dynamical semigroups. Commun Math Phys 48(2):119–130
49. Liu Z, Wang L, Wang J, Dong D, Hu X (2017) Distributed sampled-data control of nonholonomic multi-robot systems with proximity networks. Automatica 77:170–179
50. Lloyd S, Mohseni M, Rebentrost P (2014) Quantum principal component analysis. Nat Phys 10:631–633
51. Ma H, Chen C, Dong D (2015) Differential evolution with equally-mixed strategies for robust control of open quantum systems. In: IEEE international conference on systems, man and cybernetics, Hong Kong, 9–12 Oct 2015, pp 2055–2060
52. Ma H, Dong D, Ding SX, Chen C (2022) Curriculum-based deep reinforcement learning for quantum control. IEEE Trans Neural Netw Learn Syst. https://doi.org/10.1109/TNNLS.2022.3153502

53. Ma H, Dong D, Shu C-C, Zhu Z, Chen C (2017) Differential evolution with equally-mixed strategies for robust control of open quantum systems. Control Theory Technol 15:226–241
54. Ma H, Huang CJ, Chen C, Dong D, Wang Y, Wu RB, Xiang G-Y (2023) On compression rate of quantum autoencoders: control design, numerical and experimental realization. Automatica, 57:110659
55. Malandrino G, Fragala IL (2006) Lanthanide 'second-generation' precursors for MOCVD applications: effects of the metal ionic radius and polyether length on coordination spheres and mass-transport properties. Coord Chem Rev 250:1605–1620
56. Malandrino G, Incontro O, Castelli F, Fragala IL, Benelli C (1996) Synthesis, characterization, and mass-transport properties of two novel Gadolinium(III) hexafluoroacetylacetonate polyether adducts: promising precursors for MOCVD of GdF_3 films. Chem Mater 8:1292–1297
57. Mallipeddi R, Suganthan PN, Pan QK, Tasgetirenc MF (2011) Differential evolution algorithm with ensemble of parameters and mutation strategies. Appl Soft Comput 11(2):1679–1696
58. Mazzarella L, Sarlette A, Ticozzi F (2015) Consensus for quantum networks: from symmetry to gossip iterations. IEEE Trans Autom Control 60(1):158–172
59. Mazzarella L, Ticozzi F, Sarlette A (2015) Extending robustness and randomization from consensus to symmetrization algorithms. SIAM J Control Optim 53(4):2076–2099
60. Meng Q, Witte RJ, May PS, Berry MT (2009) Photodissociation and photoionization mechanisms in lanthanide-based fluorinated β-diketonate metal-organic chemical-vapor deposition precursors. Chem Mater 21:5801–5808
61. Mesbahi M, Egerstedt M (2010) Graph theoretic methods in multiagent networks. Princeton University Press, New Jersey
62. Neri F, Tirronen V (2010) Recent advances in differential evolution: a survey and experimental analysis. Artif Intell Rev 33(1–2):61–106
63. Nielsen MA, Chuang IL (2010) Quantum computation and quantum information. Cambridge University Press, Cambridge
64. Niu MY, Boixo S, Smelyanskiy VN, Neven H (2019) Universal quantum control through deep reinforcement learning. npj Quantum Inf 5:33
65. Pan Y, Amini H, Miao Z, Gough J, Ugrinovskii V, James MR (2014) Heisenberg picture approach to the stability of quantum Markov systems. J Math Phys 55:062701
66. Pan Y, Nguyen T (2017) Stabilizing quantum states and automatic error correction by dissipation control. IEEE Trans Autom Control 62:4625–4630
67. Pan Y, Ugrinovskii V, James MR (2016) Ground-state stabilization of quantum finite-level systems by dissipation. Automatica 65:147–159
68. Paparo GD, Dunjko V, Makmal A, Martin-Delgado MA, Briegel HJ (2014) Quantum speedup for active learning agents. Phys Rev X 4:031002
69. Pennebaker WB, Mitchell JL (1992) JPEG: still image data compression standard. Springer, New York
70. Pepper A, Tischler N, Pryde GJ (2019) Experimental realization of a quantum autoencoder: the compression of qutrits via machine learning. Phys Rev Lett 122(6):060501
71. Plesch M, Băek V (2010) Efficient compression of quantum information. Phys Rev A 81:032317
72. Pollard KD, Jenkins HA, Puddephatt R (2000) Chemical vapor deposition of cerium oxide using the precursors [Ce(hfac)$_3$(glyme)]. Chem Mater 12:701–710
73. Qin AK, Huang VL, Suganthan PN (2009) Differential evolution algorithm with strategy adaptation for global numerical optimization. IEEE Trans Evol Comput 13(2):398–417
74. Rabitz H, De Vivie-Riedle R, Motzkus M, Kompa K (2000) Whither the future of controlling quantum phenomena? Science 288(5467):824–828
75. Radovic A, Williams M, Rousseau D, Kagan M, Bonacorsi D, Himmel A, Aurisano A, Terao K, Wongjirad T (2018) Machine learning at the energy and intensity frontiers of particle physics. Nature 560:41–48
76. Richardson MF, Wagner WF, Sands DE (1968) Rare-earth trishexafluoroacetylacetonates and related compounds. J Inorg Nucl Chem 30:1275–1289

77. Romero J, Olson JP, Aspuru-Guzik A (2017) Quantum autoencoders for efficient compression of quantum data. Quantum Sci Technol 2(4):045001
78. Rozema LA, Mahler DH, Hayat A, Turner PS, Steinberg AM (2014) Quantum data compression of a qubit ensemble. Phys Rev Lett 113(16):160504
79. Salimans T, Ho J, Chen X, Sidor S, Sutskever I (2017) Evolution strategies as a scalable alternative to reinforcement learning. arXiv preprint arXiv:1703.03864
80. Sepulchre R, Sarlette A, Rouchon P (2010) Consensus in non-commutative spaces. In: Proceedings of the 49th IEEE conference on decision and control, Atlanta, GA, 15–17 Dec 2010, pp 6596–6601
81. Shannon CE (1948) A mathematical theory of communication. Bell Syst Tech J 27(3):379–423
82. Shi G, Dong D, Petersen IR, Johansson KH (2016) Reaching a quantum consensus: master equations that generate symmetrization and synchronization. IEEE Trans Autom Control 59(2):374–387
83. Shir OM, Bäck T (2009) Niching with derandomized evolution strategies in artificial and real-world landscapes. Nat Comput 8(1):171–196
84. Springer CS, Meek DW, Sievers RE (1967) Rare earth chelates of 1,1,1,2,2,3,3-heptafluoro-7,7-dimethyl-4,6-octanedione. Inorg Chem 6:1105–1110
85. Steinbrecher GR, Olson JP, Englund D, Carolan J (2019) Quantum optical neural networks. npj Quantum Inf 5(60)
86. Storn R, Price K (1997) Differential evolution—a simple and efficient heuristic for global optimization over continuous spaces. J Glob Optim 11(4):341–359
87. Sutton R (1988) Learning to predict by the methods of temporal difference. Mach Learn 3:9–44
88. Sutton R, Barto AG (2018) Reinforcement learning: an introduction. MIT Press, Cambridge, MA
89. Talaga DS, Hanna SD, Zink JI (1998) Luminescent photofragments of (1,1,1,5,5,5-hexafluoro-2,4-pentanedionato) metal complexes in the gas phase. Inorg Chem 37:2880–2887
90. Tibbetts KM, Xing X, Rabitz H (2013) Systematic trends in photonic reagent induced reactions in a homologous chemical family. J Phys Chem A 117:8025–8215
91. Tibbetts KM, Xing X, Rabitz H (2013) Optimal control of molecular fragmentation with homologous families of photonic reagents and chemical substrates. Phys Chem Chem Phys 15:18012–18022
92. Tibbetts KM, Xing X, Rabitz H (2014) Laboratory transferability of optimally shaped laser pulses for quantum control. J Chem Phys 140:074302
93. Ticozzi F (2016) Symmetrizing quantum dynamics beyond gossip-type algorithms. Automatica 74:38–46
94. Tsubouchi M, Momose T (2008) Rovibrational wave-packet manipulation using shaped mid-infrared femtosecond pulses toward quantum computation: optimization of pulse shape by a genetic algorithm. Phys Rev A 77(5):052326
95. Wan KH, Dahlsten O, Kristjánsson H, Gardner R, Kim M (2017) Quantum generalisation of feedforward neural networks. npj Quantum Inf 3(1):36
96. Wang D, Liu D, Wei Q, Zhao D, Jin N (2012) Optimal control of unknown nonaffine nonlinear discrete-time systems based on adaptive dynamic programming. Automatica 48(8):1825–1832
97. Wang Y, Yin Q, Dong D, Qi B, Petersen IR, Hou Z, Yonezawa H, Xiang G-Y (2019) Quantum gate identification: error analysis, numerical results and optical experiment. Automatica 101:269–279
98. Watkins C, Dayan P (1992) Q-learning. Mach Learn 8:279–292
99. Wiseman HM, Milburn GJ (2010) Quantum measurement and control. Cambridge University Press, Cambridge
100. Witten IH, Neal RM, Cleary JG (1987) Arithmetic coding for data compression. Commun ACM 30(6):520–540

101. Wu RB, Chu B, Owens DH, Rabitz H (2018) Data-driven gradient algorithm for high-precision quantum control. Phys Rev A 97:042122
102. Xing X, Rey-de-Castro R, Rabitz H (2014) Assessment of optimal control mechanism complexity by experimental landscape Hessian analysis: fragmentation of CH_2BrI. New J Phys 16:125004
103. Yang Y, Chiribella G, Ebler D (2016) Efficient quantum compression for ensembles of identically prepared mixed states. Phys Rev Lett 116:080501
104. Yang Y, Chiribella G, Hayashi M (2016) Optimal compression for identically prepared qubit states. Phys Rev Lett 117:090502
105. Yang F, Cong S, Long R, Ho TS, Wu RB, Rabitz H (2013) Exploring the transition-probability-control landscape of open quantum systems: application to a two-level case. Phys Rev A 88(3):033420
106. You JQ, Nori F (2005) Superconducting circuits and quantum information. Phys Today 58:42–47
107. Zahedinejad E, Ghosh J, Sanders BC (2015) High-fidelity single-shot Toffoli gate via quantum control. Phys Rev Lett 114(20):200502
108. Zahedinejad E, Ghosh J, Sanders BC (2016) Designing high-fidelity single-shot three-qubit gates: a machine-learning approach. Phys Rev Appl 6:054005
109. Zahedinejad E, Schirmer S, Sanders BC (2014) Evolutionary algorithms for hard quantum control. Phys Rev A 90(3):032310
110. Zare RN (1998) Laser control of chemical reactions. Science 279:1875–1879
111. Zeidler D, Frey S, Kompa K, Motzkus M (2001) Evolutionary algorithms and their application to optimal control studies. Phys Rev A 64(2):023420
112. Zhang X, Wei Z, Asad R, Yang X, Wang X (2019) When does reinforcement learning stand out in quantum control? A comparative study on state preparation. npj Quantum Inf 5:85
113. Ziv J, Lempel A (1977) A universal algorithm for sequential data compression. IEEE Trans Inf Theory 23(3):337–343

Chapter 6
Sliding Mode Control of Quantum Systems

Abstract Sliding mode control is a powerful variable structure control method for classical control systems. The basic idea of sliding mode control can also be extended to the quantum domain for robust control. This chapter presents sliding mode control results for quantum systems. In Sect. 6.1, the concept of sliding mode control is introduced to the quantum domain and sliding modes are defined for quantum systems. Section 6.2 presents results on sliding mode control for two-level quantum systems.[a] Sliding mode control methods are extended to multi-level quantum systems in Sect. 6.3[b] and then to open quantum systems in Sect. 6.4.[c] A summary and further reading are given in Sect. 6.5.

6.1 Sliding Mode Control

The variable structure control strategy is a widely used design method in classical control theory and industrial applications where one can change the controller structure according to a specified switching logic in order to obtain desired closed-loop properties [1, 43]. Sliding modes play an important role in variable structure control [43], and sliding mode control (SMC) is a useful approach to robust controller design. In SMC, a sliding surface (sliding mode) and a switching function are involved [19]. A sliding mode is generally defined as a specified state region where the system has a desired dynamic behavior. A switching function is designed

[a] Section 6.2 contains materials reprinted, with permission, from Automatica [15] © 2012 Elsevier.
[b] Section 6.3 contains materials reprinted from [13], https://doi.org/10.1088/1367-2630/11/10/105033, © Deutsche Physikalische Gesellschaft. Reproduced by permission of IOP Publishing. CC BY-NC-SA.
[c] Section 6.4 contains materials reprinted, with permission, from IEEE Transactions on Automatic Control [17] © 2013 IEEE.

D. Dong and I. R. Petersen, *Learning and Robust Control in Quantum Technology*, Communications and Control Engineering,
https://doi.org/10.1007/978-3-031-20245-2_6

to control the system to the sliding mode. SMC has two main advantages [19]: (i) the dynamic behavior of the system may be determined by the particular choice of switching function; (ii) the closed-loop response becomes totally insensitive to a particular class of uncertainties. It is the order reduction property and low sensitivity to uncertainty that make SMC an efficient tool for controlling complex high-order systems operating under uncertain conditions [44]. In this chapter, we aim to extend SMC to the quantum domain. However, the direct generalization is extremely challenging (if not impossible) due to the unique characteristics of quantum systems. For example, it is usually difficult to acquire information for designing feedback control laws for SMC of quantum systems since quantum measurement usually changes the system state under measurement.

To apply the idea of SMC to quantum systems, we first need to define a sliding mode where the quantum system has desired dynamics. A sliding mode can be represented as a functional S of the state ρ and Hamiltonian H; i.e., $S(\rho, H) = 0$. For example, an eigenstate $\rho_s = |\phi_j\rangle\langle\phi_j|$ of H_0 can be selected as a sliding surface. We may define

$$S(\rho, H) = 1 - \langle\phi_j|\rho|\phi_j\rangle = 0.$$

If the initial state ρ_0 is in the sliding mode; i.e.,

$$S(\rho_0, H) = 1 - \langle\phi_j|\rho_0|\phi_j\rangle = 0,$$

we can prove that the quantum system will maintain its state in this surface under only the action of the free Hamiltonian H_0. In fact,

$$\rho(t) = \mathrm{e}^{-\mathrm{i}H_0 t}\rho_0 \mathrm{e}^{\mathrm{i}H_0 t}.$$

Hence,

$$S(\rho(t), H) = 1 - \langle\phi_j|\rho(t)|\phi_j\rangle = 1 - \langle\phi_j|\rho_0|\phi_j\rangle = 0.$$

That is, an eigenstate of H_0 can be identified as a sliding mode.

More generally, an invariant state subspace of a quantum system satisfying $S(\rho, H) = 0$ can be defined as a sliding mode. For example, the wavefunction controllable subspace in [12] can be identified as a sliding mode. Some other state subspaces such as decoherence-free subspaces [31, 32] can also be used as a sliding mode.

If a quantum system is driven into a sliding mode, its state will be maintained in the sliding surface under the action of some classes of Hamiltonians determined by the sliding mode. However, in practical applications, it is inevitable that there exist noises and uncertainties in the system Hamiltonian, initial states and/or control fields. An important advantage of SMC is its robustness against uncertainties. Our main motivation for introducing SMC to quantum systems is to deal with uncertainties. The control problem is converted as: for a given initial state, design a control law to steer the quantum system's state into and then maintain its state in a sliding mode domain (a neighborhood region containing the sliding mode) even with uncertainties. However, it is significantly different from traditional SMC. Once the uncertainties take the state slightly away from the sliding mode, it always has a probability (we call it the probability of failure) to make the system's state collapse out of the sliding mode domain. If the allowed probability of failure is p_0, we may define the sliding mode domain as

$$\mathcal{D}_{\text{SMC}} = \{\rho : \text{Tr}(\rho\rho_{\text{s}}) \geq 1 - p_0, \rho_{\text{s}} \in \{\rho_{\text{s}} : S(\rho_{\text{s}}, H) = 0\}\}.$$

We expect that the control law can ensure the system's state in the sliding mode domain $\mathcal{D}_{\text{SMC}}$ except that the measurement operation may take it away from $\mathcal{D}_{\text{SMC}}$ with a small probability (not greater than p_0). The problem includes three main subtasks:

(i) for any initial state (assumed known), design a control law to drive the system into a specified sliding mode domain $\mathcal{D}_{\text{SMC}}$;
(ii) design a control law to maintain the system's state in $\mathcal{D}_{\text{SMC}}$;
(iii) design a control law to drive the system back to $\mathcal{D}_{\text{SMC}}$ once a measurement takes it away from $\mathcal{D}_{\text{SMC}}$.

6.2 Sliding Mode Control of Two-Level Quantum Systems

6.2.1 SMC Design Using Periodic Measurement

In this section, we focus on two-level pure-state quantum systems. We denote the Pauli matrices $\sigma = (\sigma_x, \sigma_y, \sigma_z)$. One may select the free Hamiltonian of the two-level quantum system as $H_0 = \frac{1}{2}\sigma_z$. Its two eigenstates are denoted as $|0\rangle$ and $|1\rangle$. To control a quantum system, we introduce the control Hamiltonian $H_u = \sum_k u_k(t)H_k$, where $u_k(t) \in \mathbb{R}$ and $\{H_k\}$ is a set of time-independent Hamiltonians. Without loss of generality, the control Hamiltonian for two-level systems can be written as

$$H_u = \frac{1}{2}u_x(t)\sigma_x + \frac{1}{2}u_y(t)\sigma_y + \frac{1}{2}u_z(t)\sigma_z.$$

The controlled dynamical equation can be described as

$$\dot{\rho}(t) = -\mathrm{i}\left[H_0 + \frac{1}{2}\sum_{k=x,y,z} u_k(t)\sigma_k, \rho(t)\right]. \tag{6.1}$$

This control problem is converted into the following problem: given an initial state and a target state, find a set of controls $\{u_k(t)\}$ in (6.1) to drive the controlled system from the initial state to the target state.

For a two-level quantum system, the state ρ can be represented in terms of the Bloch vector $\mathbf{r} = [x\ y\ z] = [\text{Tr}\{\rho\sigma_x\}\ \text{Tr}\{\rho\sigma_y\}\ \text{Tr}\{\rho\sigma_z\}]$:

$$\rho = \frac{1}{2}(I + \mathbf{r} \cdot \sigma). \tag{6.2}$$

With the representation of the Bloch vector, the pure states of a two-level quantum system correspond to the surface of the Bloch sphere, where

$$(x, y, z) = (\sin\theta \cos\varphi, \sin\theta \sin\varphi, \cos\theta),$$

$\theta \in [0, \pi], \varphi \in [0, 2\pi]$.

For two-level quantum systems, we may select either $\rho_s = |0\rangle\langle 0|$ or $\rho_s = |1\rangle\langle 1|$ as a sliding mode. Without loss of generality, we identify the eigenstate $|0\rangle$ of a two-level quantum system as the sliding mode. This means that if a quantum system is driven into the sliding mode, its state will be maintained in the sliding surface under the action of the free Hamiltonian. However, in practical applications, it is inevitable that there exist noises and uncertainties. Here, we suppose that the uncertainties can be approximately described as perturbations in the Hamiltonian. That is, the uncertainties can be denoted as

$$H_\Delta = \frac{1}{2}\varepsilon_x(t)\sigma_x + \frac{1}{2}\varepsilon_y(t)\sigma_y + \frac{1}{2}\varepsilon_z(t)\sigma_z.$$

The unitary errors in [39] belong to this class of uncertainties, and uncertainties in a one-qubit gate also correspond to this class of uncertainties [13]. For a spin system in solid-state NMR, external noisy magnetic fields and unwanted coupling with other spins may lead to uncertainties in this class. Further, we assume the uncertainties are bounded; i.e.,

$$\sqrt{\varepsilon_x^2(t) + \varepsilon_y^2(t) + \varepsilon_z^2(t)} \leq \varepsilon \ \ (\varepsilon \geq 0). \tag{6.3}$$

When $\varepsilon = 0$, $H_\Delta = 0$. That is, there exist no uncertainties, which is trivial for our problem. Hence, we assume $\varepsilon > 0$. We further suppose that the corresponding system without uncertainties is completely controllable and arbitrary unitary control operations can be generated. This assumption can be guaranteed for a two-level quantum system if we can realize arbitrary rotations along the z-axis and ζ-axis ($\zeta = x$ or y) (e.g., see [8] for details).

The control problem under consideration is stated as follows [15]: design a control law to drive and then maintain the quantum system's state in a sliding mode domain even when bounded uncertainties exist in the system Hamiltonian. Here a sliding mode domain may be defined as

$$\mathcal{D}_{\text{SMC}} = \{\rho : \langle 0|\rho|0\rangle \geq 1 - p_0, 0 < p_0 < 1\},$$

where p_0 is a given constant. Here we assume $p_0 \neq 0, 1$ since the case $p_0 = 0$ only occurs in the sliding mode surface and the case $p_0 = 1$ is always true. Hence, the two cases with $p_0 = 0$ and $p_0 = 1$ are trivial for our problem. The definition of the sliding mode domain implies that the system has a probability of at most p_0 to collapse out of $\mathcal{D}_{SMC}$ when making a measurement. This behavior is quite different from that which occurs in traditional SMC. Hence, we expect that our control laws will guarantee that the system's state remains in $\mathcal{D}_{SMC}$ except that a measurement operation may take it away from $\mathcal{D}_{SMC}$ with a small probability (not greater than p_0). Then we can design a control law to complete the three main subtasks (i)–(iii).

Although quantum measurement often has deleterious effects in quantum control tasks, some existing results have shown that it can be combined with unitary transformations to complete specific quantum manipulation tasks and enhance the capability of quantum control [2, 10, 11, 18, 38, 40, 45]. For example, some nonunitarily controllable systems may become controllable by using "measurement plus evolution" [45]. Quantum measurement can be used as a control tool as well as a method of information acquisition. It is worth mentioning that the effect of measurement on a quantum system as a control tool can be achieved through the interaction between the system and measurement apparatus. Here, we combine the Lyapunov methodology and projective measurements (with the measurement operator σ_z) to accomplish the SMC task for two-level quantum systems. The projective measurement with σ_z on a two-level system makes the system's state collapse into $|0\rangle$ (corresponding to eigenvalue 1 of σ_z) or $|1\rangle$ (corresponding to eigenvalue -1 of σ_z).

The steps in the control algorithm are as follows (see Fig. 6.1):

1. Select an eigenstate $|0\rangle$ of H_0 as a sliding mode $S(\rho, H) = 0$, and define the sliding mode domain as $\mathcal{D}_{SMC} = \{\rho : \langle 0|\rho|0\rangle \geq 1 - p_0\}$.
2. For a known initial state ρ_0, design a Lyapunov control law to drive ρ_0 into the sliding mode S.
3. For a specified probability of failure p_0, construct the control period T_0 so that the control law can drive the system's state into $\mathcal{D}_{SMC}$ in a time period T_0.
4. For an initial condition $\rho = |1\rangle\langle 1|$, design a Lyapunov control law and construct the period T_1 by using a similar method to that in Step 3.
5. According to p_0 and ε, design the period T for periodic projective measurements.
6. Use the designed control law to drive the system's state into $\mathcal{D}_{SMC}$ in T_0 and make a projective measurement at $t = T_0$. Then repeat the following operations: make periodic projective measurements with the period T to maintain the system's state in $\mathcal{D}_{SMC}$; if the measurement result corresponds to $|1\rangle$, we use the corresponding control law to drive the state back into $\mathcal{D}_{SMC}$.

In the SMC control algorithm, an important task is to design the measurement period T. We can estimate a bound according to the bound ε on the uncertainties and the allowed probability of failure p_0. Then, we construct a period T to guarantee control performance according to the estimated bound. An extreme case is $T \to 0$. That is, after the quantum system's state is driven into the sliding mode, we make frequent measurements. This corresponds to the quantum Zeno effect [23], which is the inhibition of transitions between quantum states by frequent measurement of the

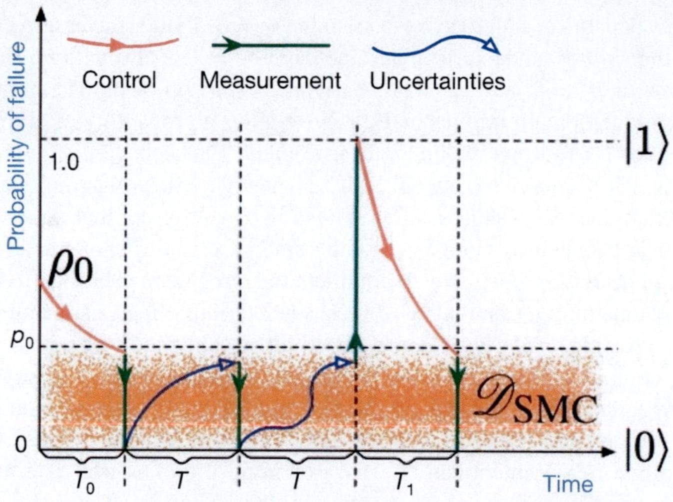

Fig. 6.1 SMC scheme for a two-level quantum system. In this figure, "Control," "Measurement" and "Uncertainties" mean the evolution processes of the quantum system under the control law, the projective measurement and uncertainties in the system Hamiltonian, respectively

state (see, e.g., [23, 36]). Frequent measurements (i.e., $T \to 0$) can guarantee that the state is maintained in the sliding mode in spite of the presence of uncertainties. However, it is usually a difficult task to make such frequent measurements. We may conclude that the smaller T is, the bigger the cost of accomplishing the periodic measurements becomes. Hence, in contrast to the quantum Zeno effect, we wish to design a measurement period T as large as possible. In the following section, we propose two approaches of designing T for different situations.

6.2.2 Design of the Measurement Period

We select the sliding mode as $S(\rho, H) = 1 - \langle 0|\rho|0\rangle = 0$. If there exist no uncertainties and we have driven the system's state to the sliding mode at time t_0, it will be maintained in the sliding mode using only the free Hamiltonian H_0; i.e., $S(\rho(t \geq t_0), H_0) = 0$. That is, if the quantum system's state is driven into the sliding mode, it will evolve in the sliding surface. However, uncertainties may drive the system's state away from the sliding mode. We expect to design a control law to ensure the desired robustness in the presence of uncertainties. Assume that the state at time t is $\rho(t)$. If we make measurements on this system, the probability of failure p that it will collapse into $|1\rangle$ is

$$p = \langle 1|\rho_t|1\rangle = \frac{1 - z_t}{2}, \tag{6.4}$$

where $z_t = \mathrm{Tr}(\rho(t)\sigma_z)$. We now give a detailed discussion on the design of the measurement period T for possible uncertainties.

First we consider the special case $H_\Delta = \frac{1}{2}\varepsilon(t)\sigma_z$ ($|\varepsilon(t)| \leq \varepsilon$). This case corresponds to phase-flip type bounded uncertainties. For any $H_\Delta = \frac{1}{2}\varepsilon_z\sigma_z$ (where $|\varepsilon_z| \leq \varepsilon$), if $S(\rho_0, H) = 0$, we have $S(\rho(t), H) = 0$. This type of uncertainty does not drive the system's state away from the sliding mode. Hence, we ignore this type of uncertainty in our method.

Now we assume that the uncertainty is $H_\Delta = \frac{1}{2}\varepsilon(t)\sigma_\zeta$ ($\zeta = x$ or y) and $|\varepsilon(t)| \leq \varepsilon$. We have the following theorem:

Theorem 6.1 *For a two-level quantum system with the initial state* $\rho_0 = |0\rangle\langle 0|$ *at the time* $t = 0$*, the system evolves to* $\rho(t)$ *under the action of*

$$H(t) = \frac{1}{2}\sigma_z + \frac{1}{2}\varepsilon(t)\sigma_\zeta,$$

where $\zeta = x$ *or* y, $|\varepsilon(t)| \leq \varepsilon$ *and* $\varepsilon > 0$. *If* $p_0 \in (0, \frac{\varepsilon^2}{1+\varepsilon^2}]$ *and* $t \in [0, T^{(1)}]$, *where*

$$T^{(1)} = \frac{\arccos\left[1 - 2(1 + \frac{1}{\varepsilon^2})p_0\right]}{\sqrt{1+\varepsilon^2}}, \tag{6.5}$$

the system's state will remain in

$$\mathcal{D}_{SMC} = \{\rho(t) : \langle 0|\rho(t)|0\rangle \geq 1 - p_0\},$$

where $0 < p_0 < 1$. *When one makes a projective measurement with the measurement operator* σ_z *at the time* t, *the probability of failure* $p = \langle 1|\rho(t)|1\rangle$ *is not greater than* p_0.

The proof of Theorem 6.1 is presented in Appendix. Using Theorem 6.1, we may try to maintain the system's state in $\mathcal{D}_{\mathrm{SMC}}$ (i.e., the subtask (ii)) by implementing periodic projective measurements with the measurement period $T = T^{(1)}$.

Now we consider the unknown uncertainties

$$H_\Delta = \frac{1}{2}\varepsilon_x(t)\sigma_x + \frac{1}{2}\varepsilon_y(t)\sigma_y$$

satisfying the bounded condition $\sqrt{\varepsilon_x^2(t) + \varepsilon_y^2(t)} \leq \varepsilon$ and have the following theorem:

Theorem 6.2 *For a two-level quantum system with the initial state* $\rho_0 = |0\rangle\langle 0|$ *at the time* $t = 0$*, the system evolves to* $\rho(t)$ *under the action of*

$$H(t) = \frac{1}{2}\sigma_z + \frac{1}{2}\varepsilon_x(t)\sigma_x + \frac{1}{2}\varepsilon_y(t)\sigma_y,$$

where $\sqrt{\varepsilon_x^2(t)+\varepsilon_y^2(t)} \leq \varepsilon$ *and* $\varepsilon > 0$. *If* $t \in [0, T^{(2)}]$, *where*

$$T^{(2)} = \frac{\arccos(1-2p_0)}{\varepsilon}, \tag{6.6}$$

the system's state will remain in

$$\mathcal{D}_{SMC} = \{\rho(t) : \langle 0|\rho(t)|0\rangle \geq 1 - p_0\}$$

with $0 < p_0 < 1$. *When one makes a projective measurement with the measurement operator* σ_z *at the time* t, *the probability of failure* $p = \langle 1|\rho(t)|1\rangle$ *is not greater than* p_0.

The proof of Theorem 6.2 can be found in [15]. The two theorems mean the following fact: For a two-level quantum system with unknown uncertainties

$$H_\Delta = \frac{1}{2}\varepsilon_x(t)\sigma_x + \frac{1}{2}\varepsilon_y(t)\sigma_y,$$

where $\sqrt{\varepsilon_x^2(t)+\varepsilon_y^2(t)} \leq \varepsilon$, if its initial state is in the sliding mode $|0\rangle$, we can ensure that the probability of failure is not greater than a given constant p_0 $(0 < p_0 < 1)$ through implementing periodic projective measurements with the measurement period $T = T^{(2)}$ using (6.6). Further, if we know that p_0 and ε satisfy the relationship

$$0 < p_0 \leq \frac{\varepsilon^2}{1+\varepsilon^2}$$

and there exists only one type of uncertainty (i.e., $H_\Delta = \frac{1}{2}\varepsilon(t)\sigma_x$ or $H_\Delta = \frac{1}{2}\varepsilon(t)\sigma_y$, where $|\varepsilon(t)| \leq \varepsilon$), we can design a measurement period $T = T^{(1)}$ using (6.5) which is larger than $T^{(1)}$. The proof of Theorem 6.1 also shows that $T^{(1)}$ is an optimal measurement period. This measurement period will guarantee the required robustness. It is easy to prove the relationship $T^{(1)} \geq T^{(2)}$ for arbitrary $p_0 \in (0, \frac{\varepsilon^2}{1+\varepsilon^2}]$. Based on the above analysis, the selection rule for T is summarized in Table 6.1. Moreover, from (6.5) and (6.6), it is clear that for a constant ε, $T^{(1)} \to 0$ and $T^{(2)} \to 0$ when $p_0 \to 0$. That is, for a given bound ε on the uncertainties, if we expect to guarantee the probability of failure $p_0 \to 0$, it requires us to implement frequent measurements such that the measurement period $T \to 0$. Another special case is $\varepsilon \to +\infty$, which leads to $T^{(1)} \to 0$ and $T^{(2)} \to 0$. That is, to deal with very large uncertainties, we need to make frequent measurements ($T \to 0$) to guarantee the desired robustness. From (6.6), we also know that for a given p_0, $T^{(2)}$ monotonically decreases with increasing ε. This means that we need to employ a smaller measurement period to deal with uncertainties with a larger bound ε.

Table 6.1 A summary on the selection rule of the measurement period T

Type of uncertainties	$H_\Delta = \frac{1}{2}\varepsilon(t)\sigma_\zeta$ ($\zeta = x$ or y)		$H_\Delta = \frac{1}{2}\varepsilon_x(t)\sigma_x + \frac{1}{2}\varepsilon_y(t)\sigma_y$
Allowed probability of failure p_0	$0 < p_0 \leq \frac{\varepsilon^2}{1+\varepsilon^2}$	$\frac{\varepsilon^2}{1+\varepsilon^2} < p_0 < 1$	$0 < p_0 < 1$
Measurement period T	$T = T^{(1)}$	$T = T^{(2)}$	$T = T^{(2)}$

6.2.3 Rapid Lyapunov Control Design

In the above SMC method, the design of Lyapunov control is another important task which aims to design control laws to drive the controlled system to the chosen sliding mode domain $\mathcal{D}_{\text{SMC}}$ in subtasks (i) and (iii). For simplicity, we suppose that there exist no uncertainties during the control processes (i) and (iii).

Lyapunov control methods are widely used in traditional SMC. If the gradient of a Lyapunov function is negative in the neighborhood of the sliding surface, then the controlled system's state will be attracted to and maintained in $\mathcal{D}_{\text{SMC}}$. The Lyapunov methodology has also been used to design control laws for quantum systems [29, 30, 35]. Since the measurement of a quantum system will inevitably destroy the measured state, most existing results on Lyapunov control for quantum systems in fact use a feedback design to construct an open-loop control. That is, Lyapunov control methodology can be used to first design a feedback control law which is then used to find the open-loop control by simulating the closed-loop system. Then the control can be applied to the quantum system in an open-loop way. Hence, the traditional SMC methods using Lyapunov control cannot be directly applied to our SMC problem. Here, we employ rapid Lyapunov control to achieve the subtasks (i) and (iii) in the SMC method.

Before presenting a specific rapid Lyapunov control approach for SMC of two-level systems, we first consider Lyapunov control of finite-dimensional quantum systems. Assume that the quantum system under consideration is an N-dimensional and controllable closed system [8], described by the following Liouville–von Neumann equation:

$$\dot{\rho}(t) = -\mathrm{i}\left[H_0 + \sum_{k=1}^{m} H_k u_k(t),\ \rho(t)\right], \tag{6.7}$$

where $\rho(t) \in \mathbb{C}^{N\times N}$ is a density matrix describing the state of the system; H_0 is the free Hamiltonian, and H_k is the control Hamiltonian that describes the interaction between the external control fields and the system (H_0 and H_k are time-independent Hermitian matrices); and $u_k(t)$ ($k = 1, \ldots, m$) are external real-valued control fields. In the energy representation, H_0 has a diagonal form, i.e., $H_0 = \mathrm{diag}(\lambda_1, \lambda_2, \ldots, \lambda_N)$. We call $\omega_{ab} \triangleq \lambda_a - \lambda_b$ ($a, b \in \{1, 2, \ldots, N\}$) the transition frequency between the energy levels λ_a and λ_b. Denote $|\lambda_j\rangle$ as the eigenvector of H_0 corresponding to the eigenvalue λ_j, i.e., $|\lambda_j\rangle = [0, \ldots, 0, 1, 0, \ldots, 0]^{\mathrm{T}}$ where the jth element is 1 and other elements are 0. All $|\lambda_j\rangle$ ($j = 1, \ldots, N$) form an orthogonal basis of the N-dimensional complex Hilbert space $\mathbb{H} \cong \mathbb{C}^N$.

For the quantum system in (6.7), its evolution can also be described by a unitary operator $U(t)$ as [14]

$$\dot{U}(t) = -\mathrm{i}\left(H_0 + \sum_{k=1}^{m} H_k u_k(t)\right) U(t) = -\mathrm{i}H(t)U(t), \tag{6.8}$$

where $U(t)U^{\dagger}(t) = U^{\dagger}(t)U(t) = I$ and $U(0) = I$. Given an initial state ρ_0, the quantum state at time t, $\rho(t)$, can always be written as

$$\rho(t) = U(t)\rho_0 U^{\dagger}(t), t \geq 0. \tag{6.9}$$

Equation (6.9) indicates that the system state $\rho(t)$ at any time always has the same spectrum with the initial state ρ_0.

We assume that the control objective is to steer the system to an eigenstate of H_0, $\rho_f \triangleq |\lambda_f\rangle\langle\lambda_f|$, $(f \in \{1, 2, \ldots, N\})$. Also, the following conditions are assumed on the system:

$$\omega_{af} \neq \omega_{bf} \ (a \neq b;\ a, b \neq f); \tag{6.10}$$

$$(H_{k'})_{jf} \neq 0 \ \left(j \neq f;\ \exists k' \in \{1, 2, \ldots, m\}\right). \tag{6.11}$$

Condition (6.10) means that the transition frequencies between the target eigenstate and other eigenstates are distinguishable, and that H_0 is non-degenerate; i.e., its all eigenvalues are mutually different. Condition (6.11) implies that there exists a direct coupling between the target eigenstate and any other eigenstate. The two conditions are helpful for providing strict theoretical results. On the other hand, they could also be relaxed in practical applications for achieving good control performance.

We can use the Lyapunov method to design a control law for the model (6.7) and then apply the control law to the real quantum system in an open-loop way. Consider the following Lyapunov function:

$$V = \mathrm{Tr}\left(P\rho(t)\right), \tag{6.12}$$

where P is a positive semidefinite Hermitian operator that needs to be constructed for completing a given control task.

The time derivative of Lyapunov function (6.12) is calculated as

$$\dot{V} = \mathrm{Tr}(P\dot{\rho}) = \mathrm{Tr}\left(-\mathrm{i}\rho\left[P,\ H_0\right]\right) + \sum_{k=1}^{m} u_k(t)\mathrm{Tr}\left(-\mathrm{i}\rho\left[P,\ H_k\right]\right). \tag{6.13}$$

We design the control laws by guaranteeing $\dot{V} \leq 0$ in (6.13). Considering that $\mathrm{Tr}(-\mathrm{i}\rho[P, H_0])$ in (6.13) is independent of the control field $u_k(t)$ while P is an unknown Hermitian matrix to be constructed, we let

$$[P,\ H_0] = 0. \tag{6.14}$$

Since the diagonal matrix H_0 is non-degenerate, (6.14) implies that P is also a diagonal matrix. Such a P is easy to design since we have complete flexibility to choose P. We denote $P \triangleq \text{diag}(p_1, p_2, \ldots, p_N)$. By using (6.14), (6.13) can be written as

$$\dot{V} = \sum_{k=1}^{m} u_k(t)\tau_k(t), \tag{6.15}$$

where $\tau_k(t) \triangleq \text{Tr}(-\text{i}\rho(t)[P, H_k])$. For notational simplicity, we also denote $\tau_k(t)$ as τ_k in the sequel.

By guaranteeing $\dot{V} \leq 0$, we can design a control law with the following general form:

$$u_k(t) = f_k(\tau_k), (k = 1, 2, \ldots, m), \tag{6.16}$$

where the control function $f_k(\cdot)$ satisfies:

(1) $f_k(x)$ $(x \in \mathbb{R})$ is continuously differentiable with respect to x.
(2) $f_k(0) = 0$.
(3) $f_k(x) \cdot x \leq 0$. Since $\tau_k = \tau_k^\dagger$ is a real number, we have $\dot{V} = \sum_{k=1}^{m} f_k(\tau_k)\tau_k \leq 0$.

In particular, the following standard Lyapunov control can be designed:

$$u_k(t) = -K_k\tau_k(t), (k = 1, 2, \ldots, m), \tag{6.17}$$

where the control gain $K_k > 0$ is used to adjust the amplitude of the control field $u_k(t)$.

Control law (6.16) means that the whole system is a nonlinear autonomous system. We use the LaSalle invariance principle to analyze the stability of the system. The LaSalle principle ensures that system (6.7) with the control fields in (6.16) necessarily converges to the largest invariant set E contained in $M \triangleq \{\rho : \dot{V}(\rho) = 0\}$. Assume $\bar{\rho} \in E$ and let $\rho_0 = \bar{\rho}$. We have the following stability results whose proof can be found in [30]:

Theorem 6.3 *[30] Given an arbitrary initial pure ρ_0, and under the action of the control fields in (6.16), the system (6.7) converges to the invariant set*

$$E(\rho_0) = \{\bar{\rho} : \lambda(\bar{\rho}) = \lambda(\rho_0);\ \bar{\rho} = \bar{\rho}^\dagger;\ \mathcal{M}\mathcal{P}\Im(\xi_k) = 0, \mathcal{M}\tilde{\Omega}\mathcal{P}\Re(\xi_k) = 0\},$$

where $k = 1, \ldots, m$ and we denote $FN = N(N-1) - 2$,

$$\begin{aligned}\xi_k \triangleq [&(H_k)_{12}\bar{\rho}_{21}, \ldots, (H_k)_{1N}\bar{\rho}_{N1}, (H_k)_{23}\bar{\rho}_{32}, \ldots, (H_k)_{2N}\bar{\rho}_{N2},\\ &\ldots, (H_k)_{N-1,N}\bar{\rho}_{N,N-1}]^T,\end{aligned} \tag{6.18}$$

$$\mathcal{M} \triangleq \begin{bmatrix} 1 & \cdots & 1 & 1 & \cdots & 1 & \cdots & 1 \\ \omega_{12}^2 & \cdots & \omega_{1N}^2 & \omega_{23}^2 & \cdots & \omega_{2N}^2 & \cdots & \omega_{N-1,N}^2 \\ \omega_{12}^4 & \cdots & \omega_{1N}^4 & \omega_{23}^4 & \cdots & \omega_{2N}^4 & \cdots & \omega_{N-1,N}^4 \\ \vdots & \cdots & \vdots & \vdots & \cdots & \vdots & \cdots & \vdots \\ \omega_{12}^{FN} & \cdots & \omega_{1N}^{FN} & \omega_{23}^{FN} & \cdots & \omega_{2N}^{FN} & \cdots & \omega_{N-1,N}^{FN} \end{bmatrix}, \tag{6.19}$$

$$\mathcal{P} \triangleq diag(p_2 - p_1, \ldots, p_N - p_1, p_3 - p_2, \ldots, p_N - p_2, \ldots, p_N - p_{N-1}), \tag{6.20}$$

$$\tilde{\Omega} \triangleq diag(\omega_{12}, \ldots, \omega_{1N}, \omega_{23}, \ldots, \omega_{2N}, \ldots, \omega_{N-1,N}). \tag{6.21}$$

From Theorem 6.3, the invariant set that the system converges to is dependent on the transition frequencies of the system, the diagonal values of P, the initial state ρ_0 and the connectivity of H_k. We further study the construction method of P to achieve convergence to the target eigenstate ρ_f. When the initial state ρ_0 satisfies $\mathrm{Tr}(\rho_0\rho_f) \neq 0$, we have the following result whose proof can be found in [30]:

Theorem 6.4 *Consider system (6.7) satisfying conditions (6.10), (6.11) and with the control fields in (6.16). Assume that the target eigenstate ρ_f and the initial pure state ρ_0 satisfy $\mathrm{Tr}(\rho_0\rho_f) \neq 0$. If the diagonal elements of P satisfy $p_j = p > p_f \geq 0, (j = \{1, 2, \ldots, N\}/f)$, then ρ_f is isolated in the invariant set $E(\rho_0)$ and the system state starting from ρ_0 necessarily converges to ρ_f.*

For general continuously differentiable control function (6.16), the construction relation in Theorem 6.4 ensures convergence to the target eigenstate. Based on the construction relation of P, we may design Lyapunov control laws to accomplish the subtasks (i) and (iii) in our SMC method. In [30], a result has shown that rapid control can make the control law more robust to uncertainties in the model or in the control process. In [21], an optimal Lyapunov design method has been proposed to design a control law for rapid state transfer in quantum systems. Under power-type and strength-type constraints on the control fields, two kinds of Lyapunov control laws were proposed. In particular, the strength-type constraint led to a bang-bang Lyapunov control. In [47], the convergence problem for bang-bang Lyapunov control law was further discussed for two-level quantum systems. In order to achieve rapidly convergent control in state transfer, two classes of rapid Lyapunov control methods have been proposed in [30], including switching Lyapunov control and approximate bang-bang (ABB) Lyapunov control. Here we employ the following ABB Lyapunov control law for SMC of quantum systems:

$$u_k(\tau_k) = \frac{2S_k}{1 + e^{\eta_k \tau_k}} - S_k, \ (k = 1, \ldots, m), 6, \tag{6.22}$$

where $S_k > 0$ is the maximum admissible strength of the control field u_k, and $\eta_k > 0$ is a parameter used to adjust the hardness of the control function. The bigger η_k is, the harder the characteristic of $u_k(\tau_k)$ is. As $\eta_k \to +\infty$, the characteristic of $u_k(\tau_k)$ approaches a bang-bang Lyapunov control. The ABB Lyapunov control law (6.22) is a special form of the smooth control law (6.16). Therefore, the convergence results in Theorems 6.3 and 6.4 naturally hold for the ABB control law. That is to say, with the conditions in Theorem 6.4, the corresponding ABB Lyapunov control law (6.22) is always stable.

6.2.4 An Illustrative Example

Example 6.1 Now, we present an illustrative example to demonstrate the proposed method. Assume $p_0 = 0.005$. Consider two cases: (a) $\varepsilon = 0.01$; (b) $\varepsilon = 0.1$. For simplicity, we assume $\rho_0 = |1\rangle\langle 1|$. Hence, $T_0 = T_1$. We first design the control and T_1 using (6.22). Here, we consider control only using $H_u = \frac{1}{2}u(t)\sigma_x$. Using (6.22), we select

$$u(\tau) = \frac{2S}{1 + e^{\eta\tau}} - S, \tag{6.23}$$

where

$$\tau = \operatorname{Tr}\left(-i\rho(t)\left[P, \frac{1}{2}\sigma_x\right]\right),$$

$S = 40$, $\eta = 100$ and $P = \operatorname{diag}(0.5, 1)$. Let the time step size be given by $\delta t = 10^{-5}$. We can obtain the probability curve of $|0\rangle$ shown in Fig. 6.2a, the control value shown in Fig. 6.2b and $T_1 = 0.085$. For $\varepsilon = 0.01$, we can design the measurement period $T = T^{(2)} = 14.154$ using (6.6). For $\varepsilon = 0.1$, we can design the measurement period $T = T^{(2)} = 1.415$ using (6.6). Since

$$p' = \frac{\varepsilon^2}{1 + \varepsilon^2} = 0.010 > p_0$$

when $\varepsilon = 0.1$, if the uncertainties take the form of $H_\Delta = \frac{1}{2}\varepsilon(t)\sigma_\zeta$ ($\zeta = x$ or y), we can improve the measurement period to $T = T^{(1)} = 1.573$ using (6.5). It is clear that $T \gg T_1$ in these two cases. For some practical quantum systems such as spin systems in NMR, we can use strong control actions (e.g., $S = 10^4$) to drive the system from $|1\rangle$ into $\mathcal{D}_{\mathrm{SMC}}$ within a short time period T_1 [25]. These facts make the assumption of no uncertainties in the control process reasonable. Moreover, the fact that the measurement period T is much greater than the control time required to go to $|0\rangle$ from $|1\rangle$ indicates the possibility of realizing such a periodic measurement on a practical quantum system.

Remark 6.1 In the process of designing the Lyapunov control for driving the system's state from $|1\rangle$ to $|0\rangle$, we ignore possible uncertainties. By simulation, we find that small uncertainties can also be tolerated in this process. For example, if $\varepsilon = 0.01$ and the uncertainty $\varepsilon(t)$ is the noise with a uniform distribution on the interval $[-0.01, 0.01]$, the probability curves of $|0\rangle$ are shown in Fig. 6.3 when we

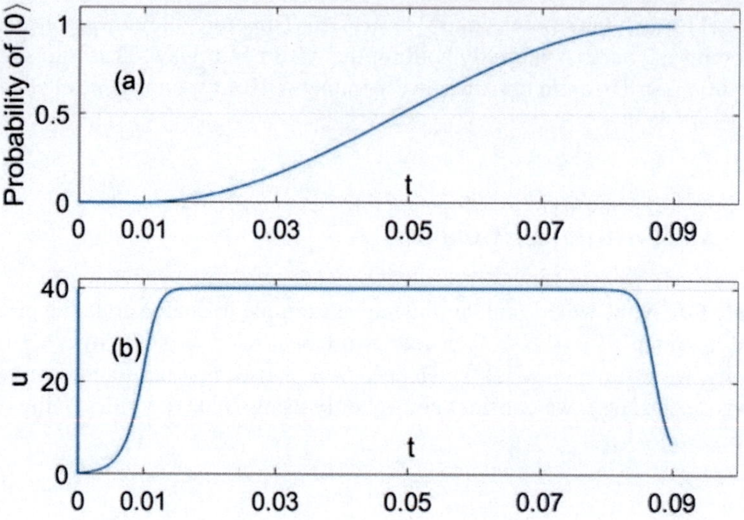

Fig. 6.2 Results under rapid Lyapunov control. **a** The probability of $|0\rangle$ under rapid Lyapunov control. **b** The control value $u(t)$

apply the control obtained from Fig. 6.2b to the quantum system. The probabilities of $|0\rangle$ for the cases with uncertainties are very close to the probability of $|0\rangle$ for the case without uncertainties. By more simulation, we find that the final probability of $|0\rangle$ is $(99.50 \pm 0.12)\%$ for $\frac{1}{2}\varepsilon(t)\sigma_x$ or $\varepsilon(t)\frac{1}{2}\sigma_y$ ($|\varepsilon(t)| \leq \varepsilon$ where $\varepsilon = 0.01$ or 0.1). The comparison results for $\varepsilon = 0.01$ are presented in Fig. 6.3 where the uncertainties are generated randomly according to a uniform distribution in every interval and each noise case is obtained by using the average of fifty runs. If we use a smaller probability of failure $\tilde{p}_0$ (e.g., $\tilde{p}_0 = 0.6p_0$) as the terminal condition of the Lyapunov control or employ a longer T_1, additional simulations suggest that it is possible to ensure that the Lyapunov control will drive the system's state into the sliding mode domain even when there exist small uncertainties.

6.3 Sliding Mode Control of Multi-level Quantum Systems

6.3.1 *SMC Based on Amplitude Amplification and Periodic Measurements*

For a multi-level quantum system, we may define a subspace as a sliding mode. For example, we consider the following five-level system:

$$\mathrm{i}\dot{\rho}(t) = [H_0 + u(t)H_1, \rho(t)], \tag{6.24}$$

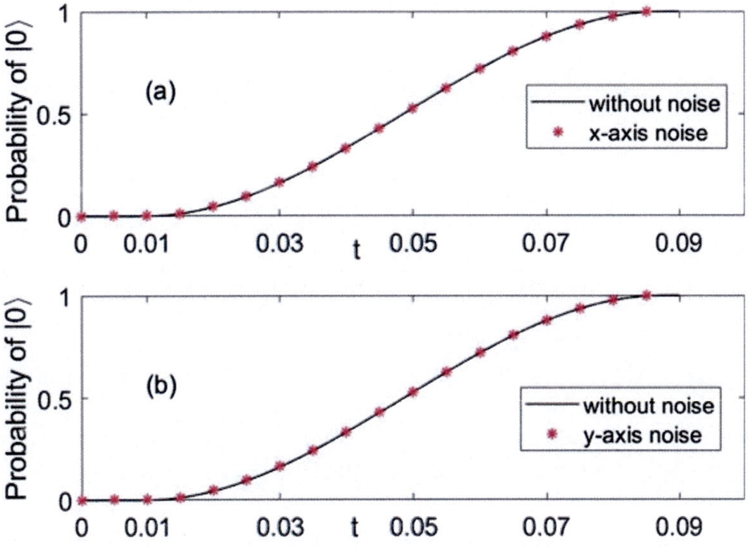

Fig. 6.3 Probability curves of $|0\rangle$ for the case without uncertainties (without noise) and the cases with uncertainties (noise) when we apply the control in Fig. 6.2b to the quantum system. **a** Comparison between the case without noise and the case with x-axis noise. **b** Comparison between the case without noise and the case with y-axis noise. The ξ-axis noise ($\xi = x$ or y) means the existence of $\frac{1}{2}\varepsilon(t)\sigma_\xi$ where $\varepsilon(t)$ is the noise with a uniform distribution on the interval $[-0.01, 0.01]$

where [42]

$$H_0 = \begin{bmatrix} 1.0 & 0 & 0 & 0 & 0 \\ 0 & 1.2 & 0 & 0 & 0 \\ 0 & 0 & 1.3 & 0 & 0 \\ 0 & 0 & 0 & 2.0 & 0 \\ 0 & 0 & 0 & 0 & 2.15 \end{bmatrix}, \quad H_1 = \begin{bmatrix} 0 & 0 & 0 & 1 & 1 \\ 0 & 0 & 0 & 0 & 0 \\ 0 & 0 & 0 & 0 & 0 \\ 1 & 0 & 0 & 0 & 1 \\ 1 & 0 & 0 & 1 & 0 \end{bmatrix}. \tag{6.25}$$

Denote the five eigenstates of H_0 as $\{|\phi_1\rangle, |\phi_2\rangle, |\phi_3\rangle, |\phi_4\rangle, |\phi_5\rangle\}$. In [12], it has been proven that the subspace Ω spanned by $\{|\phi_1\rangle, |\phi_4\rangle, |\phi_5\rangle\}$ is a wavefunction controllable subspace [10]. We may select Ω as a sliding surface. We can prove that if the initial state of this system is in this sliding surface, its state will be maintained in this surface under the action of the Hamiltonian $H = H_0 + u(t)H_1$. In fact, we can express the sliding mode as follows:

$$S(\rho, H) = 1 - (\langle\phi_1|\rho|\phi_1\rangle + \langle\phi_4|\rho|\phi_4\rangle + \langle\phi_5|\rho|\phi_5\rangle) = 0.$$

If $S(\rho_0, H) = 0$, we can obtain $S(\rho(t), H) = 0$. We denote $p_n(t) = \langle\phi_n|\rho(t)|\phi_n\rangle$ for $n = 1, \ldots, 5$. From $S(\rho_0, H) = 0$, it is clear that we can obtain $p_1(0) + p_4(0) + p_5(0) = 1$. On the other hand, we may calculate the derivative of $p_n(t)$ as

$$\frac{\mathrm{d}}{\mathrm{d}t}p_n(t) = \langle \phi_n | \dot{\rho}(t) | \phi_n \rangle.$$

Using (6.24), we have the following relationship after a detailed calculation:

$$\frac{\mathrm{d}}{\mathrm{d}t}(p_1(t) + p_4(t) + p_5(t)) = 0. \tag{6.26}$$

It is clear that

$$p_1(t) + p_4(t) + p_5(t) = p_1(0) + p_4(0) + p_5(0) = 1.$$

That is, $S(\rho(t), H) = 0$; i.e., the state of this system will be maintained in this surface under the action of the Hamiltonian $H = H_0 + u(t)H_1$.

Now, we consider the following quantum control problem: For a multi-level quantum system with uncertainties where a wavefunction controllable subspace Ω is used as a sliding mode; i.e.,

$$S(\rho, H) = 1 - \langle \Phi | \rho | \Phi \rangle = 0$$

with $|\Phi\rangle \in \Omega$,

(i) drive the controlled quantum system into the sliding mode domain

$$\mathcal{D}_{\mathrm{SMC}} = \{\rho : \langle \Phi | \rho | \Phi \rangle \geq 1 - p_0, |\Phi\rangle \in \Omega\};$$

(ii) maintain its state in $\mathcal{D}_{\mathrm{SMC}}$ except that measurements may take it away from $\mathcal{D}_{\mathrm{SMC}}$ with at most probability p_0;

(iii) once the system's state is away from $\mathcal{D}_{\mathrm{SMC}}$, design a control to drive it back to $\mathcal{D}_{\mathrm{SMC}}$.

We may employ quantum amplitude amplification to help the design of control laws. The quantum amplitude amplification technology is a powerful ingredient in quantum algorithms [5, 20, 22, 34]. The central task of quantum amplitude amplification is to find a suitable operator $\mathbb{Q}$ whose iterative action on the initial state can increase the probability of pointed eigenstates. If we denote $\mathbb{X} = \{|1\rangle, \ldots, |x\rangle, \ldots, |N\rangle\}$ as a set of orthonormal basis in the N-dimensional complex Hilbert space $\mathbb{H}$, a pure state $|\psi\rangle$ of an N-level quantum system can be represented as $|\psi\rangle = \sum_{x=1}^{N} c_x |x\rangle$, where $\sum_{x=1}^{N} |c_x|^2 = 1$. A Boolean function $\chi : \mathbb{X} \to \{0, 1\}$ induces two orthogonal subspaces of $\mathbb{H}$: the "good" subspace and the "bad" subspace. The good subspace is spanned by the set of basis states $|x\rangle \in \mathbb{X}$ satisfying $\chi(x) = 1$, and the bad subspace is its orthogonal complement in $\mathbb{H}$. We may decompose $|\psi\rangle$ as $|\psi\rangle = |\psi_{\mathrm{g}}\rangle + |\psi_{\mathrm{b}}\rangle$, where $|\psi_{\mathrm{g}}\rangle = P_{\mathrm{g}}|\psi\rangle$ denotes the projection of $|\psi\rangle$ onto the good subspace with the corresponding projector P_{g}, and $|\psi_{\mathrm{b}}\rangle = (I - P_{\mathrm{g}})|\psi\rangle$ denotes the projection of $|\psi\rangle$ onto the bad subspace. It is clear that the occurrence probabilities of the "good" state $|x\rangle$ $[\chi(x) = 1]$ and the "bad" state $|x\rangle$ $[\chi(x) = 0]$ upon measuring $|\psi\rangle$ are $g = \langle \psi_{\mathrm{g}} | \psi_{\mathrm{g}} \rangle$ and $b = \langle \psi_{\mathrm{b}} | \psi_{\mathrm{b}} \rangle = 1 - g$, respectively.

Let $|\psi\rangle = U|1\rangle$. Given two angles $0 \leq \varphi_1, \varphi_2 \leq \pi$, a general quantum amplitude amplification can be realized by the following operator [5]:

$$\mathbb{Q} = \mathbb{Q}(U, \chi, \varphi_1, \varphi_2) = -U\mathcal{P}_0^{\varphi_1}U^{-1}\mathcal{P}_\chi^{\varphi_2}. \tag{6.27}$$

The operators $\mathcal{P}_0^{\varphi_1}$ and $\mathcal{P}_\chi^{\varphi_2}$ conditionally change the phases of state $|1\rangle$ and the good states, respectively [5], and they can be expressed as [18]:

$$\mathcal{P}_0^{\varphi_1} = I - (1 - \mathrm{e}^{\mathrm{i}\varphi_1})|1\rangle\langle 1| \tag{6.28}$$

$$\mathcal{P}_\chi^{\varphi_2} = I - (1 - \mathrm{e}^{\mathrm{i}\varphi_2}) \sum_{\chi(x)=1} |x\rangle\langle x|. \tag{6.29}$$

The action of $\mathbb{Q}$ can be described by the following relationship [18]:

$$\mathbb{Q}|\psi\rangle = [(1 - \mathrm{e}^{\mathrm{i}\varphi_1})(1 - g + g\mathrm{e}^{\mathrm{i}\varphi_2}) - \mathrm{e}^{\mathrm{i}\varphi_2}]|\psi_\mathrm{g}\rangle + [g(1 - \mathrm{e}^{\mathrm{i}\varphi_1})(\mathrm{e}^{\mathrm{i}\varphi_2} - 1) - \mathrm{e}^{\mathrm{i}\varphi_1}]|\psi_\mathrm{b}\rangle. \tag{6.30}$$

Thus, we can amplify (or shrink) the amplitude of $|\psi_\mathrm{g}\rangle$ (or $|\psi_\mathrm{b}\rangle$) by a suitable selection of the parameters φ_1, φ_2 in $\mathbb{Q}$.

The main steps of the SMC algorithm based on amplitude amplification and periodic projective measurements are as follows:

1. Select a state subspace Ω as a sliding mode $S(\rho, H) = 0$.
2. For a known initial state ρ_0, identify Ω as a "good" subspace and construct an amplitude amplification operator $\mathbb{Q}(S)$ to amplify the probability of projecting $|\psi_0\rangle$ into Ω.
3. According to the probability p_0 and $\mathbb{Q}(S)$, determine a number L_0 of iterations; i.e., it guarantees that the control law can drive the system into the sliding mode domain $\mathcal{D}_{\mathrm{SMC}} = \{\rho : \langle\Phi|\rho|\Phi\rangle \geq 1 - p_0, |\Phi\rangle \in \Omega\}$.
4. For other eigenstates $|\phi_k\rangle$ out of the "good" subspace, first apply a unitary transformation U_k to $|\phi_k\rangle$ to obtain a superposition state $|\psi_k\rangle$, then construct the corresponding amplitude amplification operator $\mathbb{Q}(|\psi_k\rangle, S)$ and design the number L_k of iterations using a similar method as in Steps 2 and 3.
5. According to p_0 and ε, design the period T for periodic projective measurements.
6. Use the designed control law to drive the system's state into $\mathcal{D}_{\mathrm{SMC}}$, then implement periodic projective measurements with the period T to maintain the system's state in $\mathcal{D}_{\mathrm{SMC}}$. If the measurement result is $|\phi_k\rangle$, we use the corresponding control law to drive it into $\mathcal{D}_{\mathrm{SMC}}$.

Remark 6.2 In the control algorithm, the construction of the amplitude amplification operator $\mathbb{Q}$ is dependent on the initial state; i.e., the initial state must be known. When the initial state is an eigenstate $|\phi_k\rangle$, we usually need to apply a unitary transformation U_k to drive $|\phi_k\rangle$ to a superposition state before constructing $\mathbb{Q}$. Here U_k may be a small perturbation or an easily-realized unitary transformation. If the initial state is unknown, we make a measurement to prepare a known initial state.

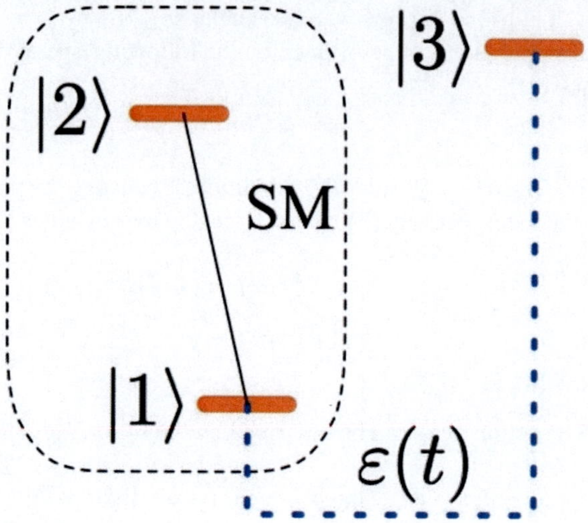

Fig. 6.4 A three-level system. The subspace spanned by $\{|1\rangle, |2\rangle\}$ corresponds to the sliding mode (SM). The uncertainty $\varepsilon(t)$ exists between $|1\rangle$ and $|3\rangle$

6.3.2 An Illustrative Example: Three-Level Systems

Example 6.2 Now consider a three-level quantum system. Consider

$$i\dot{\rho}(t) = [H_0 + H_1 + H_\Delta, \rho(t)], \tag{6.31}$$

where $H_0 = \text{diag}(-1, 0, 1)$, and H_1 and the uncertainty H_Δ are as follows:

$$H_1 = \begin{bmatrix} 0 & 1 & 0 \\ 1 & 0 & 0 \\ 0 & 0 & 0 \end{bmatrix}, \quad H_\Delta = \begin{bmatrix} 0 & 0 & \varepsilon(t) \\ 0 & 0 & 0 \\ \varepsilon(t) & 0 & 0 \end{bmatrix}, \tag{6.32}$$

where $|\varepsilon(t)| \leq \varepsilon$. The three eigenstates $\{|\phi_1\rangle, |\phi_2\rangle, |\phi_3\rangle\}$ of H_0 are denoted as $\{|1\rangle, |2\rangle, |3\rangle\}$, respectively. Suppose that the initial state is $|1\rangle$. The three-level system corresponding to (6.32) can be schematically shown in Fig. 6.4. The H_1 matrix implies that there is a direct coupling between $|1\rangle$ and $|2\rangle$. The uncertainty matrix H_Δ implies that a disturbance coupling $\varepsilon(t)$ between $|1\rangle$ and $|3\rangle$. It is not difficult to check that the subspace spanned by $\{|1\rangle, |2\rangle\}$ can be defined as a sliding mode; i.e.,

$$S(\rho, H) = 1 - (\langle 1|\rho|1\rangle + \langle 2|\rho|2\rangle) = 0.$$

The control problem may correspond to some practical tasks in quantum information. For example, the two-level system with levels $|1\rangle$ and $|2\rangle$ can be used as a qubit, and the uncertainty H_Δ can be taken as a possible leakage into states outside the qubit subspace [9].

Now we consider the design of the period T for periodic projective measurements. Expand the state of the system $|\psi\rangle$ as

$$|\psi\rangle = \sum_{j=1}^{3} c_j(t)|\phi_j\rangle.$$

Let $c_j(t) = x_j(t) + \mathrm{i}y_j(t)(j = 1, 2, 3)$, where $x_j(t), y_j(t) \in \mathbb{R}$. Let the vector $Z(t)$ denote

$$Z(t) = [x_1(t)\ y_1(t)\ x_2(t)\ y_2(t)\ x_3(t)\ y_3(t)]^T.$$

We have

$$\dot{Z}(t) = F(\varepsilon(t))Z(t), \tag{6.33}$$

where

$$F(\varepsilon(t)) = \begin{bmatrix} 0 & -1 & 0 & 1 & 0 & \varepsilon(t) \\ 1 & 0 & -1 & 0 & -\varepsilon(t) & 0 \\ 0 & 1 & 0 & 0 & 0 & 0 \\ -1 & 0 & 0 & 0 & 0 & 0 \\ 0 & \varepsilon(t) & 0 & 0 & 0 & 1 \\ -\varepsilon(t) & 0 & 0 & 0 & -1 & 0 \end{bmatrix}, \tag{6.34}$$

and $Z(0) = [x_1(0)\ y_1(0)\ x_2(0)\ y_2(0)\ x_3(0)\ y_3(0)]^{\mathrm{T}} = [1\ 0\ 0\ 0\ 0\ 0]^T$.

We now introduce the Lagrange multiplier vector

$$\lambda(t) = [\lambda_1(t)\ \lambda_2(t)\ \lambda_3(t)\ \lambda_4(t)\ \lambda_5(t)\ \lambda_6(t)]^T,$$

and obtain the corresponding Hamiltonian function as follows:

$$H(Z(t), \varepsilon(t), \lambda(t), t) = \lambda^{\mathrm{T}}(t)F(\varepsilon(t))\eta(t) = \varepsilon(t)M(t) + N(t), \tag{6.35}$$

where

$$M(t) = \lambda_5(t)y_1(t) - \lambda_6(t)x_1(t) - \lambda_2(t)x_3(t) + \lambda_1(t)y_3(t),$$

$$N(t) = \lambda_2(t)x_1(t) - \lambda_4(t)x_1(t) - \lambda_1(t)y_1(t) + \lambda_3(t)y_1(t)$$

$$- \lambda_2(t)x_2(t) + \lambda_1(t)y_2(t) - \lambda_6(t)x_3(t) + \lambda_5(t)y_3(t).$$

According to Pontryagin's minimum principle [27], a necessary condition for $\tilde{\varepsilon}(t)$ to minimize the functional $J(\varepsilon) = x_3^2(t_{\mathrm{f}}) + y_3^2(t_{\mathrm{f}})$ is

$$H(\tilde{\eta}(t), \tilde{\varepsilon}(t), \tilde{\lambda}(t), t) \le H(\tilde{\eta}(t), \varepsilon(t), \tilde{\lambda}(t), t). \tag{6.36}$$

Hence,

$$\tilde{\varepsilon}(t) = -\varepsilon \mathrm{sgn} M(t). \tag{6.37}$$

That is, the optimal control is a bang-bang control strategy; i.e., $\tilde{\varepsilon}(t) = \bar{\varepsilon} = +\varepsilon$ or $-\varepsilon$.

Without loss of generality, now we let $\tilde{\varepsilon}(t) = \varepsilon$ and focus on

$$\dot{Z}(t) = F(\varepsilon)Z(t), \tag{6.38}$$

where $Z(0) = [1\ 0\ 0\ 0\ 0\ 0]^T$.

Consider the optimal control with a fixed final time t_f and a free final state

$$Z(t_f) = [x_1(t_f)\ y_1(t_f)\ x_2(t_f)\ y_2(t_f)\ x_3(t_f)\ y_3(t_f)]^T.$$

Let $J(t) = x_3^2(t) + y_3^2(t)$. According to Pontryagin's minimum principle, $\tilde{\lambda}(t_f) = \frac{\partial}{\partial Z} J(t)|_{t=t_f}$. From this, we obtain

$$(\lambda_1(t_f), \lambda_2(t_f), \lambda_3(t_f), \lambda_4(t_f), \lambda_5(t_f), \lambda_6(t_f)) = (0, 0, 0, 0, 2x_3(t_f), 2y_3(t_f)).$$

Now we consider another necessary condition

$$\dot{\lambda}(t) = -\frac{\partial H(Z(t), \varepsilon(t), \lambda(t), t)}{\partial Z}$$

which leads to the following relationships:

$$\dot{\lambda}(t) = F(\varepsilon)\lambda(t). \tag{6.39}$$

By simulation, we can estimate the period T by considering the first monotonic interval of $J(t)$ where the sign of $M(t)$ does not change. As an example, if we consider $\varepsilon = 0.1$, by simulation we obtain $t_f = 1.1160$ where the sign of $M(t)$ does not change in $t \in [0, t_f]$. We may design the period T using the following relationship:

$$T = \begin{cases} t_1 \text{ (where } t_1 \in [0, t_f) \text{ and } J(t_1) = p_0) & \text{if } p_0 \leq J(t_f); \\ t_f & \text{otherwise.} \end{cases} \tag{6.40}$$

For example, if $p_0 = 0.5\%$, we have $p_0 \leq J(t_f)$ and may choose $T = 1.1160$.

Now we design a control law for the subtask (iii) using quantum amplitude amplification. For simplicity, we use a simple $\mathbb{Q}$ where $\varphi_1 = \pi$, $\varphi_2 = \pi$ and the subspace spanned by $\{|1\rangle, |2\rangle\}$ corresponds to the "good" subspace. When the measurement result is $|\phi_k\rangle = |3\rangle$, we first apply a unitary transformation (a perturbation) U_3 on $|3\rangle$ to obtain

$$|\psi_3\rangle = U_k|3\rangle = 0.0600|1\rangle + 0.0800|2\rangle + 0.9950|3\rangle.$$

Using a similar method as in the example of [18], we obtain the state

$$|\psi_3'\rangle = 0.5986|1\rangle + 0.7981|2\rangle + 0.0682|3\rangle$$

after 7 iterations of $\mathbb{Q}$. Now make a measurement on $|\psi_3'\rangle$, and the probability of failure is $p_0' = 0.47\%$ ($\leq 0.5\%$).

6.4 Sliding Mode Control of Open Quantum Systems

6.4.1 Control of Open Quantum Systems

In the previous sections, SMC is developed to enhance the robustness of quantum systems with uncertainties in the system Hamiltonian. Now, we further extend SMC to enhance the performance of open quantum systems with uncertainties in the system Hamiltonian as well as in the system–environment interaction and develop robust decoherence control schemes for Markovian open quantum systems [17, 24]. Decoherence occurs when a quantum system interacts with an uncontrollable environment [3, 6, 7], and it has been recognized as a bottleneck for the development of practical quantum information technology [37]. Various methods have been proposed for decoherence control including quantum error-avoiding codes [32], quantum error-correction codes [28] and dynamical decoupling [26, 46]. Here we consider a robust decoherence control scheme for quantum systems subject to Markovian decoherence. In particular, we focus on a single qubit subject to amplitude damping decoherence, phase damping decoherence and depolarizing decoherence [37]. We propose an SMC design approach to guarantee the robustness of a single qubit system with uncertainties in the system Hamiltonian and the coupling strength of the system–environment interaction. In this approach, control actions and measurement (sampling) operations are applied at discrete time instants, which makes controller design using digital computers possible for robust control of open quantum systems.

We consider a two-level quantum system (qubit) with Markovian dynamics whose evolution can be described by the following Lindblad equation [4, 49]:

$$\dot{\rho}(t) = -\mathrm{i}[H(t), \rho(t)] + \sum_{k=1}^{K} \gamma_k \mathfrak{D}[L_k]\rho(t), \tag{6.41}$$

where

$$\mathfrak{D}[L_k]\rho = L_k \rho L_k^\dagger - \frac{1}{2} L_k^\dagger L_k \rho - \frac{1}{2} \rho L_k^\dagger L_k.$$

For such a single qubit system, we can divide $H(t)$ into three parts $H(t) = H_0 + H_\Delta + H_u$, where the free Hamiltonian is $H_0 = \frac{1}{2}\sigma_z$, the control Hamiltonian is

$$H_u = \frac{1}{2} \sum_{j=x,y,z} u_j(t)\sigma_j, \quad (u_j(t) \in \mathbb{R}),$$

and the uncertainties in the system Hamiltonian are

$$H_\Delta = \frac{1}{2}\omega(t)\sigma_z + \frac{1}{2}\varepsilon_x(t)\sigma_x + \frac{1}{2}\varepsilon_y(t)\sigma_y \quad (\omega(t), \varepsilon_x(t), \varepsilon_y(t) \in \mathbb{R}).$$

H_Δ is the first class of uncertainties we consider. A second class of uncertainties involves uncertainties $\delta\gamma_k$ residing in the coupling strength γ_k. Since the Lindblad equation is an approximate equation for the open quantum system coupling with its environment, this class of uncertainties may come from inaccurate modeling as well as time-varying coupling between the system and environment. We assume that all the uncertainties are bounded, i.e., $|\omega(t)| \leq \omega$, $\sqrt{\varepsilon_x^2(t) + \varepsilon_y^2(t)} \leq \varepsilon$ and $|\delta\gamma_k| \leq \gamma$, where constants $\omega \geq 0$, $\varepsilon > 0$ and $\gamma \geq 0$ are given.

For an open qubit system, the purity of ρ is defined as $P = \mathrm{Tr}(\rho^2)$. Another useful quantity is the coherence which can be defined as $C = x^2 + y^2$, where $x = \mathrm{Tr}(\rho\sigma_x)$ and $y = \mathrm{Tr}(\rho\sigma_y)$ (see, e.g., [33]). A decoherence process due to the interaction of a quantum system with its environment may reduce its purity or coherence.

We consider the following three cases and define the required robustness using the concept of a sliding mode domain [17]:

(A) *Amplitude damping decoherence.* In this case, the population of the quantum system can change (e.g., through loss of energy by spontaneous emission). The evolution of ρ_t (i.e., $\rho(t)$) can be described by the following equation:

$$\dot{\rho}_t = -\mathrm{i}[H(t), \rho_t] + \gamma_t \left(\sigma_- \rho_t \sigma_+ - \frac{1}{2}\sigma_+\sigma_-\rho_t - \frac{1}{2}\rho_t\sigma_+\sigma_- \right), \qquad (6.42)$$

where $\sigma_- = \frac{1}{2}(\sigma_x - \mathrm{i}\sigma_y)$, $\sigma_+ = \frac{1}{2}(\sigma_x + \mathrm{i}\sigma_y)$, $\gamma_t = \gamma_0 + \delta\gamma_t$ and $|\delta\gamma_t| \leq \gamma$. We also assume that $\gamma_0 \geq \gamma$, which guarantees the coupling strength $\gamma_t \geq 0$. The corresponding sliding mode domain is defined as

$$\mathcal{D}_\mathrm{a} = \{\rho : \langle 0|\rho|0\rangle \geq 1 - p_0, 0 < p_0 < 1\}.$$

The definition of $\mathcal{D}_\mathrm{a}$ implies that the system's state has a probability of at most p_0 to collapse out of $\mathcal{D}_\mathrm{a}$ when making a projective measurement with the operator σ_z. We aim to drive and then maintain a single qubit's state in the sliding mode domain $\mathcal{D}_\mathrm{a}$. However, the uncertainties H_Δ and $\delta\gamma_t$ may take the system's state away from $\mathcal{D}_\mathrm{a}$. The measurement operation unavoidably makes the system's state change. Thus, we expect that the control law will guarantee that the system's state remains in $\mathcal{D}_\mathrm{a}$, except that the measurement process may take it away from $\mathcal{D}_\mathrm{a}$ with a small probability (not greater than p_0).

(B) *Phase damping decoherence.* In this case, a loss of quantum coherence can occur without loss of energy in the quantum system. The evolution of the state may be described by the following equation:

$$\dot{\rho}_t = -\mathrm{i}[H(t), \rho_t] + \gamma_t(\sigma_z\rho_t\sigma_z - \rho_t). \qquad (6.43)$$

The corresponding sliding mode domain is defined as

$$\mathcal{D}_{\mathrm{p}} = \{\rho : x^2 + y^2 \geq \bar{C}, x = \mathrm{Tr}(\rho\sigma_x), y = \mathrm{Tr}(\rho\sigma_y), 0 < \bar{C} \leq 1\}.$$

From the definition, it is clear that all states in $\mathcal{D}_{\mathrm{p}}$ have coherence of at least $\bar{C}$.

(C) *Depolarizing decoherence.* This decoherence maps pure states into mixed states. The dynamics can be described by the following equation:

$$\dot{\rho}_t = -\mathrm{i}[H(t), \rho_t] + \gamma_t(\sigma_x\rho_t\sigma_x - \rho_t) + \gamma_t(\sigma_y\rho_t\sigma_y - \rho_t) + \gamma_t(\sigma_z\rho_t\sigma_z - \rho_t). \tag{6.44}$$

The corresponding sliding mode domain is defined as

$$\mathcal{D}_{\mathrm{d}} = \{\rho : \mathrm{Tr}(\rho^2) \geq \bar{P}, 0.5 < \bar{P} \leq 1\}.$$

From the definition, it is clear that any state in $\mathcal{D}_{\mathrm{d}}$ has a purity of at least $\bar{P}$.

6.4.2 SMC Methods and Results

We aim to develop an SMC design method for robust control of open quantum systems with uncertainties. A key task is to design a measurement period as large as possible to guarantee the required robustness defined using a sliding mode domain. The measurement process is taken as an important control tool to modify the system dynamics nonunitarily. The sequel provides the main methods and results for the three cases of uncertain quantum systems.

6.4.2.1 Amplitude Damping Decoherence

For single qubit systems with amplitude damping decoherence, if the initial state is the excited state $|0\rangle$, the decoherence will drive this excited state to the ground state $|1\rangle$. The objective is to design a control law to guarantee the required robustness defined by $\mathcal{D}_{\mathrm{a}}$. We use the following design method: If the state is $|0\rangle$ at $t = nT_{\mathrm{a}}$ ($n = 0, 1, 2, \ldots$), we design a measurement period to maintain the system's state in $\mathcal{D}_{\mathrm{a}}$ by implementing periodic measurement with period T_{a}; if the measurement makes the state collapse into $|1\rangle$ (with a probability $p \leq p_0$), we design a unitary control to drive the state back into a subset $\mathcal{E}_{\mathrm{a}}$ of $\mathcal{D}_{\mathrm{a}}$ from $t = nT_{\mathrm{a}}$ to $(n + \beta)T_{\mathrm{a}}$, and then measure the system again at $t = (n + 1)T_{\mathrm{a}}$. In order to determine the required measurement period, we have the following result:

Theorem 6.5 *For a single qubit with initial state $\rho_0 = |0\rangle\langle 0|$ at time $t = 0$, the system evolves to ρ_t subject to (6.42) where*

$$H(t) = \frac{1}{2}[1 + \omega(t)]\sigma_z + \frac{1}{2}\varepsilon_x(t)\sigma_x + \frac{1}{2}\varepsilon_y(t)\sigma_y$$

with $\sqrt{\varepsilon_x^2(t)+\varepsilon_y^2(t)} \le \varepsilon$, $\varepsilon > 0$, $|\omega(t)| \le \omega$ *and* $\omega \ge 0$, *and the coupling strength of amplitude damping decoherence is* $\gamma_t = \gamma_0 + \delta\gamma_t$ $(|\delta\gamma_t| \le \gamma)$. *If* $t \in [0, T_a]$ *with*

$$T_a = \frac{2p_0}{\sqrt{4\varepsilon^2 + (\gamma_0+\gamma)^2} + (\gamma_0+\gamma)}, \tag{6.45}$$

the state will remain in

$$\mathcal{D}_a = \{\rho : \langle 0|\rho|0\rangle \ge 1 - p_0, 0 < p_0 < 1\}.$$

When a projective measurement is made with σ_z *at time* t, *the probability of failure* $p = \langle 1|\rho|1\rangle$ *is not greater than* p_0.

When there exist no uncertainties in the system Hamiltonian (i.e., $H_\Delta \equiv 0$), the measurement period can be designed using the following proposition:

Proposition 6.1 *For a single qubit with initial state* $\rho_0 = |0\rangle\langle 0|$ *at time* $t = 0$, *the system evolves to* ρ_t *subject to (6.42) where* $H(t) = \frac{1}{2}\sigma_z$ *and the coupling strength of amplitude damping decoherence is* $\gamma_t = \gamma_0 + \delta\gamma_t$ $(|\delta\gamma_t| \le \gamma)$. *If* $t \in [0, T_a']$ *with*

$$T_a' = -\frac{\ln(1-p_0)}{\gamma_0+\gamma}, \tag{6.46}$$

the state will remain in

$$\mathcal{D}_a = \{\rho : \langle 0|\rho|0\rangle \ge 1 - p_0, 0 < p_0 < 1\}.$$

When a projective measurement is made with σ_z *at time* t, *the probability of failure* $p = \langle 1|\rho|1\rangle$ *is not greater than* p_0.

The proofs of Theorem 6.5 and Proposition 6.1 are presented in Appendix. From the proof of Proposition 6.1, it is clear that $T_\mathrm{a}' = -\frac{\ln(1-p_0)}{\gamma_0+\gamma}$ exactly corresponds to the case $\delta\gamma_t \equiv \gamma$ when $H_\Delta \equiv 0$. In this sense, the measurement period T_a' is optimal to guarantee the required robustness. The relationship $T_\mathrm{a}' \ge T_\mathrm{a}$ for arbitrary p_0 can also be proved (see details in [17]).

For different situations we may use T_a or T_a' as the measurement period to guarantee the required performance. If the measurement result corresponds to $|1\rangle$, a unitary control can be used to drive the state back to a subset $\mathcal{E}_\mathrm{a}$ of $\mathcal{D}_\mathrm{a}$. The subset $\mathcal{E}_\mathrm{a}$ may be defined as

$$\mathcal{E}_\mathrm{a} = \{\rho : \langle 0|\rho|0\rangle \ge 1 - \alpha p_0, 0 < p_0 < 1, 0 \le \alpha \le 1\}.$$

The following theorem gives a sufficient condition on the relationships between α, p_0 and β to guarantee the required robustness: (The proof can be found in [17].)

Theorem 6.6 *For a single qubit with initial state* ρ_0 *satisfying* $\langle 1|\rho_0|1\rangle \le \alpha p_0$ $(0 \le \alpha \le 1)$ *at time* $t = 0$, *the system evolves to* $\rho(t)$ *subject to (6.42) where*

$$H(t) = \frac{1}{2}[1 + \omega(t)]\sigma_z + \frac{1}{2}\varepsilon_x(t)\sigma_x + \frac{1}{2}\varepsilon_y(t)\sigma_y$$

with $\sqrt{\varepsilon_x^2(t) + \varepsilon_y^2(t)} \leq \varepsilon$, $\varepsilon > 0$, $|\omega(t)| \leq \omega$ *and* $\omega \geq 0$, *and the coupling strength of the amplitude damping decoherence is* $\gamma_t = \gamma_0 + \delta\gamma_t$ $(|\delta\gamma_t| \leq \gamma)$. *If* $t \in [0, (1-\beta)T_a]$ *and* $\alpha \leq \beta$ *where* $0 < \beta \leq 1$ *and* T_a *satisfies (6.45), the state will remain in*

$$\mathcal{D}_a = \{\rho : \langle 0|\rho|0\rangle \geq 1 - p_0, 0 < p_0 < 1\}.$$

When a projective measurement is made with the operator σ_z *at time* t, *the probability of failure* $p = \langle 1|\rho|1\rangle$ *is not greater than* p_0.

6.4.2.2 Phase Damping Decoherence

For a single qubit, phase damping decoherence will reduce the coherence of the system. The objective is to guarantee that the state has coherence not less than $\bar{C}$ by periodic measurement when there exist uncertainties in the coupling strength of system–environment interaction and in the system Hamiltonian. To determine the required measurement period, we have the following results whose proof can be found in [17]:

Theorem 6.7 *For a single qubit with initial state* ρ_0 *satisfying* $C_0 = [Tr(\rho_0\sigma_x)]^2 + [Tr(\rho_0\sigma_y)]^2 = 1$ *at time* $t = 0$, *the system evolves to* ρ_t *subject to (6.43) where*

$$H(t) = \frac{1}{2}[1 + \omega(t)]\sigma_z + \frac{1}{2}\varepsilon_x(t)\sigma_x + \frac{1}{2}\varepsilon_y(t)\sigma_y$$

with $\sqrt{\varepsilon_x^2(t) + \varepsilon_y^2(t)} \leq \varepsilon$, $\varepsilon > 0$, $|\omega(t)| \leq \omega$ *and* $\omega \geq 0$, *and the coupling strength of the phase damping decoherence is* $\gamma_t = \gamma_0 + \delta\gamma_t$ $(|\delta\gamma_t| \leq \gamma)$. *If* $t \in [0, T_p]$ *with*

$$T_p = \begin{cases} \frac{1-\bar{C}}{4\sqrt{2}(\gamma_0+\gamma)}, & \text{when } 4(\gamma_0+\gamma)^2 \geq \varepsilon^2; \\ \frac{(1-\bar{C})\sqrt{\varepsilon^2-2(\gamma_0+\gamma)^2}}{2\varepsilon^2}, & \text{when } 4(\gamma_0+\gamma)^2 < \varepsilon^2, \end{cases} \tag{6.47}$$

the state will remain in

$$\mathcal{D}_p = \{\rho_t : [Tr(\rho_t\sigma_x)]^2 + [Tr(\rho_t\sigma_y)]^2 \geq \bar{C}, 0 < \bar{C} \leq 1\}.$$

When a periodic projective measurement is made with the operator σ_x *on the system, the measurement period* T_p *will guarantee that the state remains in* $\mathcal{D}_p$.

If $H_\Delta \equiv 0$, we can design the sampling period using the following proposition whose proof can be found in [17]:

Proposition 6.2 *For a single qubit with initial state ρ_0 satisfying $C_0 = [Tr(\rho_0\sigma_x)]^2 + [Tr(\rho_0\sigma_y)]^2 = 1$ at time $t = 0$, the system evolves to ρ_t subject to (6.43) where $H(t) = \frac{1}{2}\sigma_z$ and the coupling strength of the phase damping decoherence is $\gamma_t = \gamma_0 + \delta\gamma_t$ ($|\delta\gamma_t| \leq \gamma$). If $t \in [0, T_p']$ with*

$$T_p' = -\frac{\ln \bar{C}}{4(\gamma_0 + \gamma)}, \tag{6.48}$$

the state will remain in

$$\mathcal{D}_p = \{\rho_t : [Tr(\rho_t\sigma_x)]^2 + [Tr(\rho_t\sigma_y)]^2 \geq \bar{C}, 0 < \bar{C} \leq 1\}.$$

If a periodic projective measurement is made with the operator σ_x, the measurement period T_p' will guarantee that the system's state remains in $\mathcal{D}_p$.

In Proposition 6.2, the measurement period T_p' is optimal to guarantee the required robustness when $H_\Delta \equiv 0$. For this case with phase damping decoherence, we can also make projective measurements with the operator σ_y, which does not affect the conclusions. Moreover, in this case, no unitary control is required and measurement is the only tool needed for guaranteeing the required robustness.

6.4.2.3 Depolarizing Decoherence

For a single qubit, depolarizing decoherence will reduce the purity $P = \text{Tr}(\rho^2)$ of the system's state. The objective is to guarantee that the purity of the state is not less than $\bar{P}$ by periodic measurement when there exist uncertainties in the coupling strength of system–environment interaction and in the system Hamiltonian. For the measurement period, we have the following results whose proof can be found in [17]:

Theorem 6.8 *For a single qubit with initial state ρ_0 satisfying $Tr(\rho_0^2) = 1$ at time $t = 0$, the system evolves to ρ_t subject to (6.44) where*

$$H(t) = \frac{1}{2}[1 + \omega(t)]\sigma_z + \frac{1}{2}\varepsilon_x(t)\sigma_x + \frac{1}{2}\varepsilon_y(t)\sigma_y$$

with $\sqrt{\varepsilon_x^2(t) + \varepsilon_y^2(t)} \leq \varepsilon$, $\varepsilon > 0$, $|\omega(t)| \leq \omega$ and $\omega \geq 0$, and the coupling strength of depolarizing decoherence is $\gamma_t = \gamma_0 + \delta\gamma_t$ ($|\delta\gamma_t| \leq \gamma$). If $t \in [0, T_d]$ with

$$T_d = -\frac{\ln(2\bar{P} - 1)}{8(\gamma_0 + \gamma)}, \tag{6.49}$$

the state remains in

$$\mathcal{D}_d = \{\rho_t : Tr(\rho_t^2) \geq \bar{P}, 0.5 < \bar{P} \leq 1\}.$$

Table 6.2 Summary of measurement periods for different cases

Cases		Measurement period	$\varepsilon = 0.1$	$\varepsilon = 1.0$
Amplitude damping decoherence	$H_\Delta(x, y, z)$	$T_a = \frac{2p_0}{\sqrt{4\varepsilon^2+(\gamma_0+\gamma)^2}+(\gamma_0+\gamma)}$	$T_a = 0.0212$	$T_a = 0.0080$
	$H_\Delta \equiv 0$	$T_a' = -\frac{\ln(1-p_0)}{\gamma_0+\gamma}$	$T_a' = 0.0223$	$T_a' = 0.0223$
Phase damping decoherence	$H_\Delta(x, y, z)$	$T = T_p$ in (6.47)	$T_p = 0.0039$	$T_p = 0.0039$
	$H_\Delta \equiv 0$	$T_p' = -\frac{\ln C}{4(\gamma_0+\gamma)}$	$T_p' = 0.0056$	$T_p' = 0.0056$
Depolarizing decoherence	$H_\Delta(x, y, z)$	$T_d = -\frac{\ln(2\bar{P}-1)}{8(\gamma_0+\gamma)}$	$T_d = 0.0056$	$T_d = 0.0056$

$H_\Delta(x, y, z) = \frac{1}{2}\omega(t)\sigma_z + \frac{1}{2}\varepsilon_x(t)\sigma_x + \frac{1}{2}\varepsilon_y(t)\sigma_y$ (where $|\omega(t)| \leq \omega$ and $\sqrt{\varepsilon_x^2(t) + \varepsilon_y^2(t)} \leq \varepsilon$) and $\gamma_t = \gamma_0 + \delta\gamma_t$ ($|\delta\gamma_t| \leq \gamma$). $p_0 = 0.01$, $\gamma_0 = 0.40$, $\gamma = 0.05$ and $\bar{C} = \bar{P} = 0.99$ are assumed for the calculation of the rightmost two columns

If periodic projective measurements are made with the operator σ_z, the measurement period T_d can guarantee that the state remains in $\mathcal{D}_d$.

The selection of measurement operators (i.e., σ_x, σ_y or σ_z) and uncertainties in the system Hamiltonian ($H_\Delta \neq 0$ or $H_\Delta \equiv 0$) do not affect the conclusion in Theorem 6.8. The measurement period T_d is also optimal to guarantee the required robustness.

The measurement periods for the different cases considered above are summarized in Table 6.2.

6.4.3 Illustrative Examples

Example 6.3 The values of measurement periods are shown in the rightmost two columns of Table 6.2 for several specific cases, where we have assumed $p_0 = 0.01$, $\gamma_0 = 0.40$, $\gamma = 0.05$ and $\bar{C} = \bar{P} = 0.99$.

Example 6.4 We now consider a real quantum system consisting of a superconducting charge qubit strongly coupled to a nonresonant microwave field in an on-chip cavity. In this system, a split Cooper pair box (qubit) is coupled capacitively to the electromagnetic field of a transmission line resonator which can be described by a harmonic oscillator (for details, see, e.g., [41, 48]). We let the resonance frequency be $\tilde{\omega}_0 = 2\pi \times 100$ MHz and the cavity decay rate be $\tilde{\gamma}_0 = 2\pi \times 0.8$ MHz. This leads to the cavity decay time $T_\gamma = 198.9$ ns. Now, we assume that the bounds on the uncertainties in the decoherence strength and in the system Hamiltonian are, respectively, $\tilde{\gamma} = 2\pi \times 0.08$ MHz and $\tilde{\varepsilon} = 2\pi \times 2.0$ MHz. Using the results in Table 6.2, we obtain the real measurement periods as $\tilde{T}_a = 0.64$ ns, $\tilde{T}_a' = 1.82$ ns, $\tilde{T}_p = 0.31$ ns, $\tilde{T}_p' = 0.45$ ns and $\tilde{T}_d = 0.46$ ns.

6.5 Summary and Further Reading

In this chapter, we introduced SMC methods for quantum systems. Several SMC results on two-level and multi-level quantum systems, and open quantum systems were presented. It is worth mentioning that although the basic idea of SMC has been used in these results, a feedback-based SMC method analogous to classical control systems is still lacking for quantum systems. Future research may contribute such an SMC method as well as extend SMC approaches to robust control of high-dimensional and complex quantum systems.

Further reading may include [15, 16] for a more comprehensive treatment of SMC of two-level quantum systems, [13] for SMC of multi-level quantum systems, [17] for more detailed discussion on SMC of open quantum systems, and [30] for other rapid Lyapunov control approaches that can be used in SMC of quantum systems.

Appendix

Proof of Theorem 6.1 [15]

For $H^A = \frac{1}{2}\sigma_z + \frac{1}{2}\varepsilon(t)\sigma_x$, using $\dot{\rho} = -\mathrm{i}[H^A, \rho]$ and (6.2), we have

$$\begin{bmatrix} \dot{x}_t \\ \dot{y}_t \\ \dot{z}_t \end{bmatrix} = \begin{bmatrix} 0 & -1 & 0 \\ 1 & 0 & -\varepsilon(t) \\ 0 & \varepsilon(t) & 0 \end{bmatrix} \begin{bmatrix} x_t \\ y_t \\ z_t \end{bmatrix}, \tag{6.50}$$

where $(x_0, y_0, z_0) = (0, 0, 1)$.

We now consider $\varepsilon(t)$ as a control input and select the performance measure as

$$J(\varepsilon) = z_{\mathrm{f}}. \tag{6.51}$$

From (6.4), we know that the "worst" case (i.e., the case maximizing the probability of failure) corresponds to minimizing z_{f}. Also, we introduce the Lagrange multiplier vector $\lambda(t) = [\lambda_1(t)\ \lambda_2(t)\ \lambda_3(t)]^{\mathrm{T}}$ and obtain the corresponding Hamiltonian function as follows:

$$H(\mathbf{r}(t), \varepsilon(t), \lambda(t), t) \equiv \lambda^{\mathrm{T}}(t) \begin{bmatrix} 0 & -1 & 0 \\ 1 & 0 & -\varepsilon(t) \\ 0 & \varepsilon(t) & 0 \end{bmatrix} \begin{bmatrix} x_t \\ y_t \\ z_t \end{bmatrix}, \tag{6.52}$$

where $\mathbf{r}(t) = (x_t, y_t, z_t)$. That is

$$H = -\lambda_1(t)y_t + \lambda_2(t)x_t + \varepsilon(t)(\lambda_3(t)y_t - \lambda_2(t)z_t), \tag{6.53}$$

where $H = H(\mathbf{r}(t), \varepsilon(t), \lambda(t), t)$. According to Pontryagin's minimum principle [27], a necessary condition for $\varepsilon^*(t)$ to minimize $J(\varepsilon)$ is

$$H(\mathbf{r}^*(t), \varepsilon^*(t), \lambda^*(t), t) \leq H(\mathbf{r}^*(t), \varepsilon(t), \lambda^*(t), t). \tag{6.54}$$

The necessary condition provides a relationship to determine the optimal control $\varepsilon^*(t)$. If there exists a time interval $[t_1, t_2]$ of finite duration during which the necessary condition (6.54) provides no information about the relationship between $\mathbf{r}^*(t)$, $\varepsilon^*(t)$, $\lambda^*(t)$, we call the interval $[t_1, t_2]$ a singular interval [27]. If we do not consider singular cases (i.e., $\lambda_3(t)y_t - \lambda_2(t)z_t \equiv 0$), the optimal control $\varepsilon^*(t)$ should be chosen as follows:

$$\varepsilon^*(t) = -\varepsilon \operatorname{sgn}(\lambda_3(t)y_t - \lambda_2(t)z_t). \tag{6.55}$$

That is, the optimal control strategy for $\varepsilon(t)$ is bang-bang control; i.e., $\varepsilon^*(t) = \bar{\varepsilon} = +\varepsilon$ or $-\varepsilon$. Now we consider $H^B = \frac{1}{2}\sigma_z + \frac{1}{2}\bar{\varepsilon}\sigma_x$ which leads to the state equation

$$\begin{bmatrix} \dot{x}_t \\ \dot{y}_t \\ \dot{z}_t \end{bmatrix} = \begin{bmatrix} 0 & -1 & 0 \\ 1 & 0 & -\bar{\varepsilon} \\ 0 & \bar{\varepsilon} & 0 \end{bmatrix} \begin{bmatrix} x_t \\ y_t \\ z_t \end{bmatrix}, \tag{6.56}$$

where $(x_0, y_0, z_0) = (0, 0, 1)$. The corresponding solution is

$$\begin{bmatrix} x_t \\ y_t \\ z_t \end{bmatrix} = \begin{bmatrix} -\frac{\bar{\varepsilon}}{1+\varepsilon^2}\cos\omega t + \frac{\bar{\varepsilon}}{1+\varepsilon^2} \\ -\frac{\bar{\varepsilon}}{\sqrt{1+\varepsilon^2}}\sin\omega t \\ \frac{\varepsilon^2}{1+\varepsilon^2}\cos\omega t + \frac{1}{1+\varepsilon^2} \end{bmatrix}, \tag{6.57}$$

where $\omega = \sqrt{1+\varepsilon^2}$. From (6.57), we know that z_t is a monotonically decreasing function in t when $t \in [0, \frac{\pi}{\sqrt{1+\varepsilon^2}}]$. Hence, we only consider the case $t \in [0, t_f]$ where $t_f \in [0, \frac{\pi}{\sqrt{1+\varepsilon^2}}]$.

Now, consider the optimal control problem with a fixed final time t_f and a free final state $\mathbf{r}_f = (x_f, y_f, z_f)$. According to Pontryagin's minimum principle, $\lambda^*(t_f) = \frac{\partial}{\partial \mathbf{r}}\mathbf{r}^*(t_f)$. From this, it is straightforward to verify that $(\lambda_1(t_f), \lambda_2(t_f), \lambda_3(t_f)) = (0, 0, 1)$. Now let us consider another necessary condition

$$\dot{\lambda}(t) = -\frac{\partial H(\mathbf{r}(t), \varepsilon(t), \lambda(t), t)}{\partial \mathbf{r}}$$

which leads to the following relationships:

$$\dot{\lambda}(t) = \begin{bmatrix} \dot{\lambda}_1(t) \\ \dot{\lambda}_2(t) \\ \dot{\lambda}_3(t) \end{bmatrix} = \begin{bmatrix} 0 & -1 & 0 \\ 1 & 0 & -\bar{\varepsilon} \\ 0 & \bar{\varepsilon} & 0 \end{bmatrix} \begin{bmatrix} \lambda_1(t) \\ \lambda_2(t) \\ \lambda_3(t) \end{bmatrix}, \tag{6.58}$$

where $(\lambda_1(t_f), \lambda_2(t_f), \lambda_3(t_f)) = (0, 0, 1)$. The corresponding solution is

$$\begin{bmatrix} \lambda_1(t) \\ \lambda_2(t) \\ \lambda_3(t) \end{bmatrix} = \begin{bmatrix} -\frac{\bar{\varepsilon}}{1+\varepsilon^2} \cos \omega(t_f - t) + \frac{\bar{\varepsilon}}{1+\varepsilon^2} \\ \frac{\bar{\varepsilon}}{\sqrt{1+\varepsilon^2}} \sin \omega(t_f - t) \\ \frac{\varepsilon^2}{1+\varepsilon^2} \cos \omega(t_f - t) + \frac{1}{1+\varepsilon^2} \end{bmatrix}. \tag{6.59}$$

We obtain

$$\lambda_3(t)y_t - \lambda_2(t)z_t = -\frac{\bar{\varepsilon}}{\omega^3}[\sin \omega t + \varepsilon^2 \sin \omega t_f + \sin \omega(t_f - t)]. \tag{6.60}$$

It is easy to show that the quantity $(\lambda_3(t)y_t - \lambda_2(t)z_t)$ occurring in (6.55) does not change its sign when $t_f \in [0, \frac{\pi}{\sqrt{1+\varepsilon^2}}]$ and $t \in [0, t_f]$.

Now, we further exclude the possibility that there exists a singular case. Suppose that there exists a singular interval $[t_0, t_1]$ (where $t_0 \geq 0$) such that when $t \in [t_0, t_1]$

$$h(t) = \lambda_3(t)y_t - \lambda_2(t)z_t \equiv 0. \tag{6.61}$$

We also have the following relationship:

$$\dot{h}(t) = \lambda_3(t)x_t - \lambda_1(t)z_t \equiv 0, \tag{6.62}$$

where we have used (6.50) and the following costate equation:

$$\dot{\lambda}(t) = \begin{bmatrix} \dot{\lambda}_1(t) \\ \dot{\lambda}_2(t) \\ \dot{\lambda}_3(t) \end{bmatrix} = \begin{bmatrix} 0 & -1 & 0 \\ 1 & 0 & -\varepsilon(t) \\ 0 & \varepsilon(t) & 0 \end{bmatrix} \begin{bmatrix} \lambda_1(t) \\ \lambda_2(t) \\ \lambda_3(t) \end{bmatrix}. \tag{6.63}$$

If $t_0 = 0$, we have $(x_0, y_0, z_0) = (0, 0, 1)$. By the principle of optimality [27], we may consider the case $t_f = t_1$. Using (6.61), (6.62) and $(\lambda_1(t_1), \lambda_2(t_1), \lambda_3(t_1)) = (0, 0, 1)$, we have $x_{t_1} = 0$ and $y_{t_1} = 0$. Using the relationship of $x_t^2 + y_t^2 + z_t^2 = 1$, we obtain $z_{t_1} = 1$ or -1. If $z_{t_1} = 1$, the initial state and the final state are the same state $|0\rangle$. However, if we use the control $\varepsilon(t) = \bar{\varepsilon}$, from (6.57) we have

$$z_{t_1}(\bar{\varepsilon}) = \frac{\varepsilon^2}{1+\varepsilon^2} \cos \omega t_1 + \frac{1}{1+\varepsilon^2} < z_{t_1} = 1.$$

Hence, this contradicts the fact that we are considering the optimal case $\min z_f$. If $z_{t_1} = -1$, there exists $0 < \tilde{t}_1 < t_1$ such that $z_{\tilde{t}_1} = 0$. By the principle of optimality [27], we may consider the case $t_f = \tilde{t}_1$. From the two Eqs. (6.61) and (6.62), we know that $z_{\tilde{t}_1}^2 = 1$ which contradicts $z_{\tilde{t}_1} = 0$. Hence, no singular condition can exist if $t_0 = 0$.

If $t_0 > 0$, using (6.55) we must select $\varepsilon(t) = \bar{\varepsilon}$ when $t \in [0, t_0]$. From (6.60), we know that there exist no $t_0 \in (0, t_f)$ satisfying $\lambda_3(t_0)y_{t_0} - \lambda_2(t_0)z_{t_0} = 0$. Hence, there exist no singular cases for our problem.

From the above analysis, $\varepsilon(t) = \bar{\varepsilon}$ is the optimal control when $t \in [0, \frac{\pi}{\sqrt{1+\varepsilon^2}}]$. Hence,

$$z_t^A = z_t(\varepsilon(t)) \geq z_t(\bar{\varepsilon}) = z_t^B.$$

From (6.4), it is clear that the probabilities of failure satisfy

$$p_t^A = \frac{1 - z_t^A}{2} \leq p_t^B = \frac{1 - z_t^B}{2}.$$

That is, the probability of failure p_t^A is not greater than p_t^B for $t \in [0, \frac{\pi}{\sqrt{1+\varepsilon^2}}]$. When $t \in [0, \frac{\pi}{\sqrt{1+\varepsilon^2}}]$, z_t^B is monotonically decreasing and p_t^B is monotonically increasing. When $t = \frac{\pi}{\sqrt{1+\varepsilon^2}}$, using (6.57) we have $z_t^B = \frac{1-\varepsilon^2}{1+\varepsilon^2}$. That is, the probability of failure $p' = \frac{\varepsilon^2}{1+\varepsilon^2}$. Hence, we can design the measurement period T using the case of H^B when $0 < p_0 \leq \frac{\varepsilon^2}{1+\varepsilon^2}$.

Using (6.4) and (6.57), for $t \in [0, \frac{\pi}{\sqrt{1+\varepsilon^2}}]$ we obtain the probability of failure

$$p_t^B = \frac{\varepsilon^2}{1+\varepsilon^2}\frac{1 - \cos \omega t}{2}. \tag{6.64}$$

Hence, we can design the maximum measurement period as follows:

$$T^{(1)} = \frac{\arccos[1 - 2(1 + \frac{1}{\varepsilon^2})p_0]}{\sqrt{1+\varepsilon^2}}. \tag{6.65}$$

For $H_\Delta = \frac{1}{2}\varepsilon(t)\sigma_y$ (where $|\varepsilon(t)| \leq \varepsilon$), we can obtain the same conclusion as that in the case $H_\Delta = \frac{1}{2}\varepsilon(t)\sigma_x$ (where $|\varepsilon(t)| \leq \varepsilon$).

Proof of Theorem 6.5 [17]

For the open qubit system subject to (6.42), when

$$H(t) = \frac{1}{2}[1 + \omega(t)]\sigma_z + \frac{1}{2}\varepsilon_x(t)\sigma_x + \frac{1}{2}\varepsilon_y(t)\sigma_y,$$

where $\sqrt{\varepsilon_x^2(t) + \varepsilon_y^2(t)} \leq \varepsilon, \varepsilon > 0, |\omega(t)| \leq \omega$ and $\omega \geq 0, \gamma_t = \gamma_0 + \delta\gamma_t$ $(|\delta\gamma_t| \leq \gamma)$, using (6.2), we have

$$\begin{bmatrix} \dot{x}_t \\ \dot{y}_t \\ \dot{z}_t \end{bmatrix} = \begin{bmatrix} -\frac{1}{2}(\gamma_0 + \delta\gamma_t) & -(1+\omega(t)) & \varepsilon_y(t) \\ 1+\omega(t) & -\frac{1}{2}(\gamma_0 + \delta\gamma_t) & -\varepsilon_x(t) \\ -\varepsilon_y(t) & \varepsilon_x(t) & -(\gamma_0 + \delta\gamma_t) \end{bmatrix} \begin{bmatrix} x_t \\ y_t \\ z_t \end{bmatrix} + \begin{bmatrix} 0 \\ 0 \\ -(\gamma_0 + \delta\gamma_t) \end{bmatrix}, \tag{6.66}$$

where $(x_0, y_0, z_0) = (0, 0, 1)$. From (6.66), we have

$$\begin{aligned}\dot{z}_t &= -\varepsilon_y(t)x_t + \varepsilon_x(t)y_t - (\gamma_0 + \delta\gamma_t)(z_t + 1) \\ &\geq -2\varepsilon\sqrt{1 - z_t^2} - (\gamma_0 + \gamma)(z_t + 1).\end{aligned} \tag{6.67}$$

Letting

$$f(z) = 2\varepsilon\sqrt{1 - z_t^2} + (\gamma_0 + \gamma)(1 + z_t), \tag{6.68}$$

we have

$$\frac{\mathrm{d}f(z)}{\mathrm{d}z} = (\gamma_0 + \gamma) - 2\varepsilon\frac{z_t}{\sqrt{1 - z_t^2}}. \tag{6.69}$$

Use $\frac{\mathrm{d}f(z)}{\mathrm{d}z} = 0$ to find the solution

$$z = \frac{\gamma_0 + \gamma}{\sqrt{4\varepsilon^2 + (\gamma_0 + \gamma)^2}}.$$

Hence,

$$\max f(z) = f\left(\frac{\gamma_0 + \gamma}{\sqrt{4\varepsilon^2 + (\gamma_0 + \gamma)^2}}\right) = \sqrt{4\varepsilon^2 + (\gamma_0 + \gamma)^2} + (\gamma_0 + \gamma). \tag{6.70}$$

Therefore,

$$\dot{z}_t \geq -\max f(z) = -\sqrt{4\varepsilon^2 + (\gamma_0 + \gamma)^2} - (\gamma_0 + \gamma). \tag{6.71}$$

When $t \in [0, T_\mathrm{a}]$ where

$$T_\mathrm{a} = \frac{2p_0}{\sqrt{4\varepsilon^2 + (\gamma_0 + \gamma)^2} + (\gamma_0 + \gamma)}, \tag{6.72}$$

we have

$$z_t \geq 1 - (\max f(z))t \geq 1 - 2p_0. \tag{6.73}$$

Therefore, if one makes a measurement on the system with σ_z, the probability of failure $\langle 1|\rho_t|1\rangle = \frac{1-z_t}{2} \leq p_0$.

Proof of Proposition 6.1 [17]

When $H(t) = \frac{1}{2}\sigma_z$ and $\gamma_t = \gamma_0 + \delta\gamma_t$, the state equation of the system in (6.42) is

$$\begin{bmatrix}\dot{x}_t \\ \dot{y}_t \\ \dot{z}_t\end{bmatrix} = \begin{bmatrix}-\frac{1}{2}(\gamma_0 + \delta\gamma_t) & -1 & 0 \\ 1 & -\frac{1}{2}(\gamma_0 + \delta\gamma_t) & 0 \\ 0 & 0 & -(\gamma_0 + \delta\gamma_t)\end{bmatrix}\begin{bmatrix}x_t \\ y_t \\ z_t\end{bmatrix} + \begin{bmatrix}0 \\ 0 \\ -(\gamma_0 + \delta\gamma_t)\end{bmatrix}, \tag{6.74}$$

where $(x_0, y_0, z_0) = (0, 0, 1)$. It is clear that

$$\dot{z}_t = -(\gamma_0 + \delta\gamma_t)(1 + z_t) \geq -(\gamma_0 + \gamma)(1 + z_t). \tag{6.75}$$

From (6.75), we have

$$z_t \geq 2e^{-(\gamma_0+\gamma)t} - 1. \tag{6.76}$$

If $t \in [0, T_a']$ where

$$T_a' = -\frac{\ln(1 - p_0)}{\gamma_0 + \gamma}, \tag{6.77}$$

we have

$$z_t \geq 1 - 2p_0. \tag{6.78}$$

That is, the probability of failure is $\langle 1|\rho_t|1\rangle \leq p_0$.

References

1. Adamy J, Flemming A (2004) Soft variable structure controls: a survey. Automatica 40:1822–1844
2. Ashhab S, Nori F (2010) Control-free control: manipulating a quantum system using only a limited set of measurements. Phys Rev A 82:062103
3. Bacon D, Lidar DA, Whaley KB (1999) Robustness of decoherence-free subspaces for quantum computation. Phys Rev A 60:1944–1955
4. Bolognani S, Ticozzi F (2010) Engineering stable discrete-time quantum dynamics via a canonical QR decomposition. IEEE Trans Autom Control 55:2721–2734
5. Brassard G, Høyer P, Tapp A (1998) Quantum counting. Lect Notes Comput Sci 1443:820–831
6. Breuer H-P, Petruccione F (2002) The theory of open quantum systems, 1st edn. Oxford University Press
7. Cui W, Xi ZR, Pan Y (2008) Optimal decoherence control in non-Markovian open dissipative quantum systems. Phys Rev A 77:032117
8. D'Alessandro D (2007) Introduction to quantum control and dynamics. Chapman & Hall/CRC
9. Devitt SJ, Schirmer SG, Qi DKL, Cole JH, Hollenberg LCL (2007) Subspace confinement: how good is your qubit? New J Phys 9(10):384–384
10. Dong D, Chen C, Tarn TJ, Pechen A, Rabitz H (2008) Incoherent control of quantum systems with wavefunction controllable subspaces via quantum reinforcement learning. IEEE Trans Syst Man Cybern Part B Cybern 38:957–962
11. Dong D, Lam J, Petersen IR (2010) Robust incoherent control of of qubit systems via switching and optimisation. Int J Control 83(1):206–217
12. Dong D, Petersen IR (2009) Variable structure control of uncontrollable quantum systems. IFAC Proc Vol 42(6):237–242
13. Dong D, Petersen IR (2009) Sliding mode control of quantum systems. New J Phys 11:105033
14. Dong D, Petersen IR (2010) Quantum control theory and applications: a survey. IET Control Theory Appl 4:2651–2671
15. Dong D, Petersen IR (2012) Sliding mode control of two-level quantum systems. Automatica 48:725–735
16. Dong D, Petersen IR (2012) Notes on sliding mode control of two-level quantum systems. Automatica 48(12):3089–3097

17. Dong D, Petersen IR, Rabitz H (2013) Sampled-data design for robust control of a single qubit. IEEE Trans Autom Control 58:2654–2659
18. Dong D, Zhang C, Rabitz H, Pechen A, Tarn TJ (2008) Incoherent control of locally controllable quantum systems. J Chem Phys 129(15):154103
19. Edwards C, Spurgeon SK (1998) Sliding mode control: theory and applications. Taylor & Francis, Florida
20. Grover LK (1998) Quantum computers can search rapidly by using almost any transformation. Phys Rev Lett 80:4329–4332
21. Hou SC, Khan MA, Yi XX, Dong D, Petersen IR (2012) Optimal Lyapunov-based quantum control for quantum systems. Phys Rev A 86(2):022321
22. Høyer P (2000) Arbitrary phases in quantum amplitude amplification. Phys Rev A 62:052304
23. Itano WM, Heinzen DJ, Bollinger JJ, Wineland DJ (1990) Quantum Zeno effect. Phys Rev A 41:2295–2300
24. Ji Y, Hu J-J, Huang J-H, Ke Q (2018) Sampled-data design for sliding mode control based on various robust specifications in open quantum system. Int J Quantum Inf 16:1850004
25. Khaneja N, Brockett R, Glaser SJ (2001) Time optimal control in spin systems. Phys Rev A 63:032308
26. Khodjasteh K, Lidar DA, Viola L (2010) Arbitrarily accurate dynamical control in open quantum systems. Phys Rev Lett 104:090501
27. Kirk DE (1970) Optimal control theory: an introduction. Prentice-Hall Inc., Englewood Cliffs, NJ
28. Knill E, Laflamme R, Viola L (2000) Theory of quantum error correction for general noise. Phys Rev Lett 84:2525–2528
29. Kuang S, Cong S (2008) Lyapunov control methods of closed quantum systems. Automatica 44:98–108
30. Kuang S, Dong D, Petersen IR (2017) Rapid Lyapunov control of finite-dimensional quantum systems. Automatica 81:164–175
31. Kwiat PG, Berglund AJ, Altepeter JB, White AG (2000) Experimental verification of decoherence-free subspaces. Science 290:498–501
32. Lidar DA, Chuang IL, Whaley KB (1998) Decoherence-free subspaces for quantum computation. Phys Rev Lett 81(12):2594–2597
33. Lidar DA, Schneider S (2005) Stabilizing qubit coherence via tracking-control. Quantum Inf Comput 5:350–363
34. Long GL (2001) Grover algorithm with zero theoretical failure rate. Phys Rev A 64:022307
35. Mirrahimi M, Rouchon P, Turinici G (2005) Lyapunov control of bilinear Schrödinger equations. Automatica 41:1987–1994
36. Misra B, Sudarshan ECG (1977) The Zeno's paradox in quantum theory. J Math Phys 18:756–763
37. Nielsen MA, Chuang IL (2010) Quantum computation and quantum information. Cambridge University Press, Cambridge
38. Pechen A, Il'in N, Shuang F, Rabitz H (2006) Quantum control by von Neumann measurements. Phys Rev A 74:052102
39. Pravia MA, Boulant N, Emerson J, Fortunato EM, Havel TF, Cory DG, Farid A (2003) Robust control of quantum information. J Chem Phys 119:9993–10001
40. Romano R, D'Alessandro D (2006) Environment-mediated control of a quantum system. Phys Rev Lett 97:080402
41. Schuster DI, Wallraff A, Blais A, Frunzio L, Huang R-S, Majer J, Girvin SM, Schoelkopf RJ (2005) AC Stark shift and dephasing of a superconducting qubit strongly coupled to a cavity field. Phys Rev Lett 94:123602
42. Tersigni SH, Gaspard P, Rice SA (1990) On using shaped light pulses to control the selectivity of product formation in a chemical reaction: an application to a multiple level system. J Chem Phys 93:1670–1680
43. Utkin VI (1977) Variable structure systems with sliding modes. IEEE Trans Autom Control 22(2):212–222

44. Utkin VI, Guldner J, Shi J (1999) Sliding mode control in electromechanical systems. Taylor & Francis, London
45. Vilela Mendes R, Man'ko VI (2003) Quantum control and the Strocchi map. Phys Rev A 67:053404
46. Viola L, Knill E, Lloyd S (1999) Dynamical decoupling of open quantum systems. Phys Rev Lett 82:2417–2421
47. Wang LC, Hou SC, Yi XX, Dong D, Petersen IR (2014) Optimal Lyapunov control of two-level systems: convergence and extended techniques. Phys Lett A 378:1074–1080
48. Wiseman HM, Milburn GJ (2010) Quantum measurement and control. Cambridge University Press, Cambridge
49. Zhang J, Li CW, Wu RB, Tarn TJ, Liu XS (2005) Maximal suppression of decoherence in Markovian quantum systems. J Phys A Math Gen 38:6587–6601

Chapter 7
Robust Stability and Performance Analysis of Quantum Systems

Abstract This chapter presents several results on robust stability and performance analysis for uncertain quantum systems using the small gain theorem and a Popov approach. In Sect. 7.2, the problem of robust stability for linear quantum systems is presented. Section 7.3 discusses the robust stability of several classes of nonlinear quantum systems including cases with non-quadratic Hamiltonian perturbations, uncertainties in nonlinear coupling operators and nonlinear dynamic uncertainties. In Sect. 7.4, performance analysis and guaranteed cost control are investigated for uncertain linear quantum systems[a]. Section 7.5 presents several results on performance analysis for nonlinear quantum systems.

7.1 Introduction

Linear quantum systems are a class of quantum systems whose dynamics can be described by linear differential equations [9, 13, 32]. Many quantum optical systems, quantum superconducting circuits and optomechanical systems can be described or can be approximately characterized as linear quantum systems [17]. In this chapter, we consider several classes of uncertain quantum systems where the nominal systems are linear quantum systems. Robust stability and performance analysis are investigated for these classes of quantum systems. The notion of robust stability is an important concept for systems with an uncertain block which satisfies a sector bound condition; e.g., see [11]. This enables a frequency domain condition for robust stability to be given. This characterization of robust stability enables robust feedback controller synthesis to be carried out using H^∞ control theory; e.g., see [33]. This section extends classical results on robust stability to several classes of quantum

[a] Section 7.4 contains materials reprinted, with permission, from IEEE Transactions on Automatic Control [30] © 2017 IEEE.

D. Dong and I. R. Petersen, *Learning and Robust Control in Quantum Technology*, Communications and Control Engineering,
https://doi.org/10.1007/978-3-031-20245-2_7

systems. We build on the results of [8] to obtain robust stability results for uncertain quantum systems where the system Hamiltonian is decomposed as $H = H_1 + H_2$. Here, H_1 is a self-adjoint operator on the underlying Hilbert space referred to as the nominal Hamiltonian and H_2 is a self-adjoint operator on the underlying Hilbert space referred to as the perturbation Hamiltonian, which is contained in a specified set of Hamiltonians $\mathbb{W}$. In this chapter, we mainly use the (S, L, H) notation (as mentioned in Chap. 2) to describe the quantum systems and discuss various cases with quadratic or non-quadratic H_2 and with coupling perturbation in L.

In particular, the small gain theorem and the Popov approach are used to investigate robust stability and performance analysis for uncertain quantum systems. The small gain theorem and the Popov approach are two of the most important methods for the analysis of robust stability in classical control [11]. Both of these stability criteria consider Lur'e systems which involve the feedback interconnection between a linear time-invariant system and a sector bounded nonlinearity or uncertainty [10]. The key difference between the small gain theorem and the Popov criterion is that the former result guarantees robustness in the presence of time-varying uncertainties and nonlinearities while the latter approach only applies to static time-invariant nonlinearities. The reason for this is that the small gain theorem is based on conventional or fixed quadratic Lyapunov functions which are used to establish stability with respect to time-varying perturbations. In contrast, the Popov theorem relies on a Lyapunov function of the Lur'e Postnikov form, which is a function of sector bounded nonlinearity, that is, a parameter-dependent quadratic Lyapunov function. The Popov method guarantees stability by means of a family of Lyapunov functions and is less conservative than the small gain theorem. First, we employ the small gain theorem and the Popov approach to provide concrete conditions on robust stability for uncertain linear and nonlinear quantum systems. Second, the small gain theorem and the Popov approach are further applied to performance analysis and guaranteed cost controller design for uncertain linear quantum systems. Then, the performance analysis of uncertain nonlinear quantum systems is presented.

7.2 Robust Stability of Linear Quantum Systems

Here we consider the problem of robust stability for uncertain linear quantum systems with quadratic perturbations to the system Hamiltonian [17]. In particular, we consider open quantum systems defined by parameters (S, L, H) where $H = H_1 + H_2$; e.g., see [6, 8]. The corresponding generator for this quantum system is given by

$$\mathcal{G}(X) = -\mathrm{i}[X, H] + \mathcal{L}(X), \tag{7.1}$$

where

$$\mathcal{L}(X) = \frac{1}{2}L^\dagger[X, L] + \frac{1}{2}[L^\dagger, X]L.$$

The triple (S, L, H), along with the corresponding generators defines the Heisenberg evolution $X(t)$ of an operator X according to a QSDE; e.g., see [8].

The problem considered involves establishing robust stability properties for an uncertain open quantum system for the case $H_2 \in \mathbb{W}$. Using the notation of [8], the set $\mathbb{W}$ defines a set of exosystems. The main robust stability results presented in this chapter will build on the following result from [8]:

Lemma 7.1 *(See Lemma 3.4 of [8]) Consider an open quantum system defined by (S, L, H) and suppose there exists a non-negative self-adjoint operator V on the underlying Hilbert space such that*

$$\mathcal{G}(V) + cV \leq \lambda, \tag{7.2}$$

where $c > 0$ and λ are real numbers. Then for any plant state, we have

$$\langle V(t)\rangle \leq e^{-ct}\langle V\rangle + \frac{\lambda}{c}, \quad \forall t \geq 0.$$

Here $V(t)$ denotes the Heisenberg evolution of the operator V and $\langle\cdot\rangle$ denotes quantum expectation; e.g., see [8].

Given a set of non-negative self-adjoint operators $\mathcal{P}$ and real parameters $\gamma > 0$, $\delta \geq 0$, we now define a particular set of perturbation Hamiltonians $\mathbb{W}_1$. This set $\mathbb{W}_1$ is defined in terms of the commutator decomposition

$$[V, H_2] = [V, z^\dagger]w - w^\dagger[z, V] \tag{7.3}$$

for $V \in \mathcal{P}$ where w and z are vectors of operators. Here, the notation $[z, V]$ for a vector of operators z and a scalar operator V denotes the corresponding vector of commutators. This set will be defined in terms of the sector bound condition:

$$w^\dagger w \leq \frac{1}{\gamma^2}z^\dagger z + \delta. \tag{7.4}$$

We define

$$\mathbb{W}_1 = \left\{ \begin{array}{l} H_2 = H_2^\dagger : \exists w,\ z \text{ such that (7.4) is satisfied} \\ \text{and (7.3) is satisfied } \forall V \in \mathcal{P} \end{array} \right\}. \tag{7.5}$$

Using this definition, we obtain the following theorem [22] whose proof is presented in the Appendix of this chapter:

Theorem 7.1 *Consider a set of non-negative self-adjoint operators $\mathcal{P}$ and an open quantum system (S, L, H) where $H = H_1 + H_2$, $H_1 = H_1^\dagger$ and $H_2 \in \mathbb{W}_1$ defined in (7.5). If there exists a $V \in \mathcal{P}$ and real constants $c > 0$, $\tilde{\lambda} \geq 0$ such that*

$$-i[V, H_1] + \mathscr{L}(V) + [V, z^\dagger][z, V] + \frac{1}{\gamma^2} z^\dagger z + cV \leq \tilde{\lambda}, \tag{7.6}$$

then

$$\langle V(t) \rangle \leq e^{-ct} \langle V \rangle + \frac{\tilde{\lambda} + \delta}{c}, \quad \forall t \geq 0. \tag{7.7}$$

Now we consider a specific set $\mathbb{W}_2$ of quadratic perturbation Hamiltonians of the form

$$H_2 = \frac{1}{2} \begin{bmatrix} \zeta^\dagger & \zeta^T \end{bmatrix} \Delta \begin{bmatrix} \zeta \\ \zeta^\# \end{bmatrix}, \tag{7.8}$$

where $\Delta \in \mathbb{C}^{2m \times 2m}$ is a Hermitian matrix of the form

$$\Delta = \begin{bmatrix} \Delta_1 & \Delta_2 \\ \Delta_2^\# & \Delta_1^\# \end{bmatrix} \tag{7.9}$$

and $\Delta_1 = \Delta_1^\dagger$, $\Delta_2 = \Delta_2^T$. Also, $\zeta = E_1 a + E_2 a^\#$. Here a is a vector of annihilation operators on the underlying Hilbert space and $a^\#$ is the corresponding vector of creation operators. In the case of vectors of operators, the notation $^\#$ refers to the vector of adjoint operators, and in the case of complex matrices, this notation refers to the complex conjugate matrix.

We assume a and $a^\#$ satisfy the canonical commutation relations:

$$\left[\begin{bmatrix} a \\ a^\# \end{bmatrix}, \begin{bmatrix} a \\ a^\# \end{bmatrix}^\dagger \right] = \begin{bmatrix} a \\ a^\# \end{bmatrix} \begin{bmatrix} a \\ a^\# \end{bmatrix}^\dagger - \left(\begin{bmatrix} a \\ a^\# \end{bmatrix}^\# \begin{bmatrix} a \\ a^\# \end{bmatrix}^T \right)^T = J, \tag{7.10}$$

where $J = \begin{bmatrix} I & 0 \\ 0 & -I \end{bmatrix}$; e.g., see [5, 7, 13].

The matrix Δ is subject to the norm bound

$$\|\Delta\| \leq \frac{2}{\gamma}, \tag{7.11}$$

where $\|\cdot\|$ denotes the matrix-induced norm (maximum singular value). Then

$$\mathbb{W}_2 = \left\{ \begin{array}{l} H_2 \text{ of the form (7.8), (7.9) such that} \\ \text{condition (7.11) is satisfied} \end{array} \right\}. \tag{7.12}$$

For any set of self-adjoint operators $\mathscr{P}$, we know that $\mathbb{W}_2 \subset \mathbb{W}_1$. The proof is given in the Appendix of this chapter [22].

We now consider the case in which the nominal quantum system is a linear quantum system; e.g., see [13]. In this case, H_1 is of the form

$$H_1 = \frac{1}{2} \begin{bmatrix} a^\dagger & a^T \end{bmatrix} M \begin{bmatrix} a \\ a^\# \end{bmatrix}, \tag{7.13}$$

where $M \in \mathbb{C}^{2n \times 2n}$ is a Hermitian matrix of the form

$$M = \begin{bmatrix} M_1 & M_2 \\ M_2^\# & M_1^\# \end{bmatrix}, \tag{7.14}$$

and $M_1 = M_1^\dagger$, $M_2 = M_2^T$. Also, we assume L is of the form

$$L = \begin{bmatrix} N_1 & N_2 \end{bmatrix} \begin{bmatrix} a \\ a^\# \end{bmatrix} = \tilde{N} \begin{bmatrix} a \\ a^\# \end{bmatrix}, \tag{7.15}$$

where $N_1 \in \mathbb{C}^{m \times n}$ and $N_2 \in \mathbb{C}^{m \times n}$. Also, we write

$$\begin{bmatrix} L \\ L^\# \end{bmatrix} = N \begin{bmatrix} a \\ a^\# \end{bmatrix} = \begin{bmatrix} N_1 & N_2 \\ N_2^\# & N_1^\# \end{bmatrix} \begin{bmatrix} a \\ a^\# \end{bmatrix}.$$

In addition, we assume that V is of the form

$$V = \begin{bmatrix} a^\dagger & a^T \end{bmatrix} P \begin{bmatrix} a \\ a^\# \end{bmatrix}, \tag{7.16}$$

where $P \in \mathbb{C}^{2n \times 2n}$ is a positive-definite Hermitian matrix of the form

$$P = \begin{bmatrix} P_1 & P_2 \\ P_2^\# & P_1^\# \end{bmatrix}. \tag{7.17}$$

Hence, we consider the set of non-negative self-adjoint operators $\mathscr{P}_1$ defined as

$$\mathscr{P}_1 = \left\{ \begin{array}{l} V \text{ of the form (7.16) such that P} > 0 \text{ is a} \\ \text{Hermitian matrix of the form (7.17)} \end{array} \right\}. \tag{7.18}$$

In the linear case, we consider a notion of robust mean square stability as follows:

Definition 7.1 An uncertain open quantum system defined by (S, L, H) where $H = H_1 + H_2$ with H_1 of the form (7.13), and L of the form (7.15), is said to be *robustly mean square stable* if for any $H_2 \in \mathbb{W}$, there exist constants $c_1 > 0$, $c_2 > 0$ and $c_3 \geq 0$ such that

$$\left\langle \begin{bmatrix} a(t) \\ a^\#(t) \end{bmatrix}^\dagger \begin{bmatrix} a(t) \\ a^\#(t) \end{bmatrix} \right\rangle \leq c_1 e^{-c_2 t} \left\langle \begin{bmatrix} a \\ a^\# \end{bmatrix}^\dagger \begin{bmatrix} a \\ a^\# \end{bmatrix} \right\rangle + c_3 \quad \forall t \geq 0. \tag{7.19}$$

Here $\begin{bmatrix} a(t) \\ a^{\#}(t) \end{bmatrix}$ denotes the Heisenberg evolution of the vector of operators $\begin{bmatrix} a \\ a^{\#} \end{bmatrix}$; e.g., see [8].

Now we present concrete conditions for robust mean square stability of a linear nominal system and show that a sufficient condition for robust mean square stability can be characterized by a strict bounded real condition. We use the relationship:

$$z = \begin{bmatrix} \zeta \\ \zeta^{\#} \end{bmatrix} = \begin{bmatrix} E_1 & E_2 \\ E_2^{\#} & E_1^{\#} \end{bmatrix} \begin{bmatrix} a \\ a^{\#} \end{bmatrix} = E \begin{bmatrix} a \\ a^{\#} \end{bmatrix}. \tag{7.20}$$

We have the following robust stability results:

Theorem 7.2 *Consider an uncertain open quantum system defined by* (S, L, H) *such that* $H = H_1 + H_2$ *where* H_1 *is of the form (7.13),* L *is of the form (7.15) and* $H_2 \in \mathbb{W}_2$. *Furthermore, assume that the following strict bounded real condition is satisfied:*

(1) The matrix

$$F = -iJM - \frac{1}{2} JN^{\dagger} JN \, is \; Hurwitz. \tag{7.21}$$

(2)

$$\left\| E\,(sI - F)^{-1}\, D \right\|_{\infty} < \frac{\gamma}{2}, \tag{7.22}$$

where $D = iJE^{\dagger}$.

Then the uncertain quantum system is robustly mean square stable.

The proof can be found in [22]. This robust stability result can be regarded as an extension of the classical small gain theorem for robust stability to the case of quantum systems. As mentioned in Sect. 7.1, the small gain theorem and the Popov criterion are two of the most useful tests for robust stability and nonlinear system stability in classical control theory. The Popov criterion is less conservative than the small gain theorem. Here, we present a quantum Popov stability criterion for linear quantum systems aiming to obtain less conservative results [10].

We now consider a set of self-adjoint perturbation Hamiltonians $H_2 \in \mathbb{W}_3$. For a given set of non-negative self-adjoint operators $\mathscr{P}$, a set of Popov scaling parameters $\theta \subset [0, \infty]$, the nominal Hamiltonian H_1, a coupling operator L, and for the H_2 in (7.8), the matrix Δ is subject to the bounds

$$0 \leq \Delta \leq \frac{4}{\gamma} I. \tag{7.23}$$

The set $\mathbb{W}_3$ is defined as

$$\mathbb{W}_3 = \left\{ \begin{array}{l} H_2 \text{ of the form (7.8), (7.9) such that} \\ \text{condition (7.23) is satisfied} \end{array} \right\}. \tag{7.24}$$

We have the following robust stability result based on the Popov criterion whose proof can be found in [10]:

Theorem 7.3 *Consider an uncertain open quantum system defined by* (S, L, H) *such that* $H = H_1 + H_2$ *where* H_1 *is of the form (7.13),* L *is of the form (7.15) and* $H_2 \in \mathbb{W}_3$*. Furthermore, assume there exist a constant* $\theta \geq 0$ *such that:*

(1) The matrix F

$$F = -iJM - \frac{1}{2}JN^{\dagger}JN \, is \, Hurwitz. \tag{7.25}$$

(2) The transfer function

$$G(s) = -2iE\,(sI - F)^{-1}\,JE^{\dagger} \tag{7.26}$$

satisfies the strict positive real condition

$$\gamma I - (1 + \theta i\omega)G(i\omega) - (1 - \theta i\omega)G(i\omega)^{\dagger} > 0 \tag{7.27}$$

for all $\omega \in [-\infty, \infty]$.

Then the uncertain quantum system is robustly mean square stable.

Example 7.1 [10] We consider an example of an open quantum system with

$$S = I, \; L = \sqrt{\kappa}a,$$
$$H = \frac{1}{2}\mathrm{i}\left(\left(a^{\dagger}\right)^2 - a^2\right),$$

which corresponds to a linearized optical parametric amplifier; see [4]. We let

$$H_1 = 0, \; H_2 = \frac{1}{2}\mathrm{i}\left(\left(a^{\dagger}\right)^2 - a^2\right).$$

This defines a linear quantum system of the form considered in Theorem 7.2 with $M_1 = 0, M_2 = 0, N_1 = \sqrt{\kappa}, N_2 = 0, E_1 = 1, E_2 = 0, \Delta_1 = 0, \Delta_2 = \mathrm{i}$. Hence, $M = 0$,

$$N = \begin{bmatrix} \sqrt{\kappa} & 0 \\ 0 & \sqrt{\kappa} \end{bmatrix},$$
$$F = \begin{bmatrix} -\frac{\kappa}{2} & 0 \\ 0 & -\frac{\kappa}{2} \end{bmatrix}$$

which is Hurwitz, $E = I$, and $D = \mathrm{i}J$. In this case, $\Delta^2 = I$. Hence, we can choose $\gamma = 1$ to ensure that (7.11) is satisfied and $H_2 \in \mathbb{W}_2$. Also,

$$\left\| E\,(sI - F)^{-1}\,D \right\|_{\infty} = \left\| \begin{bmatrix} \frac{1}{s+\kappa/2} & 0 \\ 0 & -\frac{1}{s+\kappa/2} \end{bmatrix} \right\|_{\infty} = \frac{2}{\kappa}.$$

Hence, it follows from Theorem 7.2 that this system will be mean square stable if $\frac{2}{\kappa} < \frac{1}{2}$; i.e., $\kappa > 4$.

Now we apply the Popov criterion to investigate the stability. In order to apply the theory in Theorem 7.3 to this example, we let

$$H_1 = \frac{1}{2}[a^\dagger \ a] \begin{bmatrix} -1 & 0 \\ 0 & -1 \end{bmatrix} \begin{bmatrix} a \\ a^\dagger \end{bmatrix}$$

and

$$H_2 = \frac{1}{2}[a^\dagger \ a] \begin{bmatrix} 1 & \mathrm{i} \\ -\mathrm{i} & 1 \end{bmatrix} \begin{bmatrix} a \\ a^\dagger \end{bmatrix} \geq 0.$$

This defines a linear quantum system of the form considered in Theorem 7.3 with $M_1 = -1$, $M_2 = 0$, $N_1 = \sqrt{\kappa}$, $N_2 = 0$, $E_1 = 1$, $E_2 = 0$, and

$$\Delta = \begin{bmatrix} 1 & \mathrm{i} \\ -\mathrm{i} & 1 \end{bmatrix} \geq 0.$$

We may choose $\gamma = 2$ to ensure $H_2 \in \mathbb{W}_3$ and use Theorem 7.3 to demonstrate that when $\theta = 0.2$ this system is mean square stable for $\kappa = 2.1$. More details could be found in [10]. It is clear that the Popov criterion provides a considerable improvement over the small gain theorem approach (where the system is mean square stable when $\kappa > 4$).

7.3 Robust Stability of Nonlinear Quantum Systems

7.3.1 Robust Stability of Quantum Systems with Non-quadratic Hamiltonian Perturbation

We first consider the robust stability of quantum systems where the nominal system is a linear quantum system while the perturbation Hamiltonian is not quadratic. For a given set of non-negative self-adjoint operators $\mathscr{P}$ and real parameters $\gamma > 0$, $\delta_1 \geq 0$, $\delta_2 \geq 0$, a set of perturbation Hamiltonians $\mathbb{W}_4$ is defined in terms of the commutator decomposition

$$[V, H_2] = [V, z]w_1^\dagger - w_1[z^\dagger, V] + \frac{1}{2}[z, [V, z]]\,w_2^\dagger - \frac{1}{2}w_2\,[z, [V, z]]^\dagger \qquad (7.28)$$

for $V \in \mathscr{P}$ where w_1, w_2 and z are given scalar operators. Also, the set $\mathbb{W}_4$ will be defined in terms of the sector bound condition

$$w_1 w_1^\dagger \leq \frac{1}{\gamma^2} z z^\dagger + \delta_1 \tag{7.29}$$

and the condition

$$w_2 w_2^\dagger \leq \delta_2. \tag{7.30}$$

Then, we define

$$\mathbb{W}_4 = \left\{ \begin{array}{l} H_2 = H_2^\dagger : \exists w_1,\ w_2,\ z \text{ such that (7.29) and (7.30)} \\ \text{are satisfied and (7.28) is satisfied } \forall V \in \mathscr{P} \end{array} \right\}. \tag{7.31}$$

Using this definition, we obtain the following theorem [23]:

Theorem 7.4 *Consider a set of non-negative self-adjoint operators $\mathscr{P}$ and an open quantum system (S, L, H) where $H = H_1 + H_2$, $H_1 = H_1^\dagger$ and $H_2 \in \mathbb{W}_4$ defined in (7.31). If there exists a $V \in \mathscr{P}$ and real constants $c > 0$, $\tilde{\lambda} \geq 0$ such that $\mu = [z, [V, z]]$ is a constant and*

$$-i[V, H_1] + \mathscr{L}(V) + [V, z][z^\dagger, V] + \frac{1}{\gamma^2} z z^\dagger + cV \leq \tilde{\lambda}. \tag{7.32}$$

Then

$$\langle V(t) \rangle \leq e^{-ct} \langle V \rangle + \frac{\tilde{\lambda} + \delta_1 + \mu\mu^\dagger/4 + \delta_2}{c}, \quad \forall t \geq 0.$$

Now, we define a set of non-quadratic perturbation Hamiltonians as denoted as $\mathbb{W}_5$. For a given set of non-negative self-adjoint operators $\mathscr{P}$ and real parameters $\gamma > 0$, $\delta_1 \geq 0$, $\delta_2 \geq 0$, the set $\mathbb{W}_5$ is defined in terms of the following power series (which is assumed to converge in the sense of the induced operator norm on the underlying Hilbert space):

$$H_2 = f(\zeta, \zeta^*) = \sum_{k=0}^{\infty} \sum_{\ell=0}^{\infty} S_{k\ell} \zeta^k (\zeta^*)^\ell = \sum_{k=0}^{\infty} \sum_{\ell=0}^{\infty} S_{k\ell} H_{k\ell}. \tag{7.33}$$

Here $S_{k\ell} = S_{\ell k}^*$, $H_{k\ell} = \zeta^k (\zeta^*)^\ell$, and ζ is a scalar operator on the underlying Hilbert space. It follows from this definition that

$$H_2^* = \sum_{k=0}^{\infty} \sum_{\ell=0}^{\infty} S_{k\ell}^* \zeta^\ell (\zeta^*)^k = \sum_{\ell=0}^{\infty} \sum_{k=0}^{\infty} S_{\ell k} \zeta^\ell (\zeta^*)^k = H_2$$

and thus H_2 is a self-adjoint operator. Note that it follows from the use of Wick ordering that the form (7.33) is the most general form for a perturbation Hamiltonian defined in terms of a single scalar operator ζ [23].

Also, we let

$$f'(\zeta,\zeta^*) = \sum_{k=1}^{\infty}\sum_{\ell=0}^{\infty} k S_{k\ell}\zeta^{k-1}(\zeta^*)^{\ell}, \tag{7.34}$$

$$f''(\zeta,\zeta^*) = \sum_{k=1}^{\infty}\sum_{\ell=0}^{\infty} k(k-1) S_{k\ell}\zeta^{k-2}(\zeta^*)^{\ell} \tag{7.35}$$

and consider the sector bound condition

$$f'(\zeta,\zeta^*)^* f'(\zeta,\zeta^*) \leq \frac{1}{\gamma^2}\zeta\zeta^* + \delta_1 \tag{7.36}$$

and the condition

$$f''(\zeta,\zeta^*)^* f''(\zeta,\zeta^*) \leq \delta_2. \tag{7.37}$$

Then we define the set $\mathbb{W}_5$ as follows:

$$\mathbb{W}_5 = \left\{ \begin{array}{l} H_2 \text{ of the form (7.33) such that} \\ \text{conditions (7.36) and (7.37) are satisfied} \end{array} \right\}. \tag{7.38}$$

The condition (7.37) effectively amounts to a global Lipschitz condition on the quantum nonlinearity. Given any $V \in \mathscr{P}$, we assume that the quantity

$$\mu = [\zeta, [V, \zeta]] = \zeta[V, \zeta] - [V, \zeta]\zeta$$

is a constant. Under this assumption, from [23], we know that $\mathbb{W}_5 \subset \mathbb{W}_4$. We consider the case that the nominal quantum system is a linear quantum system described by (7.13) and (7.15). Define the vector of operators on the underlying Hilbert space $z = \begin{bmatrix} z_1 \; z_2 \; \ldots \; z_m \end{bmatrix}^T$ as

$$z = E_1 a + E_2 a^{\#} = \begin{bmatrix} E_1 \; E_2 \end{bmatrix}\begin{bmatrix} a \\ a^{\#} \end{bmatrix} = \tilde{E}\begin{bmatrix} a \\ a^{\#} \end{bmatrix}. \tag{7.39}$$

We have the following robust stability result:

Theorem 7.5 *Consider an uncertain open quantum system defined by (S, L, H) such that $H = H_1 + H_2$ where H_1 is of the form (7.13), L is of the form (7.15) and $H_2 \in \mathbb{W}_5$. Furthermore, assume that the following strict bounded real condition is satisfied:*

(1) The matrix

$$F = -iJM - \frac{1}{2}JN^{\dagger}JN \; is \; Hurwitz. \tag{7.40}$$

(2)

$$\left\| \tilde{E}^{\#}\Sigma\,(sI - F)^{-1}\,\tilde{D} \right\|_{\infty} < \frac{\gamma}{2}, \tag{7.41}$$

where $\tilde{D} = J\Sigma\tilde{E}^T$ *and* $\Sigma = \begin{bmatrix} 0 & I \\ I & 0 \end{bmatrix}$.

Then the uncertain quantum system is robustly mean square stable.

Example 7.2 [14] We consider a quantum system consisting of a small Josephson junction coupled to an electromagnetic resonant cavity. This system has been considered in [1] where a Hamiltonian for the system is derived. The paper [1] considered the system as a closed quantum system defined purely in terms of a system Hamiltonian. We modify this description of the system to consider an open quantum system model which interacts with external fields by introducing coupling operators for the system in order to apply our results given above. The coupling operator which is introduced is taken as a standard coupling operator for a resonant cavity coupled to a single field as well as a corresponding coupling operator for the Josephson junction coupled to an external heat bath.

A Josephson junction consists of a thin insulating material between two superconducting layers. As in [1], we consider a Josephson junction in a resonant cavity. The following Hamiltonian for the Josephson junction system is derived in [1]:

$$H = \frac{1}{2\hbar}\hat{U}(\hat{n} - \hat{\bar{n}})^2 - \frac{\hat{J}}{\hbar}\cos\hat{\phi} + \frac{1}{2\hbar}(\hat{p}^2 + \omega^2\hat{q}^2) - g\sqrt{\frac{\omega}{\hbar}}\hat{p}\hat{n} + \frac{U\omega\bar{n}^2 g^2}{2\hat{U}}, \quad (7.42)$$

where $\hat{\bar{n}} = \frac{U\bar{n}}{\hat{U}}$, $\hat{U} = U + \hbar\omega\hat{g}^2$, $[\hat{\phi}, \hat{n}] = \mathrm{i}$, and $[\hat{q}, \hat{p}] = \mathrm{i}\hbar$. Here

$$\hat{\phi} = \phi - g\sqrt{\frac{\omega}{\hbar}}q, \quad \hat{n} = n$$
$$\hat{p} = p + g\sqrt{\omega\hbar}n, \quad \hat{q} = q,$$

where q and p are the position and momentum operators for the resonant cavity. Also, n is an operator which represents the difference between the number of Cooper pairs on the two superconducting islands which make up the junction. Furthermore, ϕ is an operator which represents the phase difference across the junction. Note that compared to the expression for the Hamiltonian given in [1], we have normalized the expression (7.42) by dividing through by a factor of $\hbar$.

The quantities U, $\hat{J}$, $\bar{n}$, $\hat{g}$, ω are physical constants associated with the junction and the resonant cavity. In particular, U is the charging energy of the Josephson junction, $\hat{J}$ is the Josephson energy of the junction, $\bar{n}$ is an experimental parameter relating to the gate voltage applied to the superconducting islands, $\hat{g}$ is a parameter related to the dimensions of the junction, and ω is an angular frequency related to the detuning of the cavity; see [1].

By defining new variables $\hat{\hat{n}} = \hat{n} - \bar{n}$, $\hat{\hat{p}} = \hat{p} - g\sqrt{\hbar\omega}$ and neglecting the constant terms, we can rewrite the Hamiltonian as follows:

$$\hat{H} = \frac{1}{2}[\hat{q}\ \hat{\hat{p}}\ \hat{\hat{n}}\ \hat{\phi}] \begin{bmatrix} \frac{\omega^2}{\hbar} & 0 & 0 & 0 \\ 0 & \frac{1}{\hbar} & -g\sqrt{\frac{\omega}{\hbar}} & 0 \\ 0 & -\hat{g}\sqrt{\frac{\omega}{\hbar}} & \frac{\hat{U}}{\hbar} & 0 \\ 0 & 0 & 0 & 0 \end{bmatrix} \begin{bmatrix} \hat{q} \\ \hat{\hat{p}} \\ \hat{\hat{n}} \\ \hat{\phi} \end{bmatrix} - \frac{\hat{J}}{\hbar}\cos\hat{\phi}. \tag{7.43}$$

Now we define new operators

$$a_1 = \frac{\omega\hat{q} + i\hat{\hat{p}}}{\sqrt{2\hbar\omega}},$$
$$a_2 = \frac{\hat{\hat{\phi}} + i\hat{\hat{n}}}{\sqrt{2}}$$

which satisfy the canonical commutation relations $[a_1, a_1^\dagger] = 1$ and $[a_2, a_2^\dagger] = 1$. Then, the Hamiltonian can be rewritten in the form

$$H = \frac{1}{2}\left[a^\dagger\ a^T \right] M \begin{bmatrix} a \\ a^\# \end{bmatrix} - \frac{\hat{J}}{\hbar}\cos\left(\frac{a_2 + a_2^\dagger}{\sqrt{2}}\right), \tag{7.44}$$

where $a = \begin{bmatrix} a_1 \\ a_2 \end{bmatrix}$ and M is a Hermitian matrix of the form (7.14). Note that in order to write the Hamiltonian in this form with the matrix M satisfying (7.14), it is necessary to apply these canonical commutation relations and neglect further constant terms. Also, the fact that the operators a_1 and a_2 commute is used to redistribute terms within the matrix M.

We now modify the model of [1] by assuming that the cavity and Josephson modes are coupled to fields corresponding to coupling operators of the form

$$L = \begin{bmatrix} \sqrt{\kappa_1}a_1 \\ \sqrt{\kappa_2}a_2 \end{bmatrix}.$$

This modification of the quantum system is necessary in order to obtain a damped quantum system whose stability can be established and is reasonable for an experimental system which will experience damping in both the electromagnetic cavity and in the Josephson junction circuit.

In order to apply Theorem 7.5 to analyze the stability of this quantum system, we rewrite (7.44) as $H = H_1 + H_2$ where

$$H_1 = \frac{1}{2}\left[a^\dagger\ a^T \right] M \begin{bmatrix} a \\ a^\# \end{bmatrix}$$

and

$$H_2 = -\frac{\hat{J}}{\hbar}\cos(\frac{a_2 + a_2^\dagger}{\sqrt{2}}).$$

That is, we have a quadratic nominal Hamiltonian and a non-quadratic perturbation Hamiltonian. Then, we define $\zeta = a_2/\sqrt{2}$ and

$$f(\zeta, \zeta^*) = -\frac{\hat{J}}{\hbar}\cos(\zeta + \zeta^*),$$
$$f'(\zeta, \zeta^*) = \frac{\hat{J}}{\hbar}\sin(\zeta + \zeta^*),$$
$$f''(\zeta, \zeta^*) = \frac{\hat{J}}{\hbar}\cos(\zeta + \zeta^*).$$

From this it follows that

$$f'(\zeta, \zeta^*)^* f'(\zeta, \zeta^*) \leq 4\frac{\hat{J}^2}{\hbar^2}\zeta\zeta^*,$$

$$f''(\zeta, \zeta^*)^* f''(\zeta, \zeta^*) \leq \frac{\hat{J}^2}{\hbar^2}$$

and $\gamma = \frac{\hbar}{2\hat{J}}$.

The numerical values of the constants ω, $\hat{g}$, U and $\hat{J}$ are chosen as in [1] as $\frac{\omega}{2\pi} =$ 100 GHz, $g = 0.15$, $U = 2.2087 \times 10^{-22}$, $\hat{J} = 3.6652 \times 10^{11}$. Also, we calculate $\gamma/2 = 6.8209 \times 10^{-13}$. The parameters κ_1 and κ_2 will be chosen using Theorem 7.5 in order to guarantee the robust mean square stability of the quantum system. Indeed, for various values of κ_1 and κ_2, we form the transfer function $G_{\kappa_1,\kappa_2}(s) = E^\# \Sigma\,(sI - F)^{-1}\,\tilde{D}$ and calculate its H^∞ norm.

The H^∞ norm of $G_{\kappa_1,\kappa_2}(s)$ is found to be virtually independent of κ_1 and so a physically reasonable value of $\kappa_1 = 10^{11}$ is chosen. With this value of κ_1, the stability can be guaranteed for $\kappa_2 > 2.2 \times 10^{12}$. Hence, choosing a value of $\kappa_2 = 2.5 \times 10^{12}$, it follows that stability of Josephson junction system can be guaranteed using Theorem 7.5. Indeed, with this value of κ_2, we calculate the matrix $F = -\mathrm{i}JM - \frac{1}{2}JN^\dagger JN$ and find its eigenvalues to be

$$-5.0000 \times 10^{10} \pm 3.3507 \times 10^3\mathrm{i}$$

and

$$-1.2500 \times 10^{12} \pm 1.4842 \times 10^3\mathrm{i},$$

which implies that the matrix F is Hurwitz. Also, a magnitude Bode plot of the corresponding transfer function $G_{\kappa_1,\kappa_2}(s)$ implies that

$$\|G_{\kappa_1,\kappa_2}(s)\|_\infty = 5.5554 \times 10^{-13} < \gamma/2 = 6.8209 \times 10^{-13}.$$

Hence, using Theorem 7.5, we conclude that the quantum system is robustly mean square stable. More details can be found in [14].

In [15], the small gain theorem results were further extended to non-quadratic perturbation Hamiltonians which depend on multiple parameters. In particular, we assume the uncertain non-quadratic part of the system Hamiltonian H_2 is defined by a formal power series of the form

$$H_2 = f(z, z^*) = \sum_{i=1}^{p}\sum_{j=1}^{p}\sum_{k=0}^{\infty}\sum_{\ell=0}^{\infty} S_{ijk\ell} z_i^k (z_j^*)^\ell = \sum_{i=1}^{p}\sum_{j=1}^{p}\sum_{k=0}^{\infty}\sum_{\ell=0}^{\infty} S_{ijk\ell} H_{ijk\ell} \quad (7.45)$$

which is assumed to converge in some suitable sense. Here $S_{ijk\ell} = S^*_{ji\ell k}$, $H_{ijk\ell} = z_i^k (z_j^*)^\ell$. Also, we write

$$\tilde{E} = \begin{bmatrix} \tilde{E}_1 \\ \tilde{E}_2 \\ \vdots \\ \tilde{E}_p \end{bmatrix}.$$

We define the following partial derivatives:

$$\frac{\partial f(z, z^*)}{\partial z_i} \triangleq \sum_{j=1}^{p}\sum_{k=1}^{\infty}\sum_{\ell=0}^{\infty} k S_{ijk\ell} z_i^{k-1} (z_j^*)^\ell, \quad (7.46)$$

$$\frac{\partial^2 f(z, z^*)}{\partial z_i^2} \triangleq \sum_{j=1}^{p}\sum_{k=1}^{\infty}\sum_{\ell=0}^{\infty} k(k-1) S_{ijk\ell} z_i^{k-2} (z_j^*)^\ell. \quad (7.47)$$

For given constants $\gamma > 0$, $\delta_1 \geq 0$, $\delta_2 \geq 0$, we consider the sector bound condition

$$\sum_{i=1}^{p} \frac{\partial f(z, z^*)}{\partial z_i}^* \frac{\partial f(z, z^*)}{\partial z_i} \leq \frac{1}{\gamma^2} \sum_{i=1}^{p} z_i z_i^* + \delta_1 \quad (7.48)$$

and the condition

$$\sum_{i=1}^{p} \frac{\partial^2 f(z, z^*)}{\partial z_i^2}^* \frac{\partial^2 f(z, z^*)}{\partial z_i^2} \leq \delta_2. \quad (7.49)$$

Then we define the set of perturbation Hamiltonians $\mathbb{W}_6$ as follows:

$$\mathbb{W}_6 = \left\{ \begin{array}{l} H_2 = f(\cdot) \text{ of the form (7.45) such that} \\ \text{conditions (7.48) and (7.49) are satisfied} \end{array} \right\}. \quad (7.50)$$

We have the following result [15]:

Theorem 7.6 *Consider an uncertain open quantum system defined by (S, L, H) such that $H = H_1 + H_2$ where H_1 is of the form (7.13), L is of the form (7.15) and $H_2 \in \mathbb{W}_6$. Furthermore, assume that the following strict bounded real condition is satisfied:*

(1) The matrix

$$F = -iJM - \frac{1}{2}JN^{\dagger}JN \; is \; Hurwitz. \tag{7.51}$$

(2)

$$\left\| \tilde{E}^{\#}\Sigma\,(sI - F)^{-1}\,J\Sigma\tilde{E}^{T} \right\|_{\infty} < \frac{\gamma}{2} \tag{7.52}$$

where $\Sigma = \begin{bmatrix} 0 & I \\ I & 0 \end{bmatrix}$.

Then the uncertain quantum system is robustly mean square stable.

In [15], this result has been used to analyze robust stability of an optical parametric amplifier.

Now we further develop the Popov criterion to investigate the robust stability of a class of quantum systems. For given constants $\gamma > 0, \beta > 0, \delta_1 \geq 0, \delta_2 \geq 0, \delta_3 \geq 0$, we consider the sector bound condition

$$\left(\frac{\partial f(z, z^*)}{\partial z} - \frac{1}{\gamma}z^*\right)^* \left(\frac{\partial f(z, z^*)}{\partial z} - \frac{1}{\gamma}z^*\right) \leq \frac{1}{\gamma^2}zz^* + \delta_1; \tag{7.53}$$

and the smoothness conditions

$$\frac{\partial^2 f(z, z^*)}{\partial z^2}^* \frac{\partial^2 f(z, z^*)}{\partial z^2} \leq \delta_2; \tag{7.54}$$

$$\frac{\partial^2 f(z, z^*)}{\partial z \partial z^*}^* \frac{\partial^2 f(z, z^*)}{\partial z \partial z^*} \leq \delta_3. \tag{7.55}$$

Also, we consider the following upper and lower bounds on the perturbation Hamiltonian:

$$0 \leq f(z, z^*) \leq \beta zz^*. \tag{7.56}$$

Then we define the set of perturbation Hamiltonians $\mathbb{W}_7$ as follows:

$$\mathbb{W}_7 = \left\{ \begin{array}{l} H_2 \text{ of the form (7.33) such that conditions} \\ \text{(7.53), (7.54), (7.55) and (7.56) are satisfied} \end{array} \right\}. \tag{7.57}$$

We have the following robust stability results based on the Popov criterion [16]:

Theorem 7.7 *Consider an uncertain open nonlinear quantum system defined by (S, L, H) such that $H = H_1 + H_2$ where H_1 is of the form (7.13), L is of the form (7.15) and $H_2 \in \mathbb{W}_7$. Furthermore, assume that there exists a constant $\theta \geq 0$ such that the following frequency domain condition is satisfied:*

(1) The matrix

$$F = -iJM - \frac{1}{2}JN^{\dagger}JN \; is \; Hurwitz. \tag{7.58}$$

(2) The transfer function

$$G(s) = 2i\tilde{E}^{\#}\Sigma\,(sI - F)^{-1}\,\Sigma J\tilde{E}^{T} \tag{7.59}$$

satisfies the strict positive real condition

$$\gamma + (1 + \theta i\omega)G(i\omega) + (1 - \theta i\omega)G(i\omega)^{*} > 0 \tag{7.60}$$

for all $\omega \in [-\infty, \infty]$.

Then the uncertain quantum system is robustly mean square stable.

This theorem provides a means to analyze the stability of uncertain nonlinear quantum system models. In particular, the theorem provides a sufficient condition for robust mean square stability. This means that if a nonlinear quantum system can be modeled by a perturbed (S, L, H) model which satisfies the conditions of the theorem, then its dynamics are guaranteed to be mean square stable. These conditions are straightforward to check numerically and do not require the simulation of the system dynamics. However, the conditions given are only sufficient conditions and not necessary conditions which means that if the perturbed (S, L, H) model of a nonlinear quantum system does not satisfy the conditions of the theorem, then we cannot say anything about the stability of the system using this approach. Note that the strict positive real condition (7.60) can be rewritten as

$$\frac{\gamma}{2} + \Re[G(i\omega)] - \theta\omega\Im[G(i\omega)] > 0 \tag{7.61}$$

for all $\omega \in [-\infty, \infty]$. The condition (7.61) can be tested graphically by producing a plot of $\omega\Im[G(i\omega)]$ versus $\Re[G(i\omega)]$ with $\omega \in [-\infty, \infty]$ as a parameter. Such a parametric plot is referred to as the Popov plot. Then, the condition (7.61) will be satisfied if and only if the Popov plot lies below the straight line of slope $\frac{1}{\theta}$ and with x-axis intercepts $-\frac{\gamma}{2}$. This provides a graphical way to check the conditions of the theorem [20].

We consider the example of an optical cavity containing a Kerr medium [16, 20].

Example 7.3 A standard (S, L, H) model for an optical cavity containing a Kerr medium is as follows:

$$S = I, \; H = \left(a^*\right)^2 a^2, \; L = \sqrt{\kappa} a. \tag{7.62}$$

We first attempt to apply the results of Theorem 7.5 and Theorem 7.7 to this quantum system. Let

$$M = 0$$

and

$$f(z, z^*) = z^2 \left(z^*\right)^2 .$$

where $z = a^*$. This defines a nonlinear quantum system of the form considered in Theorem 7.5 and Theorem 7.7 with $M_1 = 0$, $M_2 = 0$, $N_1 = \sqrt{\kappa}$, $N_2 = 0$, $E_1 = 0$, $E_2 = 1$. In [16, 20], it has been demonstrated that the conditions required are not satisfied. In order to overcome this difficulty, we note that any physical realization of a Kerr nonlinearity will not be exactly described by the model (7.62) but rather will exhibit some saturation of the Kerr effect. In order to represent this effect, we assume that the true function $\tilde{f}(\cdot)$ describing the Hamiltonian of the Kerr medium is such that its Taylor series expansion (7.33) satisfies $S_{0,k} = S_{1,k} = 0$ for all $k = 0, 1, \ldots$ and $S_{2,2} = 1$. That is, the first nonzero term in the Taylor series expansion corresponds to the standard Kerr Hamiltonian given in (7.62). Furthermore, we assume that the function $\tilde{f}(\cdot)$ is such that the conditions (7.36), (7.53), (7.54), (7.55), (7.56) are all satisfied for suitable values of the constants $\gamma > 0$, $\beta > 0$, $\delta_1 \geq 0$, $\delta_2 \geq 0$, $\delta_3 \geq 0$. Here, the quantity $\frac{1}{\gamma}$ will be proportional to the saturation limit for the Kerr nonlinearity. Under these assumptions, we can assume $\tilde{f}(\cdot) \in \mathbb{W}_5$ and $\tilde{f}(\cdot) \in \mathbb{W}_7$.

This system has

$$F = \begin{bmatrix} -\frac{\kappa}{2} & 0 \\ 0 & -\frac{\kappa}{2} \end{bmatrix},$$

which is Hurwitz for all $\kappa > 0$ and

$$G(s) = -\frac{2\mathrm{i}}{s + \frac{\kappa}{2}}.$$

Applying Theorem 7.5 to this system, we can guarantee that the system is mean square stable provided

$$\kappa > \frac{4}{\gamma}. \tag{7.63}$$

We now apply Theorem 7.7 to further analyze the stability of the system. In fact, the frequency domain condition (7.60) will be satisfied for all $\gamma > 0$. Furthermore, we can construct the Popov plot of the system for different values of $\kappa > 0$ and verify that for a suitable value of $\theta = \frac{2}{\kappa} > 0$, the frequency domain condition (7.60) will be satisfied for all $\gamma > 0$ and all $\kappa > 0$. Thus, using Theorem 7.7, we can conclude that the optical cavity containing a saturated Kerr medium is in fact mean square stable for all $\gamma > 0$ and $\kappa > 0$. This condition is clearly less restrictive than the condition (7.63) obtained by applying Theorem 7.5. More details can be found in [16].

7.3.2 Robust Stability of Quantum Systems with Nonlinear Coupling Operators

Now, we consider robust stability for uncertain quantum systems with a nonlinear coupling operator [24]. Assume that the coupling operator is decomposed as $L = L_1 + L_2$ where L_1 is a known nominal coupling operator and L_2 is a perturbation coupling operator contained in a specified set of coupling operators $\mathbb{W}$. Given a set of non-negative self-adjoint operators $\mathscr{P}$ and real parameters $\gamma > 0$, $\delta_1 \geq 0$, $\delta_2 \geq 0$, $\delta_3 \geq 0$, we define a particular set of perturbation coupling operators $\mathbb{W}_8$. This set $\mathbb{W}_8$ is defined in terms of the commutator decomposition

$$[V, L_2] = w_1[V, \zeta] - \frac{1}{2}w_2\,[\zeta, [V, \zeta]] \tag{7.64}$$

for $V \in \mathscr{P}$ where w_1, w_2 and ζ are given scalar operators. We say $L_2 \in \mathbb{W}_8$ if the following sector bound condition holds:

$$L_2^* L_2 \leq \frac{1}{\gamma^2}\zeta^*\zeta + \delta_1 \tag{7.65}$$

and

$$w_1^* w_1 \leq \delta_2, \tag{7.66}$$

$$w_2^* w_2 \leq \delta_3. \tag{7.67}$$

Here, we use the convention that for operator inequalities, terms consisting of real constants are interpreted as that constant multiplying the identity operator.

Then, we define

$$\mathbb{W}_8 = \left\{ \begin{array}{l} L_2 : \exists w_1,\ w_2,\ \zeta \text{ such that (7.65), (7.66) and (7.67)} \\ \text{are satisfied and (7.64) is satisfied } \forall V \in \mathscr{P} \end{array} \right\}. \tag{7.68}$$

Using this definition, we obtain the following theorem whose proof can be found in [24]:

Theorem 7.8 *Consider a set of non-negative self-adjoint operators $\mathscr{P}$ and an open quantum system (S, L, H) where $L = L_1 + L_2$ and $L_2 \in \mathbb{W}_8$ defined in (7.68). If there exists a $V \in \mathscr{P}$ and real constants $c > 0$, $\tilde{\lambda} \geq 0$, $\tau_1 > 0$, $\tau_2 > 0, \ldots, \tau_5 > 0$ such that $\mu = -\frac{1}{2}[\zeta, [V, \zeta]]$ is a constant and*

$$-i[V,H]+\mathscr{L}_{L_1}(V)+\left(\frac{\tau_1^2}{2}+\frac{\tau_2^2}{2}\right)L_1^*L_1+\left(\frac{\delta_2}{2\tau_1^2}+\frac{\delta_2}{2\tau_4^2}\right)[V,\zeta]^*[V,\zeta]$$
$$+\left(\frac{\tau_3^2}{2\gamma^2}+\frac{\tau_4^2}{2\gamma^2}+\frac{\tau_5^2}{2\gamma^2}\right)\zeta^*\zeta+\frac{[V,L_1]^*[V,L_1]}{2\tau_3^2}+cV\le\tilde{\lambda}, \tag{7.69}$$

then

$$\langle V(t)\rangle\le e^{-ct}\langle V\rangle+\frac{\tilde{\lambda}+\left(\frac{\delta_3}{2\tau_2^2}+\frac{\delta_3}{2\tau_5^2}\right)\mu^*\mu+\left(\frac{\tau_3^2}{2}+\frac{\tau_4^2}{2}+\frac{\tau_5^2}{2}\right)\delta_1}{c}$$

for all $t\ge 0$.

Now, we define a set of nonlinear perturbation coupling operators denoted $\mathbb{W}_9$. For a given set of non-negative self-adjoint operators $\mathscr{P}$ and real parameters $\gamma>0$, $\delta_1\ge 0$, $\delta_2\ge 0$, $\delta_3\ge 0$, consider perturbation coupling operators defined in terms of the following power series (which is assumed to converge in some suitable sense):

$$L_2=f(\zeta)=\sum_{k=0}^{\infty}S_k\zeta^k=\sum_{k=0}^{\infty}S_kL_k. \tag{7.70}$$

Here ζ is a scalar operator on the underlying Hilbert space and $L_k=\zeta^k$.

Also, we let

$$f'(\zeta)=\sum_{k=1}^{\infty}kS_k\zeta^{k-1}, \tag{7.71}$$

$$f''(\zeta)=\sum_{k=1}^{\infty}k(k-1)S_k\zeta^{k-2} \tag{7.72}$$

and consider the sector bound condition

$$f(\zeta)^*f(\zeta)\le\frac{1}{\gamma^2}\zeta^*\zeta+\delta_1 \tag{7.73}$$

and the conditions

$$f'(\zeta)^*f'(\zeta)\le\delta_2, \tag{7.74}$$

$$f''(\zeta)^*f''(\zeta)\le\delta_3. \tag{7.75}$$

Then we define the set $\mathbb{W}_9$ as follows:

$$\mathbb{W}_9=\left\{\begin{array}{l}L_2\text{ of the form (7.70) such that}\\ \text{conditions (7.73), (7.74) and (7.75) are satisfied}\end{array}\right\}. \tag{7.76}$$

We assume the set of non-negative self-adjoint operators $\mathscr{P}$ to satisfy the following assumption:

Assumption 7.1 Given any $V \in \mathscr{P}$, the quantity

$$\mu = -\frac{1}{2}[\zeta, [V, \zeta]] = -\frac{1}{2}\zeta[V, \zeta] + \frac{1}{2}[V, \zeta]\zeta$$

is a constant.

From [24], we know that $\mathbb{W}_9 \subset \mathbb{W}_8$ if the set of self-adjoint operators $\mathscr{P}$ satisfies Assumption 7.1.

We now consider the case in which the nominal quantum system corresponds to a linear quantum system. We assume L_1 is of the form

$$L_1 = \begin{bmatrix} N_1 & N_2 \end{bmatrix} \begin{bmatrix} a \\ a^{\#} \end{bmatrix} = \tilde{N} \begin{bmatrix} a \\ a^{\#} \end{bmatrix}, \qquad (7.77)$$

where $N_1 \in \mathbb{C}^{1\times n}$ and $N_2 \in \mathbb{C}^{1\times n}$. Also, we write

$$\begin{bmatrix} L_1 \\ L_1^* \end{bmatrix} = N \begin{bmatrix} a \\ a^{\#} \end{bmatrix} = \begin{bmatrix} N_1 & N_2 \\ N_2^{\#} & N_1^{\#} \end{bmatrix} \begin{bmatrix} a \\ a^{\#} \end{bmatrix}.$$

We define

$$\zeta = E_1 a + E_2 a^{\#} = \begin{bmatrix} E_1 & E_2 \end{bmatrix} \begin{bmatrix} a \\ a^{\#} \end{bmatrix} = \tilde{E} \begin{bmatrix} a \\ a^{\#} \end{bmatrix}, \qquad (7.78)$$

where ζ is assumed to be a scalar operator. We consider a similar definition of robust mean square stability to Definition 7.1 and have the following result:

Theorem 7.9 *Consider an uncertain open quantum system defined by* (S, L, H) *such that* $L = L_1 + L_2$ *where* L_1 *is of the form (7.77),* H *is of the form (7.13) and* $L_2 \in \mathbb{W}_9$. *Furthermore, assume that there exists constants* $\tau_1 > 0$, $\tau_3 > 0$, *and* $\tau_4 > 0$ *such that the following strict bounded real condition is satisfied:*

(1) The matrix

$$F = -iJM - \frac{1}{2}JN^{\dagger}JN \; is \; Hurwitz. \qquad (7.79)$$

(2)

$$\left\| \bar{C}(sI - F)^{-1}\bar{B} \right\|_{\infty} < 1 \qquad (7.80)$$

where

$$\bar{C} = \begin{bmatrix} \frac{\sqrt{\tau_3^2 + \tau_4^2}}{\gamma}\tilde{E} \\ \tau_1 \tilde{N} \end{bmatrix} \qquad (7.81)$$

and

$$\bar{B} = \begin{bmatrix} \sqrt{\delta_2\left(\frac{1}{\tau_1^2} + \frac{1}{\tau_4^2}\right)} J\tilde{E}^{\dagger} & \frac{1}{\tau_3} J\tilde{N}^{\dagger} \end{bmatrix}. \qquad (7.82)$$

Then the uncertain quantum system is robustly mean square stable.

The proof can be found in [24]. It is clear that the robust stability condition for this class of quantum systems with a nonlinear coupling operator is given in terms of a scaled strict bounded real condition.

7.3.3 Robust Stability of Quantum Systems with Nonlinear Dynamic Uncertainties

It is well known that the classical small gain robust stability condition also applies in the case of nonlinear dynamic uncertainties. Such uncertainties can be described in terms of integral quadratic constraints; e.g., see [25]. Now we extend the quantum small gain result to allow for nonlinear dynamic uncertainties which are described by a certain quantum stochastic integral quadratic constraint [18]. This uncertainty description can be regarded as a continuous-time quantum version of the stochastic uncertainty constraint considered in [21]. Here we assume that the presence of dynamic uncertainties is represented by a perturbation Hamiltonian which depends on system variables which are in addition to those which occur in the nominal Hamiltonian. In particular, H is the system Hamiltonian operator which is assumed to be of the form

$$H = \frac{1}{2}\left[a^\dagger \; a^T\right] M \begin{bmatrix} a \\ a^\# \end{bmatrix} + f(b, b^\#, z, z^*). \tag{7.83}$$

Here b is a vector of annihilation operators on the underlying Hilbert space and $b^\#$ is the corresponding vector of creation operators. Furthermore, $M \in \mathbb{C}^{2n \times 2n}$ is a Hermitian matrix of the form (7.14). The matrix M is assumed to be known and defines the nominal quadratic part of the system Hamiltonian. Furthermore, we assume the uncertain non-quadratic part of the system Hamiltonian $f(b, b^\#, z, z^*)$ is defined by a formal power series of the form

$$H_2 = f(b, b^\#, z, z^*) = \sum_{k=0}^{\infty}\sum_{\ell=0}^{\infty} S_{k\ell}(b, b^\#) z^k (z^*)^\ell = \sum_{k=0}^{\infty}\sum_{\ell=0}^{\infty} S_{k\ell}(b, b^\#) H_{k\ell}, \tag{7.84}$$

which is assumed to converge in some suitable sense. Here $S_{k\ell}(b, b^\#) = S_{\ell k}(b, b^\#)^*$, $H_{k\ell} = z^k (z^*)^\ell$, and z is a known scalar operator as defined in (7.39).

The term $f(b, b^\#, z, z^*)$ is referred to as the perturbation Hamiltonian. It is assumed to be unknown but is contained within a known set which will be defined below. It follows from this definition that $f(b, b^\#, z, z^*)$ is a self-adjoint operator. The fact that $f(b, b^\#, z, z^*)$ depends on the quantities b and $b^\#$ which do not appear in the nominal Hamiltonian corresponds to our assumption that we allow nonlinear dynamic uncertainties in the quantum system.

We assume the coupling operator L is known and is of the form

$$L = \begin{bmatrix} L_a \\ L_b \end{bmatrix}, \tag{7.85}$$

where

$$L_a = \begin{bmatrix} N_{a1} & N_{a2} \end{bmatrix} \begin{bmatrix} a \\ a^\# \end{bmatrix} \tag{7.86}$$

and

$$L_b = \begin{bmatrix} N_{b1} & N_{b2} \end{bmatrix} \begin{bmatrix} b \\ b^\# \end{bmatrix}. \tag{7.87}$$

Here, $N_{a1} \in \mathbb{C}^{m \times n}$, $N_{a2} \in \mathbb{C}^{m \times n}$, $N_{b1} \in \mathbb{C}^{\tilde{m} \times \tilde{n}}$ and $N_{b2} \in \mathbb{C}^{\tilde{m} \times \tilde{n}}$.
Also, we write

$$\begin{bmatrix} L_a \\ L_a^\# \end{bmatrix} = N_a \begin{bmatrix} a \\ a^\# \end{bmatrix} = \begin{bmatrix} N_{a1} & N_{a2} \\ N_{a2}^\# & N_{a1}^\# \end{bmatrix} \begin{bmatrix} a \\ a^\# \end{bmatrix}$$

and

$$\begin{bmatrix} L_b \\ L_b^\# \end{bmatrix} = N_b \begin{bmatrix} b \\ b^\# \end{bmatrix} = \begin{bmatrix} N_{b1} & N_{b2} \\ N_{b2}^\# & N_{b1}^\# \end{bmatrix} \begin{bmatrix} b \\ b^\# \end{bmatrix}.$$

The annihilation and creation operators b and $b^\#$ are assumed to satisfy the canonical commutation relations:

$$\left[\begin{bmatrix} b \\ b^\# \end{bmatrix}, \begin{bmatrix} b \\ b^\# \end{bmatrix}^\dagger \right] = J. \tag{7.88}$$

Also, we assume that all of the elements of the vectors a and $a^\#$ commute with all of the elements of the vectors b and $b^\#$.

To define the set of allowable perturbation Hamiltonians $f(\cdot)$, we first define the following formal partial derivatives:

$$\frac{\partial f(b, b^\#, z, z^*)}{\partial z} \triangleq \sum_{k=1}^{\infty} \sum_{\ell=0}^{\infty} k S_{k\ell}(b, b^\#) z^{k-1} (z^*)^\ell; \tag{7.89}$$

$$\frac{\partial^2 f(b, b^\#, z, z^*)}{\partial z^2} \triangleq \sum_{k=1}^{\infty} \sum_{\ell=0}^{\infty} k(k-1) S_{k\ell}(b, b^\#) z^{k-2} (z^*)^\ell. \tag{7.90}$$

Then, we consider the following quantum stochastic differential equations describing the uncertainty dynamics (e.g., see [3] and [26]):

$$\begin{aligned}
d\begin{bmatrix} b \\ b^\# \end{bmatrix} &= -i\left[\begin{bmatrix} b \\ b^\# \end{bmatrix}, f(b, b^\#, z, z^*)\right]dt + \frac{1}{2}\left(L_b^\dagger\left[\begin{bmatrix} b \\ b^\# \end{bmatrix}, L_b^T\right]^T\right)^T dt \\
&\quad + \frac{1}{2}\left[L_b^\#, \begin{bmatrix} b \\ b^\# \end{bmatrix}\right]^T L_b dt + \left[\begin{bmatrix} b \\ b^\# \end{bmatrix}, L^T\right]d\mathscr{B}^\# - \left[\begin{bmatrix} b \\ b^\# \end{bmatrix}, L^\dagger\right]d\mathscr{B} \\
&= -i\sum_{k=0}^{\infty}\sum_{\ell=0}^{\infty}\left[\begin{bmatrix} b \\ b^\# \end{bmatrix}, S_{k\ell}(b, b^\#)\right] z^k (z^*)^\ell dt \\
&\quad - \frac{1}{2}JN_b^\dagger JN_b\begin{bmatrix} b \\ b^\# \end{bmatrix}dt - JN_b^\dagger J\begin{bmatrix} d\mathscr{B} \\ d\mathscr{B}^\# \end{bmatrix},
\end{aligned} \tag{7.91}$$

where $\mathscr{B}(t)$ is a vector of bosonic annihilation operators corresponding to the quantum fields acting on the uncertainty system and $\mathscr{B}(t)^\#$ is the corresponding vector of creation operators; e.g., see [12]. The vector $\mathscr{B}(t)$ corresponds to a vector of standard quantum Weiner processes. The set of allowable perturbation Hamiltonians will be defined in terms of quantum stochastic integral quadratic constraints for the system (7.91). These conditions are defined in a similar way to the definition of dissipativity in [8]; i.e., the given inequalities are required to hold for all interconnections between the system (7.91) and an exosystem $\tilde{W}$ contained in a suitable class of exosystems.

For given constants $\gamma > 0$ and $\delta_1 \geq 0$, we consider the quantum stochastic integral quadratic constraint

$$\limsup_{T\to\infty} \frac{1}{T}\int_0^T \left(\langle w_1(t)w_1(t)^*\rangle - \frac{1}{\gamma^2}\langle z(t)z(t)^*\rangle\right)dt \leq \delta_1, \tag{7.92}$$

where

$$w_1(t) = \frac{\partial f(b(t), b(t)^\#, z(t), z(t)^*)^*}{\partial z}$$

and $b(t)$, $b(t)^\#$, $z(t)$, $z(t)^*$ denote the Heisenberg evolutions of the operators b, $b^\#$, z, z^*, respectively, for the system formed by the interconnection between the quantum system (7.91) and an exosystem $\tilde{W}$. Similarly, for a given constant $\delta_2 \geq 0$, we consider the quantum stochastic integral quadratic constraint

$$\limsup_{T\to\infty} \frac{1}{T}\int_0^T \langle w_2(t)w_2(t)^*\rangle dt \leq \delta_2, \tag{7.93}$$

where

$$w_2(t) = \frac{\partial^2 f(z, z^*)^*}{\partial z^2}.$$

Then we define the set of perturbation Hamiltonians $\mathbb{W}_{10}$ as follows:

$$\mathbb{W}_{10} = \left\{ \begin{array}{l} H_2 \text{ of the form (7.84) such that} \\ \text{conditions (7.92) and (7.93) are satisfied} \end{array} \right\}. \tag{7.94}$$

We consider the following notion of robust mean square stability which is somewhat different from the definition considered in Definition 7.1 due to the presence of dynamic uncertainties:

Definition 7.2 An uncertain open quantum system defined by (S, L, H) where H is of the form (7.13), $H_2 \in \mathbb{W}_{10}$, and L is of the form (7.85) is said to be *robustly mean square stable* if there exists a constant $c > 0$, such that for any $H_2 \in \mathbb{W}_{10}$,

$$\limsup_{T\to\infty} \frac{1}{T} \int_0^T \left\langle \begin{bmatrix} a(t) \\ a^{\#}(t) \end{bmatrix}^{\dagger} \begin{bmatrix} a(t) \\ a^{\#}(t) \end{bmatrix} \right\rangle dt \leq c. \tag{7.95}$$

Here $\begin{bmatrix} a(t) \\ a^{\#}(t) \end{bmatrix}$ denotes the Heisenberg evolution of the vector of operators $\begin{bmatrix} a \\ a^{\#} \end{bmatrix}$.

We have the following result on robust mean square stability of the nonlinear quantum system under consideration when $H_2 \in \mathbb{W}_{10}$:

Theorem 7.10 *Consider an uncertain open nonlinear quantum system defined by (S, L, H) such that $S = I$, H is of the form (7.13), L is of the form (7.85) and $H_2 \in \mathbb{W}_{10}$. Furthermore, assume that the following strict bounded real condition is satisfied:*

(1) The matrix

$$F = -iJM - \frac{1}{2}JN_a^{\dagger}JN_a \; is \; Hurwitz. \tag{7.96}$$

(2)

$$\left\| \tilde{E}^{\#}\Sigma\,(sI - F)^{-1}\,J\Sigma\tilde{E}^{T} \right\|_{\infty} < \frac{\gamma}{2}, \tag{7.97}$$

where $\Sigma = \begin{bmatrix} 0 & I \\ I & 0 \end{bmatrix}$.

Then the uncertain quantum system is robustly mean square stable.

The proof can be found in [18]. This result has also been used to analyze a linearized model of an optical parametric amplifier in [18]. A robust stability condition for this system was obtained there and for more details, please see [18].

7.4 Performance Analysis and Guaranteed Cost Control for Linear Quantum Systems

7.4.1 Small Gain Theorem for Performance Analysis and Control Design

For the linear quantum systems with quadratic perturbation considered in Section 7.2, we may use Theorem 7.2 to investigate their robust stability. Now we provide several results on performance analysis for this class of quantum systems [28]. The results presented here are built on the following result from [10]:

Lemma 7.2 *[10] Consider an open quantum system defined by* (S, L, H) *and suppose there exist non-negative self-adjoint operators* V *and* W *on the underlying Hilbert space such that*

$$\mathscr{G}(V) + W \leq \lambda, \tag{7.98}$$

where λ *is a real number. Then for any plant state, we have*

$$\limsup_{T\to\infty} \frac{1}{T}\int_0^T \langle W(t)\rangle dt \leq \lambda. \tag{7.99}$$

Here $W(t)$ *denotes the Heisenberg evolution of the operator* W *and* $\langle\cdot\rangle$ *denotes quantum expectation; e.g., see [10].*

We denote

$$W = \begin{bmatrix} a^\dagger & a^T \end{bmatrix} R \begin{bmatrix} a \\ a^\# \end{bmatrix}, \tag{7.100}$$

and employ the following cost function:

$$J_c = \limsup_{T\to\infty} \frac{1}{T}\int_0^T \left\langle \begin{bmatrix} a^\dagger & a^T \end{bmatrix} R \begin{bmatrix} a \\ a^\# \end{bmatrix} \right\rangle \mathrm{d}t, \tag{7.101}$$

where $R > 0$.

Now we have the following conclusion which can be used to analyze the performance of a given quantum system using the quantum small gain method [30]:

Theorem 7.11 *Consider an uncertain quantum system* (S, L, H)*, where* $H = H_1 + H_2$*,* H_1 *is the form (7.13),* L *of the form (7.15) and* $H_2 \in \mathbb{W}_2$*. If*

$$F = -iJM - \frac{1}{2}JN^\dagger JN \text{ is Hurwitz},$$

$$\begin{bmatrix} F^\dagger P + PF + \frac{E^\dagger E}{\gamma^2 \tau^2} + R & 2PJE^\dagger \\ 2EJP & -I/\tau^2 \end{bmatrix} < 0 \tag{7.102}$$

has a solution $P > 0$ in the form of (7.17) and $\tau > 0$, then

$$\begin{aligned} J_c &= \limsup_{T\to\infty} \frac{1}{T} \int_0^T \langle W(t) \rangle dt \\ &= \limsup_{T\to\infty} \frac{1}{T} \int_0^T \left\langle \begin{bmatrix} a^\dagger & a^T \end{bmatrix} R \begin{bmatrix} a \\ a^\# \end{bmatrix} \right\rangle dt \\ &\le \tilde{\lambda} + \frac{\delta}{\tau^2}, \end{aligned} \tag{7.103}$$

where

$$\tilde{\lambda} = \mathrm{Tr}(PJN^\dagger \begin{bmatrix} I & 0 \\ 0 & 0 \end{bmatrix} NJ). \tag{7.104}$$

The proof can be found in [30].

Now, we design a guaranteed cost coherent controller for this class of uncertain quantum systems and aim to make the control system not only stable but also to achieve an adequate level of performance. The coherent controller is realized by adding a controller Hamiltonian H_3. H_3 is assumed to be in the form

$$H_3 = \frac{1}{2} \begin{bmatrix} a^\dagger & a^T \end{bmatrix} K \begin{bmatrix} a \\ a^\# \end{bmatrix}, \tag{7.105}$$

where $K \in \mathbb{C}^{2n \times 2n}$ is a Hermitian matrix of the form

$$K = \begin{bmatrix} K_1 & K_2 \\ K_2^\# & K_1^\# \end{bmatrix}, \tag{7.106}$$

and $K_1 = K_1^\dagger$, $K_2 = K_2^T$. The cost function J_c is defined as

$$J_c = \limsup_{T\to\infty} \frac{1}{T} \int_0^\infty \left\langle \begin{bmatrix} a \\ a^\# \end{bmatrix}^\dagger (R + \alpha K^2) \begin{bmatrix} a \\ a^\# \end{bmatrix} \right\rangle \mathrm{d}t, \tag{7.107}$$

where $\alpha \in (0, \infty)$ is a weighting factor. We let

$$W = \begin{bmatrix} a \\ a^\# \end{bmatrix}^\dagger (R + \alpha K^2) \begin{bmatrix} a \\ a^\# \end{bmatrix}. \tag{7.108}$$

The following results (Theorem 7.12) demonstrate how one can design a guaranteed cost coherent controller for the given quantum system using the quantum small gain method whose proof can be found in [30]:

Theorem 7.12 *Consider an uncertain quantum system (S, L, H), where $H = H_1 + H_2 + H_3$, H_1 is of the form (7.13), L of the form (7.15) and $H_2 \in \mathbb{W}_2$ and the controller Hamiltonian H_3 is in the form of (7.105). With $Q = P^{-1}$, $Y = KQ$ and*

$$F = -iJM - \frac{1}{2}JN^\dagger JN,$$

if there exist a matrix $Q = \tilde{q} \times I$ ($\tilde{q}$ is a constant scalar and I is the identity matrix), a Hermitian matrix Y and a constant $\tau > 0$, such that

$$\begin{bmatrix} \tilde{q}F^\dagger + F\tilde{q} + iYJ - iJY + 4\tau^2 JE^\dagger EJ & Y & \tilde{q}R^{\frac{1}{2}} & \tilde{q}E^\dagger \\ Y & -I/\alpha & 0 & 0 \\ \tilde{q}R^{\frac{1}{2}} & 0 & -I & 0 \\ \tilde{q}E & 0 & 0 & -\gamma^2\tau^2 I \end{bmatrix} < 0, \quad (7.109)$$

then the associated cost function satisfies the bound

$$\begin{aligned} \limsup_{T\to\infty} \frac{1}{T}\int_0^T \langle W(t)\rangle dt &= \limsup_{T\to\infty} \frac{1}{T}\int_0^\infty \left\langle \begin{bmatrix} a \\ a^\# \end{bmatrix}^\dagger (R + \alpha K^2) \begin{bmatrix} a \\ a^\# \end{bmatrix} \right\rangle dt \\ &\leq \tilde{\lambda} + \frac{\delta}{\tau^2}, \end{aligned} \quad (7.110)$$

where

$$\tilde{\lambda} = \mathrm{Tr}(PJN^\dagger \begin{bmatrix} I & 0 \\ 0 & 0 \end{bmatrix} NJ). \quad (7.111)$$

7.4.2 Popov Approach for Performance Analysis and Control Design

For linear quantum systems with a quadratic perturbation, we also may use the Popov approach and Theorem 7.3 to analyze the robust stability [10]. Now we consider performance analysis and have the following result [30]:

Theorem 7.13 *Consider an uncertain quantum system (S, L, H), where $H = H_1 + H_2$, H_1 is of the form (7.13), L is of the form (7.15) and $H_2 \in \mathbb{W}_3$. If*

$$F = -iJM - \frac{1}{2}JN^\dagger JN \text{ is Hurwitz,}$$

and

$$\begin{bmatrix} PF + F^{\dagger}P + R & -2iPJE^{\dagger} + E^{\dagger} + \theta F^{\dagger}E^{\dagger} \\ 2iEJP + E + \theta EF & -\gamma I \end{bmatrix} < 0 \tag{7.112}$$

has a solution $P > 0$ in the form of (7.17) for some $\theta \geq 0$, then

$$\begin{aligned} J_c &= \limsup_{T\to\infty} \frac{1}{T}\int_0^T \langle W(t)\rangle dt \\ &= \limsup_{T\to\infty} \frac{1}{T}\int_0^T \left\langle \begin{bmatrix} a^{\dagger} & a^T \end{bmatrix} R \begin{bmatrix} a \\ a^{\#} \end{bmatrix} \right\rangle dt \\ &\leq \lambda, \end{aligned} \tag{7.113}$$

where

$$\lambda = \mathrm{Tr}(PJN^{\dagger}\begin{bmatrix} I & 0 \\ 0 & 0 \end{bmatrix} NJ) + \frac{4\theta}{\gamma}\tilde{N}^{\#}\Sigma JE^{\dagger}EJ\Sigma\tilde{N}^{T}. \tag{7.114}$$

The proof can be found in [30]. In order to design a controller to guarantee a required performance, we may use the following theorem [30]:

Theorem 7.14 *Consider an uncertain quantum system (S, L, H), where $H = H_1 + H_2 + H_3$, H_1 is of the form (7.13), L is of the form (7.15), $H_2 \in \mathbb{W}_3$, the controller Hamiltonian H_3 is in the form of (7.105). With $Q = P^{-1}$, $Y = KQ$ and*

$$F = -iJM - \frac{1}{2}JN^{\dagger}JN,$$

if there exist a matrix $Q = \tilde{q} \times I$ ($\tilde{q}$ is a constant scalar and I is the identity matrix), a Hermitian matrix Y and a constant $\theta > 0$, such that

$$\begin{bmatrix} F\tilde{q} + \tilde{q}F^{\dagger} - iJY + iYJ & B^{\dagger} & Y & \tilde{q}R^{\frac{1}{2}} \\ B & -\gamma I & 0 & 0 \\ Y & 0 & -I/\alpha & 0 \\ \tilde{q}R^{\frac{1}{2}} & 0 & 0 & -I \end{bmatrix} < 0, \tag{7.115}$$

where $B = 2iEJ + E\tilde{q} + \theta EF\tilde{q} - i\theta EJY$, then the associated cost function satisfies the bound

$$\begin{aligned} \limsup_{T\to\infty} \frac{1}{T}\int_0^T \langle W(t)\rangle dt &= \limsup_{T\to\infty} \frac{1}{T}\int_0^{\infty} \left\langle \begin{bmatrix} a \\ a^{\#} \end{bmatrix}^{\dagger} (R + \alpha K^2) \begin{bmatrix} a \\ a^{\#} \end{bmatrix} \right\rangle dt, \\ &\leq \lambda \end{aligned} \tag{7.116}$$

where

$$\lambda = \mathrm{Tr}(PJN^{\dagger}\begin{bmatrix} I & 0 \\ 0 & 0 \end{bmatrix}NJ) + \frac{4\theta}{\gamma}\tilde{N}^{\#}\overline{\Sigma}JE^{\dagger}EJ\overline{\Sigma}\tilde{N}^{T}. \tag{7.117}$$

The proof can be found in [30]. For each fixed value of θ, the problem is an LMI problem. Then, we can iterate on $\theta \in [0, \infty)$ and choose the value which minimizes the cost bound (7.117) in Theorem 7.14. In order to illustrate the above small gain and Popov methods and compare their performance, an example of an optical parametric amplifier has been presented in [29, 30]. The (S, L, H) description of this system has the following form:

$$H = \frac{1}{2}\mathrm{i}((a^{\dagger})^2 - a^2),\ S = I,\ L = \sqrt{\kappa}a.$$

We let the perturbation Hamiltonian be

$$H_2 = \frac{1}{2}\begin{bmatrix} a^{\dagger} & a^{T} \end{bmatrix}\begin{bmatrix} 1 & 0.5\mathrm{i} \\ -0.5\mathrm{i} & 1 \end{bmatrix}\begin{bmatrix} a \\ a^{\#} \end{bmatrix}$$

and the nominal Hamiltonian be

$$H_1 = \frac{1}{2}\begin{bmatrix} a^{\dagger} & a^{T} \end{bmatrix}\begin{bmatrix} -1 & 0.5\mathrm{i} \\ -0.5\mathrm{i} & -1 \end{bmatrix}\begin{bmatrix} a \\ a^{\#} \end{bmatrix}$$

so that $H_1 + H_2 = H$. The corresponding parameters considered in Theorems 7.11–7.14 are as follows:

$$\begin{aligned} M &= \begin{bmatrix} -1 & 0.5\mathrm{i} \\ -0.5\mathrm{i} & -1 \end{bmatrix}, \\ N &= \begin{bmatrix} \sqrt{\kappa} & 0 \\ 0 & \sqrt{\kappa} \end{bmatrix}, \\ E &= I \\ F &= \begin{bmatrix} -\frac{\kappa}{2} + \mathrm{i} & 0.5 \\ 0.5 & -\frac{\kappa}{2} - \mathrm{i} \end{bmatrix}, \\ \Delta &= \begin{bmatrix} 1 & 0.5\mathrm{i} \\ -0.5\mathrm{i} & 1 \end{bmatrix}. \end{aligned}$$

We choose $\gamma = 1$ and use the small gain approach for this uncertain quantum system. Compared to the performance without a controller (based on Theorem 7.11), the coherent controller (based on Theorem 7.12) can guarantee that the system is stable for a larger range of the damping parameter κ and provides the system with improved performance. It is further demonstrated that the system with a controller (based on Theorem 7.14) has better performance than the case without a controller (based on Theorem 7.13) when using the Popov approach for $\gamma = 2$. Moreover, compared with the results using the small gain approach, the Popov method obtains a lower cost bound and a larger range of robust stability. More details can be found in [30].

Remark 7.1 The above control design method for uncertain quantum systems involves building a static guaranteed cost controller by adding controller Hamiltonian. In [31], the result was further extended in terms of the quadrature form instead of annihilation–creation form to more general cases where the constraint on the matrix in Lyapunov operator V (scalar multiplied by identity matrix) is eliminated. Moreover, the work in [31] also designed dynamic coherent controllers for uncertain quantum systems. The static coherent controller only focuses on adding a controller Hamiltonian with plant variables. That is, the static controller uses the same mode as the nominal system. However, the dynamic controller design method introduces another quantum system as a desired controller and the controller variables may be different from plant variables. The dynamic quantum controller is constructed by a directly coupled controller for the given system where the controller and the plant may interact by exchanging energy, and this energy exchange is often described by an interaction Hamiltonian. Although the introduction of dynamic controllers adds new complexity compared with the static controller design, it can provide improved performance as illustrated in [31].

7.5 Performance Analysis of Nonlinear Quantum Systems

In this section, we further consider a problem of robust performance analysis with a non-quadratic cost functional for the class of uncertain quantum systems of the form considered in Sect. 7.3.1. The motivation for considering this class of problems arises from the fact that the presence of nonlinearities in the quantum system allows for the possibility of a non-Gaussian system state; e.g., see [4]. Such non-Gaussian system states include important non-classical states such as the Schrödinger cat state. These non-classical quantum states are useful in areas such as quantum information and quantum communications. The presence of such non-classical states can be verified by obtaining a suitable bound on a non-quadratic cost function (such as the Wigner function, e.g., see [4, 27]). Our approach toward obtaining a bound on the non-quadratic cost function is to extend the sector bound method considered in Sect. 7.3.1 to bound both the nonlinearity and non-quadratic cost function together. It is important that these two quantities are bounded together since the non-Gaussian state only arises due to the presence of the nonlinearity in the quantum system dynamics. We aim to derive a guaranteed upper bound on the non-quadratic cost function.

We assume $H = H_1 + H_2$ with H_1 in the form of (7.13) and $H_2 = f(z, z^*)$, and L is in the form of (7.15). Here z is a known scalar operator as defined as in (7.39). $f(z, z^*)$ is defined by a formal power series of the form

$$f(z, z^*) = \sum_{k=0}^{\infty} \sum_{\ell=0}^{\infty} S_{k\ell} z^k (z^*)^\ell = \sum_{k=0}^{\infty} \sum_{\ell=0}^{\infty} S_{k\ell} H_{k\ell}, \tag{7.118}$$

which is assumed to converge in some suitable sense. Here $S_{k\ell} = S_{\ell k}^*$, $H_{k\ell} = z^k (z^*)^\ell$.

Also, we consider a non-quadratic cost defined as

$$J_c = \limsup_{T \to \infty} \frac{1}{T} \int_0^T \langle W(z(t), z(t)^*) \rangle \mathrm{d}t, \tag{7.119}$$

where $W(z, z^*)$ is a suitable non-quadratic function. Here $z(t)$ and $z(t)^*$ denote the Heisenberg evolution of the operators z and z^*, respectively. The non-quadratic function $W(z, z^*)$ is assumed to satisfy the following quadratic upper bound condition:

$$W(z, z^*) \leq \frac{1}{\gamma_0^2} z z^* + \delta_0, \tag{7.120}$$

where $\gamma_0 > 0$, $\delta_0 \geq 0$ are given constants. $W(z, z^*)$ will also be used in the definition of the set of allowable perturbation Hamiltonians $f(\cdot)$.

To define the set of allowable perturbation Hamiltonians $f(\cdot)$, we first define the following formal partial derivatives:

$$\frac{\partial f(z, z^*)}{\partial z} \triangleq \sum_{k=1}^{\infty} \sum_{\ell=0}^{\infty} k S_{k\ell} z^{k-1} (z^*)^\ell; \tag{7.121}$$

$$\frac{\partial^2 f(z, z^*)}{\partial z^2} \triangleq \sum_{k=1}^{\infty} \sum_{\ell=0}^{\infty} k(k-1) S_{k\ell} z^{k-2} (z^*)^\ell. \tag{7.122}$$

For given constants $\gamma_1 > 0$, $\gamma_2 > 0$, $\delta_1 \geq 0$, $\delta_2 \geq 0$, $\delta_3 \geq 0$, we consider the sector bound conditions

$$W(z, z^*) + \frac{\partial f(z, z^*)}{\partial z}^* \frac{\partial f(z, z^*)}{\partial z} \leq \frac{1}{\gamma_1^2} z z^* + \delta_1, \tag{7.123}$$

$$\frac{\partial f(z, z^*)}{\partial z}^* \frac{\partial f(z, z^*)}{\partial z} \leq \frac{1}{\gamma_2^2} z z^* + \delta_2 \tag{7.124}$$

and the condition

$$\frac{\partial^2 f(z, z^*)}{\partial z^2}^* \frac{\partial^2 f(z, z^*)}{\partial z^2} \leq \delta_3. \tag{7.125}$$

Then we define the set of perturbation Hamiltonians $\mathbb{W}_{11}$ as follows:

$$\mathbb{W}_{11} = \left\{ \begin{array}{l} f(\cdot) \text{ of the form (7.118) such that} \\ \text{conditions (7.123), (7.124) and (7.125) are satisfied} \end{array} \right\}. \quad (7.126)$$

Our main result, which gives an upper bound on the non-quadratic cost function (7.119), will be given in terms of the following LMI condition dependent on a parameter $\tau_1 > 0$:

$$\begin{bmatrix} F^\dagger P + PF + \kappa \Sigma \tilde{E}^T \tilde{E}^\# \Sigma & 2PJ\Sigma\tilde{E}^T \\ 2\tilde{E}^\# \Sigma J P & -\frac{I}{\tau_1^2} \end{bmatrix} < 0, \quad (7.127)$$

where $\Sigma = \begin{bmatrix} 0 & I \\ I & 0 \end{bmatrix}$, $F = -\mathrm{i}JM - \frac{1}{2}JN^\dagger JN$ and the quantity $\kappa > 0$ is defined as

$$\kappa = \left\{ \begin{array}{ll} \frac{1}{\gamma_1^2} + \left(\frac{1}{\tau_1^2} - 1\right) & for \;\; \tau_1^2 \leq 1; \\ \frac{1}{\tau_1^2\gamma_1^2} + \frac{1}{\gamma_0^2}\left(1 - \frac{1}{\tau_1^2}\right) & for \;\; \tau_1^2 > 1. \end{array} \right. \quad (7.128)$$

Theorem 7.15 *Consider an uncertain open nonlinear quantum system defined by (S, L, H) with $H = H_1 + H_2$ and a non-quadratic cost function J_c such that H_1 is of the form (7.13), L is of the form (7.15) and $H_2 \in \mathbb{W}_{11}$. Also, assume that J_c defined in (7.119) is such that (7.120) is satisfied. Furthermore, assume that there exists a constant $\tau_1 > 0$ such that the LMI (7.127) has a solution $P > 0$. Then the cost J_c satisfies the bound:*

$$J_c \leq Tr\left(PJN^\dagger \begin{bmatrix} I & 0 \\ 0 & 0 \end{bmatrix} NJ\right) + \zeta + \sqrt{\delta_3}|\mu|, \quad (7.129)$$

where

$$\zeta = \left\{ \begin{array}{ll} \delta_1 + \left(\frac{1}{\tau_1^2} - 1\right)\delta_2 & for \;\; \tau_1^2 \leq 1; \\ \frac{1}{\tau_1^2}\delta_1 + \left(1 - \frac{1}{\tau_1^2}\right)\delta_0 & for \;\; \tau_1^2 > 1 \end{array} \right. \quad (7.130)$$

and

$$\mu = -\tilde{E}\Sigma JPJ\tilde{E}^T. \quad (7.131)$$

The proof of Theorem 7.15 can be found in the Appendix of this chapter. Note that the problem of minimizing the bound on the right-hand side of (7.129) subject to the constraint (7.127) can be converted into a standard LMI optimization problem which can be solved using standard LMI software; e.g., see [2].

Example 7.4 [19] To illustrate this robust performance analysis result, we consider an illustrative example consisting of a Josephson junction in an electromagnetic resonant cavity. This system was also considered in Example 7.2 using a model derived from the model presented in [1]. Here we consider the same model as that in Example 7.2 but with simplified parameter values. That is, we consider a Hamiltonian

$H = H_1 + H_2$ with H_1 in the form of (7.13) where

$$M = \begin{bmatrix} 1 & 0 & 0 & 0 \\ 0 & 1 & -0.5 & 0 \\ 0 & -0.5 & 1 & 0 \\ 0 & 0 & 0 & 0 \end{bmatrix}$$

and

$$H_2 = f(z, z^*) = -\cos(z + z^*)$$

with $z = \frac{a_2}{\sqrt{2}}$. Hence,

$$\tilde{E} = \left[0 \; \tfrac{1}{\sqrt{2}} \; 0 \; 0 \right].$$

Also, we consider a coupling operator vector L of the form (7.15)

$$L = \begin{bmatrix} 4a_1 \\ 4a_2 \end{bmatrix}.$$

In addition, we consider a non-quadratic cost function of the form (7.119) where

$$W(z, z^*) = 4zz^* - \sin^2(z + z^*) \leq 4zz^*.$$

Hence, we can set $\gamma_0 = \frac{1}{2}$ and $\delta_0 = 0$ in (7.120).

Furthermore, we calculate

$$\begin{aligned} \frac{\partial f(z, z^*)}{\partial z} &= \sin(z + z^*) \\ \frac{\partial^2 f(z, z^*)}{\partial z^2} &= \cos(z + z^*). \end{aligned}$$

From this, it follows that

$$W(z, z^*) + \frac{\partial f(z, z^*)}{\partial z}^* \frac{\partial f(z, z^*)}{\partial z} = 4zz^*,$$

and hence, (7.123) is satisfied with $\gamma_1 = \frac{1}{2}$ and $\delta_1 = 0$. Also,

$$\frac{\partial f(z, z^*)}{\partial z}^* \frac{\partial f(z, z^*)}{\partial z} = \sin^2(z + z^*) \leq 4zz^*,$$

and hence, (7.124) is satisfied with $\gamma_2 = \frac{1}{2}$ and $\delta_2 = 0$. Moreover,

$$\frac{\partial^2 f(z, z^*)}{\partial z^2}^* \frac{\partial^2 f(z, z^*)}{\partial z^2} = \cos^2(z + z^*) \leq 1,$$

and hence (7.125) is satisfied with $\delta_3 = 1$.

We now apply Theorem 7.15 to find a bound on the cost (7.119). This is achieved by solving the corresponding LMI optimization problem. In this case, a solution to the LMI problem is found with

$$P = \begin{bmatrix} 0.012 & 0 & 0 & -0.0006 \\ 0 & 0.75 & -0.0006 & 0 \\ 0 & -0.0006 & 0.012 & 0 \\ -0.0006 & 0 & 0 & 0.75 \end{bmatrix}$$

and $\tau_1 = 0.8165$. This leads to a cost bound (7.129) of $J_c \leq 6.0965$.

7.6 Summary and Further Reading

In this chapter, we present several results on robust stability and performance analysis of uncertain quantum systems where the nominal systems are linear quantum systems. The small gain theorem and Popov approach are used to investigate robust stability, performance analysis and guaranteed cost control design. The uncertainties involve linear or nonlinear uncertainties in the Hamiltonian and coupling operators. Several examples of quantum optical systems and superconducting systems are presented to illustrate these theoretical results.

Further reading may include [22] for more details of robust stability for linear quantum systems, [20] for quantum Popov robust stability of quantum systems with non-quadratic Hamiltonian perturbations, [24] for robust stability of quantum systems with nonlinear coupling operators, [18] for robust stability of quantum systems with nonlinear dynamic uncertainties, [30] for performance analysis and guaranteed cost control for uncertain linear quantum systems, [31] for dynamic coherent controller design, and [19] for performance analysis of nonlinear quantum systems.

Appendix

Proof of Theorem 7.1

Let $V \in \mathscr{P}$ be given and consider $\mathscr{G}(V)$ defined in (7.1). Then

$$\mathscr{G}(V) = -\mathrm{i}[V, H_1] + \mathscr{L}(V) - \mathrm{i}[V, z^\dagger]w + \mathrm{i}w^\dagger[z, V] \tag{7.132}$$

using (7.3). Now since V is self-adjoint, $[V, z^\dagger]^\dagger = [z, V]$. Therefore,

$$
\begin{aligned}
0 &\le \left([V, z^\dagger] - \mathrm{i}w^\dagger\right)\left([V, z^\dagger] - \mathrm{i}w^\dagger\right)^\dagger \\
&= [V, z^\dagger][z, V] + \mathrm{i}[V, z^\dagger]w - \mathrm{i}w^\dagger[z, V] + w^\dagger w.
\end{aligned}
$$

Substituting this into (7.132), it follows that

$$
\mathscr{G}(V) \le -\mathrm{i}[V, H_1] + \mathscr{L}(V) + [V, z^\dagger][z, V] + \frac{1}{\gamma^2} z^\dagger z + \delta \tag{7.133}
$$

using (7.4). Hence, (7.6) implies (7.2) holds with $\lambda = \tilde{\lambda} + \delta$. Therefore, the result follows from Lemma 7.1.

Proof of $\mathbb{W}_2 \subset \mathbb{W}_1$

Given any $H_2 \in \mathbb{W}_2$, let

$$
w = \frac{1}{2}\begin{bmatrix} \Delta_1 & \Delta_2 \\ \Delta_2^\# & \Delta_1^\# \end{bmatrix}\begin{bmatrix} \zeta \\ \zeta^\# \end{bmatrix} = \frac{1}{2}\begin{bmatrix} \Delta_1\zeta + \Delta_2\zeta^\# \\ \Delta_2^\#\zeta + \Delta_1^\#\zeta^\# \end{bmatrix}
$$

and

$$
z = \begin{bmatrix} \zeta \\ \zeta^\# \end{bmatrix} = \begin{bmatrix} E_1 & E_2 \\ E_2^\# & E_1^\# \end{bmatrix}\begin{bmatrix} a \\ a^\# \end{bmatrix} = E\begin{bmatrix} a \\ a^\# \end{bmatrix}. \tag{7.134}
$$

Hence,

$$
H_2 = w^\dagger z = \frac{1}{2}\begin{bmatrix} a^\dagger & a^T \end{bmatrix} E^\dagger \Delta E \begin{bmatrix} a \\ a^\# \end{bmatrix}.
$$

Then, for any $V \in \mathscr{P}$,

$$
\begin{aligned}
[V, z^\dagger]w = &\frac{1}{2}(V\zeta^\dagger\Delta_1\zeta + V\zeta^\dagger\Delta_2\zeta^\# + V\zeta^T\Delta_2^\#\zeta + V\zeta^T\Delta_1^\#\zeta^\#) \\
&-\frac{1}{2}(\zeta^\dagger V\Delta_1\zeta + \zeta^\dagger V\Delta_2\zeta^\# + \zeta^T V\Delta_2^\#\zeta + \zeta^T V\Delta_1^\#\zeta^\#).
\end{aligned}
$$

Also,

$$
\begin{aligned}
w^\dagger[z, V] = &\frac{1}{2}(\zeta^\dagger\Delta_1\zeta V + \zeta^T\Delta_2^\#\zeta V + \zeta^\dagger\Delta_2\zeta^\# V + \zeta^T\Delta_1^\#\zeta^\# V) \\
&- \frac{1}{2}(\zeta^\dagger V\Delta_1\zeta + \zeta^T V\Delta_2^\#\zeta + \zeta^\dagger V\Delta_2\zeta^\# + \zeta^T V\Delta_1^\#\zeta^\#).
\end{aligned}
$$

Hence,

$$
\begin{aligned}
[V, z^\dagger]w - w^\dagger[z, V] &= \frac{1}{2}\left(V\zeta^\dagger\Delta_1\zeta + V\zeta^\dagger\Delta_2\zeta^\# + V\zeta^T\Delta_2^\#\zeta + V\zeta^T\Delta_1^\#\zeta^\#\right) \\
&- \frac{1}{2}(\zeta^\dagger\Delta_1\zeta V + \zeta^T\Delta_2^\#\zeta V + \zeta^\dagger\Delta_2\zeta^\# V + \zeta^T\Delta_1^\#\zeta^\# V) \\
&= VH_2 - H_2V = [V, H_2]
\end{aligned}
$$

and thus (7.3) is satisfied. Also, condition (7.11) implies

$$
\frac{1}{4}\left[\zeta^\dagger \ \zeta^T\right]\Delta\Delta\begin{bmatrix}\zeta \\ \zeta^\#\end{bmatrix} \leq \frac{1}{\gamma^2}\left[\zeta^\dagger \ \zeta^T\right]\begin{bmatrix}\zeta \\ \zeta^\#\end{bmatrix}
$$

which implies (7.4) for any $\delta \geq 0$. Hence, $H_2 \in \mathbb{W}_1$. Since $H_2 \in \mathbb{W}_2$ was arbitrary, we must have $\mathbb{W}_2 \subset \mathbb{W}_1$.

Proof of Theorem 7.15

In order to prove this theorem, we require the following lemmas:

Lemma 7.3 *(See Lemma 2 of [10]) Consider an open quantum system defined by* (S, L, H) *and suppose there exists a non-negative self-adjoint operator* V *on the underlying Hilbert space such that*

$$
-i[V, H] + \frac{1}{2}L^\dagger[V, L] + \frac{1}{2}[L^\dagger, V]L + W(z, z^*) \leq \lambda, \tag{7.135}
$$

where $c > 0$ *and* λ *are real numbers. Then for any system state, we have*

$$
\limsup_{T\to\infty} \frac{1}{T}\int_0^T \langle W(t)\rangle dt \leq \lambda.
$$

Lemma 7.4 *(See Lemma 2 in [18]) Given any* $V \in \mathscr{P}_1$ *defined in (7.18), then*

$$
[V, f(z, z^*)] = [V, z]w_1^* - w_1[z^*, V] + \frac{1}{2}\mu w_2^* - \frac{1}{2}w_2\mu^*, \tag{7.136}
$$

where

$$
\begin{aligned}
w_1 &= \frac{\partial f(z, z^*)^*}{\partial z}, \\
w_2 &= \frac{\partial^2 f(z, z^*)^*}{\partial z^2},
\end{aligned} \tag{7.137}
$$

and the constant μ *is defined as in (7.131).*

Lemma 7.5 *(See Lemma 4 in [15]) Given* $V \in \mathscr{P}_1$ *and* L *defined as in (7.15), then*

$$\left[V, \frac{1}{2}\left[a^\dagger \; a^T\right] M \begin{bmatrix} a \\ a^\# \end{bmatrix}\right] = \left[\left[a^\dagger \; a^T\right] P \begin{bmatrix} a \\ a^\# \end{bmatrix}, \frac{1}{2}\left[a^\dagger \; a^T\right] M \begin{bmatrix} a \\ a^\# \end{bmatrix}\right]$$
$$= \begin{bmatrix} a \\ a^\# \end{bmatrix}^\dagger [PJM - MJP] \begin{bmatrix} a \\ a^\# \end{bmatrix}.$$

Also,

$$L^\dagger[V, L] + [L^\dagger, V]L = 2Tr\left(PJN^\dagger \begin{bmatrix} I \, 0 \\ 0 \, 0 \end{bmatrix} NJ\right)$$
$$- \begin{bmatrix} a \\ a^\# \end{bmatrix}^\dagger \left(N^\dagger JNJP + PJN^\dagger JN\right) \begin{bmatrix} a \\ a^\# \end{bmatrix}.$$

Furthermore,
$$\left[\begin{bmatrix} a \\ a^\# \end{bmatrix}, \left[a^\dagger \; a^T\right] P \begin{bmatrix} a \\ a^\# \end{bmatrix}\right] = 2JP \begin{bmatrix} a \\ a^\# \end{bmatrix}.$$

Now we may prove Theorem 7.15. It follows from (7.39) that we can write

$$z^* = E_1^\# a^\# + E_2^\# a = \left[E_2^\# \; E_1^\#\right] \begin{bmatrix} a \\ a^\# \end{bmatrix} = \tilde{E}^\# \Sigma \begin{bmatrix} a \\ a^\# \end{bmatrix}.$$

Also, it follows from Lemma 7.5 that

$$[z^*, V] = 2\tilde{E}^\# \Sigma JP \begin{bmatrix} a \\ a^\# \end{bmatrix}.$$

Furthermore, $[V, z] = [z^*, V]^*$ and hence,

$$[V, z][z^*, V] = 4 \begin{bmatrix} a \\ a^\# \end{bmatrix}^\dagger PJ\Sigma\tilde{E}^T \tilde{E}^\# \Sigma JP \begin{bmatrix} a \\ a^\# \end{bmatrix}. \tag{7.138}$$

Also, we can write
$$zz^* = \begin{bmatrix} a \\ a^\# \end{bmatrix}^\dagger \Sigma\tilde{E}^T \tilde{E}^\# \Sigma \begin{bmatrix} a \\ a^\# \end{bmatrix}. \tag{7.139}$$

Hence using Lemma 7.5, we obtain

$$
\begin{aligned}
&-\mathrm{i}\left[V, \frac{1}{2}\left[a^\dagger\ a^T\right] M\left[\begin{smallmatrix} a \\ a^\# \end{smallmatrix}\right]\right] \\
&+\frac{1}{2}L^\dagger[V, L] + \frac{1}{2}[L^\dagger, V]L + \tau_1^2[V, z][z^*, V] + \kappa z z^* \\
&= \left[\begin{smallmatrix} a \\ a^\# \end{smallmatrix}\right]^\dagger \left(F^\dagger P + PF + 4\tau_1^2 PJ\Sigma\tilde{E}^T\tilde{E}^\#\Sigma JP + \kappa\Sigma\tilde{E}^T\tilde{E}^\#\Sigma\right)\left[\begin{smallmatrix} a \\ a^\# \end{smallmatrix}\right] \\
&+\mathrm{Tr}\left(PJN^\dagger\left[\begin{smallmatrix} I & 0 \\ 0 & 0 \end{smallmatrix}\right]NJ\right), \qquad (7.140)
\end{aligned}
$$

where $F = -\mathrm{i}JM - \frac{1}{2}JN^\dagger JN$.

Applying the Schur complement to the LMI (7.127) implies that the matrix inequality

$$
F^\dagger P + PF + 4\tau_1^2 PJ\Sigma\tilde{E}^T\tilde{E}^\#\Sigma JP + \kappa\Sigma\tilde{E}^T\tilde{E}^\#\Sigma < 0 \qquad (7.141)
$$

will have a solution $P > 0$ of the form (7.17). This matrix P defines a corresponding operator $V \in \mathscr{P}_1$. From this, it follows using (7.140) that

$$
\begin{aligned}
&-\mathrm{i}\left[V, \frac{1}{2}\left[a^\dagger\ a^T\right] M\left[\begin{smallmatrix} a \\ a^\# \end{smallmatrix}\right]\right] \\
&+\frac{1}{2}L^\dagger[V, L] + \frac{1}{2}[L^\dagger, V]L + \tau_1^2[V, z][z^*, V] + \kappa z z^* \le \tilde{\lambda} \qquad (7.142)
\end{aligned}
$$

with

$$
\tilde{\lambda} = \mathrm{Tr}\left(PJN^\dagger\left[\begin{smallmatrix} I & 0 \\ 0 & 0 \end{smallmatrix}\right]NJ\right) \ge 0.
$$

Also, it follows from Lemma 7.4 that

$$
\begin{aligned}
&\frac{1}{2}\left[a^\dagger\ a^T\right] M\left[\begin{smallmatrix} a \\ a^\# \end{smallmatrix}\right]] + \frac{1}{2}L^\dagger[V, L] + \frac{1}{2}[L^\dagger, V]L + W(z, z^*) \\
&= -\mathrm{i}[V, \frac{1}{2}\left[a^\dagger\ a^T\right] M\left[\begin{smallmatrix} a \\ a^\# \end{smallmatrix}\right]] + \frac{1}{2}L^\dagger[V, L] + \frac{1}{2}[L^\dagger, V]L + W(z, z^*) \\
&-\mathrm{i}[V, z]w_1^* + \mathrm{i}w_1[z^*, V] - \frac{1}{2}\mathrm{i}\mu w_2^* + \frac{1}{2}\mathrm{i}w_2\mu^*. \qquad (7.143)
\end{aligned}
$$

Furthermore, $[V, z]^* = z^*V - Vz^* = [z^*, V]$ since V is self-adjoint. Therefore, for $\tau_1 > 0$

$$
\begin{aligned}
0 &\le \left(\tau_1[V, z] - \frac{1}{\tau_1}\mathrm{i}w_1\right)\left(\tau_1[V, z] - \frac{1}{\tau_1}\mathrm{i}w_1\right)^* \\
&= \tau_1^2[V, z][z^*, V] + \mathrm{i}[V, z]w_1^* - \mathrm{i}w_1[z^*, V] + \frac{1}{\tau_1^2}w_1 w_1^*
\end{aligned}
$$

and hence

$$-\mathrm{i}[V,z]w_1^* + \mathrm{i}w_1[z^*,V] \le \tau_1^2[V,z][z^*,V] + \frac{1}{\tau_1^2}w_1w_1^*. \tag{7.144}$$

Also, for $\tau_2 > 0$

$$\begin{aligned} 0 &\le \left(\frac{\tau_2}{2}\mu - \frac{1}{\tau_2}\mathrm{i}w_2\right)\left(\frac{\tau_2}{2}\mu_i - \frac{1}{\tau_2}\mathrm{i}w_{2i}\right)^* \\ &= \frac{\tau_2^2}{4}\mu\mu^* - \frac{\mathrm{i}}{2}w_2\mu^* + \frac{\mathrm{i}}{2}\mu w_2^* + \frac{1}{\tau_2^2}w_2w_2^* \end{aligned}$$

and hence

$$\frac{\mathrm{i}}{2}w_2\mu^* - \frac{\mathrm{i}}{2}\mu w_2^* \le \frac{\tau_2^2}{4}\mu\mu^* + \frac{1}{\tau_2^2}w_2w_2^*. \tag{7.145}$$

Also, it follows from (7.125) that

$$w_2w_2^* \le \delta_3. \tag{7.146}$$

If we let $\tau_2^2 = \frac{2\sqrt{\delta_3}}{|\mu|}$, it follows from (7.145) and (7.146) that

$$\frac{\mathrm{i}}{2}w_2\mu^* - \frac{\mathrm{i}}{2}\mu w_2^* \le \frac{1}{2}\sqrt{\delta_3}|\mu| + \frac{1}{2}\sqrt{\delta_3}|\mu| = \sqrt{\delta_3}|\mu|. \tag{7.147}$$

Furthermore, it follows from (7.123) and (7.124) that

$$W(z,z^*) + w_1w_1^* \le \frac{1}{\gamma_1^2}zz^* + \delta_1 \tag{7.148}$$

and

$$w_1w_1^* \le \frac{1}{\gamma_2^2}zz^* + \delta_2. \tag{7.149}$$

Combining these equations with (7.120), it follows that

$$\begin{aligned} &W(z,z^*) + \frac{1}{\tau_1^2}w_1w_1^* \\ &\le \begin{cases} \frac{1}{\gamma_1^2}zz^* + \delta_1 + \left(\frac{1}{\tau_1^2} - 1\right)\left(\frac{1}{\gamma_2^2}zz^* + \delta_2\right) & \text{for } \tau_1^2 \le 1; \\ \frac{1}{\tau_1^2}\left(\frac{1}{\gamma_1^2}zz^* + \delta_1\right) + \left(1 - \frac{1}{\tau_1^2}\right)\left(\frac{1}{\gamma_0^2}zz^* + \delta_0\right) & \text{for } \tau_1^2 > 1. \end{cases} \end{aligned} \tag{7.150}$$

Substituting (7.144), (7.147), and (7.148) into (7.143), it follows that

$$\begin{aligned}
&-\mathrm{i}[V,H]+\frac{1}{2}L^\dagger[V,L]+\frac{1}{2}[L^\dagger,V]L+W(z,z^*)\\
&\leq -\mathrm{i}[V,\frac{1}{2}\begin{bmatrix} a^\dagger & a^T \end{bmatrix} M \begin{bmatrix} a \\ a^\# \end{bmatrix}+\frac{1}{2}L^\dagger[V,L]+\frac{1}{2}[L^\dagger,V]L\\
&+\tau_1^2[V,z][z^*,V]+W(z,z^*)+\frac{1}{\tau_1^2}w_1w_1^*+\sqrt{\delta_3}|\mu|. \qquad (7.151)
\end{aligned}$$

Hence, if $\tau_1^2 \leq 1$, it follows from (7.150) that

$$\begin{aligned}
&-\mathrm{i}[V,H]+\frac{1}{2}L^\dagger[V,L]+\frac{1}{2}[L^\dagger,V]L+W(z,z^*)\\
&\leq -\mathrm{i}[V,\frac{1}{2}\begin{bmatrix} a^\dagger & a^T \end{bmatrix} M \begin{bmatrix} a \\ a^\# \end{bmatrix}]+\frac{1}{2}L^\dagger[V,L]+\frac{1}{2}[L^\dagger,V]L+\tau_1^2[V,z][z^*,V]\\
&+\left(\frac{1}{\gamma_1^2}+\left(\frac{1}{\tau_1^2}-1\right)\right)zz^*+\delta_1+\left(\frac{1}{\tau_1^2}-1\right)\delta_2+\sqrt{\delta_3}|\mu|. \qquad (7.152)
\end{aligned}$$

Similarly, if $\tau_1^2 > 1$, it follows from (7.150) that

$$\begin{aligned}
&-\mathrm{i}[V,H]+\frac{1}{2}L^\dagger[V,L]+\frac{1}{2}[L^\dagger,V]L+W(z,z^*)\\
&\leq -\mathrm{i}[V,\frac{1}{2}\begin{bmatrix} a^\dagger & a^T \end{bmatrix} M \begin{bmatrix} a \\ a^\# \end{bmatrix}]+\frac{1}{2}L^\dagger[V,L]+\frac{1}{2}[L^\dagger,V]L+\tau_1^2[V,z][z^*,V]\\
&+\left(\frac{1}{\tau_1^2\gamma_1^2}+\frac{1}{\gamma_0^2}\left(1-\frac{1}{\tau_1^2}\right)\right)zz^*+\frac{1}{\tau_1^2}\delta_1+\left(1-\frac{1}{\tau_1^2}\right)\delta_0+\sqrt{\delta_3}|\mu|. \qquad (7.153)
\end{aligned}$$

Hence,

$$\begin{aligned}
&-\mathrm{i}[V,H]+\frac{1}{2}L^\dagger[V,L]+\frac{1}{2}[L^\dagger,V]L+W(z,z^*)\\
&\leq -\mathrm{i}[V,\frac{1}{2}\begin{bmatrix} a^\dagger & a^T \end{bmatrix} M \begin{bmatrix} a \\ a^\# \end{bmatrix}+\kappa zz^*\\
&+\frac{1}{2}L^\dagger[V,L]+\frac{1}{2}[L^\dagger,V]L+\tau_1^2[V,z][z^*,V]+\zeta+\sqrt{\delta_3}|\mu|, \qquad (7.154)
\end{aligned}$$

where $\kappa > 0$ is defined in (7.128) and $\zeta > 0$ is defined in (7.130). Then it follows from (7.142) that

$$-\mathrm{i}[V,H]+\frac{1}{2}L^\dagger[V,L]+\frac{1}{2}[L^\dagger,V]L+W(z,z^*) \leq \tilde{\lambda}+\zeta+\sqrt{\delta_3}|\mu|.$$

From this, it follows from Lemma 7.3 with $\lambda = \tilde{\lambda}+\zeta+\sqrt{\delta_3}|\mu|$ that the bound (7.129) is satisfied.

References

1. Al-Saidi WA, Stroud D (2001) Eigenstates of a small Josephson junction coupled to a resonant cavity. Phys Rev B 65:014512
2. Boyd S, El Ghaoui L, Feron E, Balakrishnan V (1994) Linear matrix inequalities in system and control theory. SIAM, Philadelphia
3. Espinosa LAD, Miao Z, Petersen IR, Ugrinovskii V, James MR (2012) Physical realizability of multi-level quantum systems. In: Proceedings of the 2012 Australian control conference, Sydney, Australia
4. Gardiner CW, Zoller P (2000) Quantum noise. Springer, Berlin
5. Gough JE, Gohm R, Yanagisawa M (2008) Linear quantum feedback networks. Phys Rev A 78:062104
6. Gough JE, James MR (2009) The series product and its application to quantum feedforward and feedback networks. IEEE Trans Autom Control 54:2530–2544
7. Gough JE, James MR, Nurdin HI (2010) Squeezing components in linear quantum feedback networks. Phys Rev A 81:023804
8. James MR, Gough JE (2010) Quantum dissipative systems and feedback control design by interconnection. IEEE Trans Autom Control 55:1806–1821
9. James MR, Nurdin HI, Petersen IR (2008) H^{∞} control of linear quantum stochastic systems. IEEE Trans Autom Control 53:1787–1803
10. James MR, Petersen IR, Ugrinovskii V (2013) A Popov stability condition for uncertain linear quantum systems. In: Proceedings of 2013 American control conference, pp 2551–2555, Washington, DC, USA, 17–19 June 2013
11. Khalil HK (2002) Nonlinear systems, 3rd edn. Prentice-Hall, Upper Saddle River
12. Parthasarathy K (1992) An introduction to quantum Stochastic calculus. Birkhauser, Berlin
13. Petersen IR (2010) Quantum linear systems theory. In: Proceedings of the 19th international symposium on mathematical theory of networks and systems, Budapest, Hungary
14. Petersen IR (2012) Quantum robust stability of a small Josephson junction in a resonant cavity. In: Proceedings of the IEEE international conference on control applications, Dubrovnik, Croatia, 3–5 Oct 2012
15. Petersen IR (2013) Robust stability analysis of an optical parametric amplifier quantum system. In: Proceedings of the 9th Asian control conference, Istanbul, Turkey, 23–26 June 2013
16. Petersen IR (2013) Quantum Popov robust stability analysis of an optical cavity containing a saturated Kerr medium. In: Proceedings of the 2013 European control conference, Zürich, Switzerland, 17–19 July 2013
17. Petersen IR (2013) Control and robustness for quantum linear systems. In: Proceedings of the 32nd Chinese control conference, Xi'an, China, 26–28 July 2013
18. Petersen IR (2013) Robust stability of quantum systems with nonlinear dynamic uncertainties. In: Proceedings of the IEEE conference on decision and control, Firenze, Italy, 10–13 Dec 2013
19. Petersen IR (2014) Guaranteed non-quadratic performance for quantum systems with nonlinear uncertainties. In: Proceedings of the 2014 American control conference, pp 3669–3673, Portland, OR, USA, 4–6 June 2014
20. Petersen IR (2017) Quantum Popov robust stability analysis of an optical cavity containing a saturated Kerr medium. Quantum Sci Technol 2:034009
21. Petersen IR, James MR (1996) Performance analysis and controller synthesis for nonlinear systems with stochastic uncertainty constraints. Automatica 32(7):959–972
22. Petersen IR, Ugrinovskii V, James MR (2012) Robust stability of uncertain linear quantum systems. Philos Trans R Soc A 370(1979):5354–5363
23. Petersen IR, Ugrinovskii V, James MR (2012) Robust stability of uncertain quantum systems. In: Proceedings of the 2012 American control conference, Montreal, Canada

24. Petersen IR, Ugrinovskii V, James MR (2012) Robust stability of quantum systems with a nonlinear coupling operator. In: Proceedings of the 51st IEEE conference on decision and control, pp 1078–1082, Maui, HI, USA, 10–13 Dec 2012
25. Petersen IR, Ugrinovskii VA, Savkin AV (2000) Robust control design using H^∞ methods. Springer, London
26. Shaiju AJ, Petersen IR (2012) A frequency domain condition for the physical realizability of linear quantum systems. IEEE Trans Autom Control 57(8):2033–2044
27. Walls DF, Milburn GJ (2008) Quantum optics, 2nd edn. Springer, Berlin
28. Xiang C, Petersen IR, Dong D (2014) Performance analysis and coherent guaranteed cost control for uncertain quantum systems. In: Proceedings of the 2014 European control conference, Strasbourg, France
29. Xiang C, Petersen IR, Dong D (2014) A Popov approach to performance analysis and coherent guaranteed cost control for uncertain quantum systems. In: Proceedings of the 2014 Australian control conference, Canberra, Australia
30. Xiang C, Petersen IR, Dong D (2017) Performance analysis and coherent guaranteed cost control for uncertain quantum systems using small gain and Popov methods. IEEE Transactions on Automatic Control 62:1524–1529
31. Xiang C, Petersen IR, Dong D (2020) Static and dynamic coherent robust control for a class of uncertain quantum systems. Syst Control Lett 141:104702
32. Zhang G, James MR (2011) Direct and indirect couplings in coherent feedback control of linear quantum systems. IEEE Trans Autom Control 57:1535–1550
33. Zhou K, Doyle J, Glover K (1996) Robust and optimal control. Prentice-Hall, Upper Saddle River

Chapter 8
H^∞ Control and Fault-Tolerant Control of Quantum Systems

Abstract This chapter presents several results on H^∞ control and fault-tolerant control of quantum systems. We present some results on H^∞ control for a class of linear quantum systems in Sect. 8.2 and further consider robust H^∞ control for uncertain linear quantum systems in Sect. 8.3[a]. In Sect. 8.4, a time-varying fault-tolerant coherent H^∞ control problem is solved for a class of linear quantum systems subject to Markovian jump faults[b]. In Sect. 8.5, an estimator-based approach is presented for fault-tolerant control of measurement-based quantum feedback systems[c].

8.1 Introduction

H^∞ control is a well-known robust control method that has been widely used in classical control systems [9, 15]. H^∞ control has also been extended to the quantum domain for robust control analysis and design of quantum systems [3]. In this chapter, we present several results on H^∞ control and robust H^∞ control for two classes of linear quantum systems which are mainly based on the results in [3, 13].

For practical quantum systems, the stochastic fluctuations in magnetic or electric fields or fault operations on the generators of quantum resources may introduce fault signals that will deteriorate the performance of quantum systems or result in instability. For example, classical (non-quantum) fault signals may originate from a change

---

[a] Section 8.3 contains materials reprinted, with permission, from Automatica [13] © 2017 Elsevier.
[b] Section 8.4 contains materials reprinted, with permission, from IEEE Transactions on Automatic Control [4] © 2021 IEEE.
[c] Section 8.5 contains materials reprinted, with permission, from IEEE Transactions on Automatic Control [10] © 2017 IEEE.

D. Dong and I. R. Petersen, *Learning and Robust Control in Quantum Technology*, Communications and Control Engineering,
https://doi.org/10.1007/978-3-031-20245-2_8

in environmental conditions (e.g., temperature) or voltage fluctuations in controlling a laser generator. The laser output can be seen as a fault signal, which results in fluctuations of the pumping field in an Optical Parametric Oscillator (OPO). It is thus expected to develop fault-tolerant control theory for quantum systems. Most classical fault-tolerant control methods cannot be straightforwardly applied to quantum control problems due to some unique characteristics of quantum systems such as measurement collapse and non-commutative observables. In this chapter, we also present several results on quantum fault-tolerant control which are mainly based on the results in [4, 10]. We solve the time-varying H^∞ coherent feedback control problem for a linear quantum system suffering from a fault signal. The strictly bounded real lemma for time-varying quantum systems is first presented, by which the H^∞ control problem is formulated in terms of several Riccati differential equations and a group of LMIs [4]. The fault under consideration is modeled as a Markov chain on a probability space [2]. A purely quantum H^∞ controller consisting of basic optical components is designed to ensure that the system has desired robust performance even when suffering from faults. We further discuss fault-tolerant control of measurement-based quantum feedback systems [10]. In particular, we present a systematic fault-tolerant control method for a class of linear quantum stochastic systems subject to classical faults and propose an estimator-based approach to fault-tolerant control design for this class of systems.

8.2 H^∞ Control of Linear Quantum Systems

In this section, we present several results on H^∞ control for linear quantum systems, which is mainly based on the reference [3]. Consider the quantum system to be controlled which is described by the following QSDEs defined in an analogous way to (2.18):

$$\begin{aligned} dx(t) &= Ax(t)dt + [B_0 \;\; B_1 \;\; B_2] \times \begin{bmatrix} dv(t) \\ dw(t) \\ du(t) \end{bmatrix}; \\ dy(t) &= C_1x(t)dt + [D_{20} \;\; D_{21} \;\; 0_{n_y \times n_u}] \times \begin{bmatrix} dv(t) \\ dw(t) \\ du(t) \end{bmatrix}; \\ dz(t) &= C_2x(t)dt + D_{12}du(t), \end{aligned} \tag{8.1}$$

where $x(t)$ represents a vector of plant variables with initial condition $x(0) = x_0$, the input $w(t)$ represents a disturbance signal. Also, $dv(t)$ represents any additional quantum noise in the system. The control input $u(t)$ and the input $w(t)$ can be written as

$$du(t) = \beta_u(t)dt + d\tilde{u}(t), \tag{8.2}$$

$$dw(t) = \beta_w(t)dt + d\tilde{w}(t), \tag{8.3}$$

where $\beta_u(t)$ is the adapted, self-adjoint part of $u(t)$ and $\tilde{u}(t)$ is the noise part of $u(t)$. The vectors $v(t)$, $\tilde{w}(t)$ and $\tilde{u}(t)$ are quantum noises with positive semidefinite Hermitian Itô matrices F_v, $F_{\tilde{w}}$ and $F_{\tilde{u}}$ (for detailed definitions, see, e.g., Chap. 2 or [3]).

The controller is assumed to be of the following form:

$$\begin{aligned}
\mathrm{d}x(t) &= A_K\xi(t)\mathrm{d}t + [B_{K1}\ \ B_K] \times \begin{bmatrix} \mathrm{d}v_K(t) \\ \mathrm{d}y(t) \end{bmatrix}; \\
\mathrm{d}u(t) &= C_K\xi(t)\mathrm{d}t + [B_{K0}\ \ 0_{n_u\times n_y}] \times \begin{bmatrix} \mathrm{d}v_K(t) \\ \mathrm{d}y(t) \end{bmatrix},
\end{aligned} \tag{8.4}$$

where

$$\xi(t) = \begin{bmatrix} \xi_1(t) \\ \cdots \\ \xi_n(t) \end{bmatrix}$$

is a vector of self-adjoint controller variables with initial condition $\xi(0) = \xi_0$. The noise

$$v_K(t) = \begin{bmatrix} v_{K1}(t) \\ \cdots \\ v_{KK_v}(t) \end{bmatrix}$$

is a vector of non-commutative Wiener processes with canonical Hermitian Itô matrix F_{v_K}.

By identifying $\beta_u = C_K\xi(t)$ and interconnecting (8.1) and (8.4), we can obtain the resulting closed-loop system (see [3])

$$\begin{aligned}
\mathrm{d}\eta(t) &= \begin{bmatrix} A & B_2C_K \\ B_KC_1 & A_K \end{bmatrix}\eta(t)\mathrm{d}t + \begin{bmatrix} B_0 & B_2B_{K0} \\ B_KD_{20} & B_{K1} \end{bmatrix}\begin{bmatrix} \mathrm{d}v(t) \\ \mathrm{d}v_K(t) \end{bmatrix} + \begin{bmatrix} B_1 \\ B_KD_{21} \end{bmatrix}\mathrm{d}w(t); \\
\mathrm{d}z(t) &= \begin{bmatrix} C_2 & D_{12}C_K \end{bmatrix}\eta(t)\mathrm{d}t + \begin{bmatrix} 0 & D_{12}B_{K0} \end{bmatrix}\begin{bmatrix} \mathrm{d}v(t) \\ \mathrm{d}v_K(t) \end{bmatrix},
\end{aligned} \tag{8.5}$$

where $\eta(t) = \begin{bmatrix} x(t) \\ \xi(t) \end{bmatrix}$. The goal of the H^∞ controller synthesis problem is to find a controller (8.4) for a given disturbance attenuation parameter $g > 0$ such that the resulting closed-loop system satisfies (for details, see [3])

$$\int_0^t \langle \beta_z(s)^T\beta_z(s) + \varepsilon\eta(s)^T\eta(s)\rangle \mathrm{d}s \le (g^2 - \varepsilon)\int_0^t \langle \beta_\omega(s)^T\beta_\omega(s)\rangle \mathrm{d}s + \mu_1 + \mu_2 t, \forall t > 0 \tag{8.6}$$

for some real constants $\varepsilon, \mu_1, \mu_2 > 0$, where

$$\beta_z(t) = [C_2\ \ D_{12}C_K]\begin{bmatrix} x(t) \\ \xi(t) \end{bmatrix}.$$

Denote $E_1 = D_{12}^T D_{12}$ and $E_2 = D_{21}D_{21}^T$. The results on quantum H^∞ control can be stated in terms of the following pair of algebraic Riccati equations:

$$(A - B_2E_1^{-1}D_{12}^TC_2)^TX + X(A - B_2E_1^{-1}D_{12}^TC_2) + X(B_1B_1^T - g^2B_2E_1^{-1}B_2^T)X \\ +g^{-2}C_2^T(I - D_{12}E_1^{-1}D_{12}^T)C_2 = 0, \tag{8.7}$$

$$(A - B_1D_{21}^TE_2^{-1}C_1)Y + Y(A - B_1D_{21}^TE_2^{-1}C_1)^T + Y(g^{-2}C_2^TC_2 - C_1^TE_2^{-1}C_1)Y \\ +B_1(I - D_{21}^TE_2^{-1}D_{21})B_1^T = 0, \tag{8.8}$$

where X and Y are positive definite symmetric matrices. The solutions to these Riccati equations will be required to satisfy the following assumption:

Assumption 8.1 (1) $A - B_2E_1^{-1}D_{12}^TC_2 + (B_1B_1^T - g^2B_2E_1^{-1}B_2^T)X$ is a stability matrix.
(2) $A - B_1D_{21}^TE_2^{-1}C_1 + Y(g^{-2}C_2^TC_2 - C_1^TE_2^{-1}C_1)$ is a stability matrix.
(3) The matrix XY has a spectral radius strictly less than one.

If the Riccati equations (8.7) and (8.8) have solutions satisfying Assumption 8.1, a controller of the form (8.4) will solve the H^∞ control problem under consideration and the system matrices of the controller can be constructed from the Riccati solutions as follows:

$$\begin{aligned} A_K &= A + B_2C_K - B_KC_1 + (B_1 - B_KD_{21})B_1^TX \\ B_K &= (I - YX)^{-1}(YC_1^T + B_1D_{21}^T)E_2^{-1} \\ C_K &= -E_1^{-1}(g^2B_2^TX + D_{12}^TC_2). \end{aligned} \tag{8.9}$$

The necessary and sufficient conditions on H^∞ controller synthesis can be described as follows:

Theorem 8.1 *[3] Necessity. Consider the system (8.1) and assume*

(1) $D_{12}^TD_{12} = E_1 > 0$.
(2) $D_{21}D_{21}^T = E_2 > 0$.
(3) The matrix $\begin{bmatrix} A - i\omega I & B_2 \\ C_1 & D_{12} \end{bmatrix}$ *is full rank for all* $\omega \geq 0$.
(4) The matrix $\begin{bmatrix} A - i\omega I & B_1 \\ C_2 & D_{21} \end{bmatrix}$ *is full rank for all* $\omega \geq 0$.

If there exists a controller of the form (8.4) such that the resulting closed-loop system (8.5) is strictly bounded real with disturbance attenuation g, then the Riccati equations (8.7) and (8.8) will have stabilizing solutions $X \geq 0$ *and* $Y \geq 0$ *satisfying Assumption 8.1.*

Sufficiency. Suppose the Riccati equations (8.7) and (8.8) have stabilizing solutions $X \geq 0$ *and* $Y \geq 0$ *satisfying Assumption 8.1. If the controller (8.4) is such that the matrices* A_K, B_K, C_K *are as defined in (8.9), then the resulting closed-loop system (8.5) will be strictly bounded real with disturbance attenuation g.*

The proof of Theorem 8.1 can be found in [3]. Several examples of quantum optical control systems were also presented there. It is worth noting that when the controller which is constructed via this approach is a quantum controller, we need to take into account the physical realizability conditions (e.g., Theorem 2.1) as discussed in Chap. 2.

8.3 Robust H^∞ Control for Quantum Systems

In this section, we consider a class of uncertain quantum systems and design a coherent robust H^∞ controller for this class of systems. We first present the closed-loop system consisting of the plant and the controller and introduce the H^∞ control objective. Then we apply the strict bounded real inequality to build a connection between the uncertain quantum system and the scaled H^∞ problem. Lastly we provide a solution to design a robust coherent H^∞ controller.

8.3.1 H^∞ Control of Uncertain Linear Quantum Systems

We consider a quantum plant subject to perturbations in both the Hamiltonian and the coupling operator and describe it by a non-commutative stochastic model in the following form [13]:

$$\begin{aligned}
\mathrm{d}x(t) &= (A+\Delta A)x(t)\mathrm{d}t + [B_0+\Delta B_0 \quad B_1+\Delta B_1 \quad B_2+\Delta B_2]\begin{bmatrix}\mathrm{d}v(t)\\ \mathrm{d}w(t)\\ \mathrm{d}u(t)\end{bmatrix};\\
\mathrm{d}z(t) &= (C_1+\Delta C_1)x(t)\mathrm{d}t + D_{12}\mathrm{d}u(t);\\
\mathrm{d}y(t) &= (C_2+\Delta C_2)x(t)\mathrm{d}t + [D_{20} \quad D_{21} \quad 0]\begin{bmatrix}\mathrm{d}v(t)\\ \mathrm{d}w(t)\\ \mathrm{d}u(t)\end{bmatrix},
\end{aligned} \tag{8.10}$$

where $x(0)=x_0, A \in \mathbb{R}^{n\times n}, B_0 \in \mathbb{R}^{n\times n_v}, B_1 \in \mathbb{R}^{n\times n_w}, B_2 \in \mathbb{R}^{n\times n_u}, C_1 \in \mathbb{R}^{n_z\times n}, C_2 \in \mathbb{R}^{n_y\times n}$, $D_{12} \in \mathbb{R}^{n_z\times n_u}$, $D_{20} \in \mathbb{R}^{n_y\times n_v}$, $D_{21} \in \mathbb{R}^{n_y\times n_w}$. Here, $w(t)$ represents a disturbance signal and $v(t)$ represents any additional quantum noise. The signal $u(t)$ is a control input of the form

$$\mathrm{d}u(t) = \beta_u(t)\mathrm{d}t + \mathrm{d}\tilde{u}(t),$$

where $\tilde{u}(t)$ is the noise part of $u(t)$ and $\beta_u(t)$ is an adapted process. The vector $x(t)$ is required to satisfy the canonical commutation relations

$$[x(t), x(t)^T] = x(t)x(t)^T - (x(t)x(t)^T)^T = 2\mathrm{i}\Theta, \tag{8.11}$$

where Θ is a real skew-symmetric matrix [13]. The quantity $z(t)$ describes the control output and the quantity $y(t)$ describes the measured output.

The uncertain perturbation in the system Hamiltonian H_2 [13] is in the form of

$$H_2 = \frac{1}{2}x^T E^T \hat{\Delta} E x, \tag{8.12}$$

where $E \in \mathbb{R}^{m\times n}$ and $\hat{\Delta} \in \mathbb{R}^{m\times m}$ is an uncertain norm bounded real matrix satisfying

$$\hat{\Delta}^T = \hat{\Delta},\ \hat{\Delta}^2 \leq I. \tag{8.13}$$

As for the coupling matrix L, we divide it as the control input part L_u, the noise part L_v and the disturbance part L_w. Uncertainties existing in the control input part, the noise part and the disturbance part of the coupling operator are described by Ω_u, Ω_v and Ω_w, respectively.

$$L = \begin{bmatrix} L_u \\ L_v \\ L_w \end{bmatrix} = \begin{bmatrix} \Lambda_u + \Omega_u \\ \Lambda_v + \Omega_v \\ \Lambda_w + \Omega_w \end{bmatrix} x = (\Lambda + \Omega)x,$$

where $\Lambda \in \mathbb{C}^{N_{uvw}\times n}$ with $N_{uvw} = \frac{n_u+n_v+n_w}{2}$. ΔA may be represented as [13]:

$$\begin{aligned} \Delta A &= \Delta A_1 + \Delta A_2; \\ \Delta A_1 &= 2\Theta E^T \hat{\Delta} E; \\ \Delta A_2 &= 2\Theta\Im[\Lambda^\dagger \Omega + \Omega^\dagger \Lambda + \Omega^\dagger \Omega]. \end{aligned} \tag{8.14}$$

Also, the uncertainties satisfy the following bound condition:

$$\Re(\Omega)^T \Re(\Omega) + \Im(\Omega)^T \Im(\Omega) \leq r I_{n\times n}. \tag{8.15}$$

The condition (8.15) also implies that the following bound conditions hold:

$$\begin{aligned} \Re(\Omega_u)^T \Re(\Omega_u) + \Im(\Omega_u)^T \Im(\Omega_u) &\leq r I_{n\times n}; \\ \Re(\Omega_v)^T \Re(\Omega_v) + \Im(\Omega_v)^T \Im(\Omega_v) &\leq r I_{n\times n}; \\ \Re(\Omega_w)^T \Re(\Omega_w) + \Im(\Omega_w)^T \Im(\Omega_w) &\leq r I_{n\times n}. \end{aligned} \tag{8.16}$$

We assume that the coherent controller is described by the following non-commutative stochastic system:

$$\begin{aligned} \mathrm{d}\xi(t) &= A_K \xi(t)\mathrm{d}t + [B_{K1} \quad B_K] \begin{bmatrix} \mathrm{d}v_K(t) \\ \mathrm{d}y(t) \end{bmatrix} \\ \mathrm{d}u(t) &= C_K \xi(t)\mathrm{d}t + [B_{K0} \quad 0] \begin{bmatrix} \mathrm{d}v_K(t) \\ \mathrm{d}y(t) \end{bmatrix}, \end{aligned} \tag{8.17}$$

where $\xi(0) = \xi_0$, $A_K \in \mathbb{R}^{n_K \times n_K}$, $B_{K0} \in \mathbb{R}^{n_u \times n_{v_K}}$, $B_{K1} \in \mathbb{R}^{n_K \times n_{v_K}}$, $B_K \in \mathbb{R}^{n_K \times n_y}$, $C_K \in \mathbb{R}^{n_u \times n_K}$ and

$$\xi(t) = [\xi_1(t) \quad \dots \quad \xi_{n_K}(t)]^T$$

is a vector of self-adjoint controller variables.

Assume that $x(0)$ commutes with $\xi(0)$ at time $t = 0$. By interconnecting (8.10) and (8.17) and identifying $\beta_u(t) = C_K \xi(t)$, the closed-loop system is of the form

$$\begin{aligned} d\eta(t) &= \begin{bmatrix} A + \Delta A & (B_2 + \Delta B_2) C_K \\ B_K (C_2 + \Delta C_2) & A_K \end{bmatrix} \eta(t) dt \\ &\quad + \begin{bmatrix} (B_0 + \Delta B_0) & (B_2 + \Delta B_2) B_{K0} \\ B_K D_{20} & B_{K1} \end{bmatrix} \begin{bmatrix} dv(t) \\ dv_K(t) \end{bmatrix} + \begin{bmatrix} B_1 + \Delta B_1 \\ B_K D_{21} \end{bmatrix} dw(t); \\ dz(t) &= \begin{bmatrix} (C_1 + \Delta C_1) & D_{12} C_K \end{bmatrix} \eta(t) dt + \begin{bmatrix} 0_{n_z \times n_v} & D_{12} B_{K0} \end{bmatrix} \begin{bmatrix} dv(t) \\ dv_K(t) \end{bmatrix}, \end{aligned} \tag{8.18}$$

where we denote $\eta(t) = \begin{bmatrix} x(t) \\ \xi(t) \end{bmatrix}$. This state equation can also be rewritten as

$$\begin{aligned} d\eta(t) &= (\tilde{A} + \Delta\tilde{A})\eta(t)dt + (\tilde{B} + \Delta\tilde{B})dw(t) + (\tilde{G} + \Delta\tilde{G})d\zeta(t); \\ dz(t) &= (\tilde{C} + \Delta\tilde{C})\eta(t)dt + \tilde{H} d\zeta(t), \end{aligned} \tag{8.19}$$

where

$$\begin{aligned} &\zeta(t) = \begin{bmatrix} v(t) \\ v_K(t) \end{bmatrix}; \tilde{A} = \begin{bmatrix} A & B_2 C_K \\ B_K C_2 & A_K \end{bmatrix}; \tilde{B} = \begin{bmatrix} B_1 \\ B_K D_{21} \end{bmatrix}; \tilde{G} = \begin{bmatrix} B_0 & B_2 B_{K0} \\ B_K D_{20} & B_{K1} \end{bmatrix}; \\ &\tilde{C} = \begin{bmatrix} C_1 & D_{12} C_K \end{bmatrix}; \tilde{H} = \begin{bmatrix} 0 & D_{12} B_{K0} \end{bmatrix}; \\ &\Delta\tilde{A} = \begin{bmatrix} \Delta A & \Delta B_2 C_K \\ B_K \Delta C_2 & 0_{n_K \times n_K} \end{bmatrix} \text{ with } \Delta A \text{ as in (8.14)}; \Delta B_2 = 2i\Theta[-\Omega_u^\dagger \quad \Omega_u^T]\Gamma_u; \\ &\Gamma_u = P_{N_u} \text{diag}_{N_u}(M); N_u = n_u/2; \Delta C_2 = 2P_{N_y}^T \Xi_{N_y N_{vw}} \begin{bmatrix} \Re(\Omega_{vw}) \\ \Im(\Omega_{vw}) \end{bmatrix}; \\ &\Omega_{vw} = \begin{bmatrix} \Omega_v \\ \Omega_w \end{bmatrix}; N_{vw} = \frac{n_v + n_w}{2}; \Delta\tilde{B} = E_1 \Delta B_1; \Delta B_1 = 2i\Theta[-\Omega_w^\dagger \quad \Omega_w^T]\Gamma_w; \\ &\Delta\tilde{G} = \begin{bmatrix} \Delta B_0 & \Delta B_2 B_{K0} \\ 0_{n_K \times n_v} & 0_{n_K \times n_{v_K}} \end{bmatrix}; \Delta B_0 = 2i\Theta[-\Omega_v^\dagger \quad \Omega_v^T]\Gamma_v; \Delta\tilde{C} = \Delta C_1 E_2; \\ &\Delta C_1 = 2P_{N_z}^T \Xi_{N_z N_u} \begin{bmatrix} \Re(\Omega_u) \\ \Im(\Omega_u) \end{bmatrix}; E_1 = \begin{bmatrix} I_{n \times n} \\ 0_{n_K \times n} \end{bmatrix}; E_2 = \begin{bmatrix} I_{n \times n} & 0_{n \times n_K} \end{bmatrix}. \end{aligned} \tag{8.20}$$

For a given disturbance attenuation parameter $g > 0$, the H^∞ control objective is to find a coherent quantum controller of the form (8.17) for the uncertain quantum system (8.10) such that the closed-loop system (8.19) satisfies

$$\int_0^t \langle \beta_z(s)^T \beta_z(s) + \tilde{\varepsilon}\eta(s)^T \eta(s) \rangle ds \leq (g^2 - \tilde{\varepsilon}) \int_0^T \langle \beta_w(s)^T \beta_w(s) \rangle ds + \mu_1 + \mu_2 t, \forall t > 0 \tag{8.21}$$

for some real constants $\tilde{\varepsilon}, \mu_1, \mu_2 > 0$. Here, $\beta_z(t)$ is controlled output operator $\beta_z(t) = \tilde{C}\eta(t)$. It is worth noting that if the closed-loop system (8.19) is robustly strict bounded real with disturbance attenuation g, it then satisfies the H^∞ control synthesis objective (8.21).

Now, we develop a relationship between robust H^∞ control for the uncertain quantum system via a coherent controller and H^∞ control for an auxiliary scaled system without parameter uncertainties via the same coherent controller.

The uncertain quantum system is described by (8.10). We call the auxiliary system to be considered a scaled H^∞ control system. In order to solve the robust H^∞ control problem, we introduce the corresponding scaled H^∞ control system for (8.10) as follows:

$$\begin{aligned} dx(t) &= Ax(t)dt + [B_0 \;\; [J_1 \;\; g^{-1}J_2] \;\; B_2] \begin{bmatrix} dv(t) \\ d\overline{w}(t) \\ du(t) \end{bmatrix}; \\ d\overline{z}(t) &= J_3 x(t)dt + \tilde{D}_{12} du(t); \\ dy(t) &= C_2 x(t)dt + [D_{20} \;\; \tilde{D}_{21} \;\; 0] \begin{bmatrix} dv(t) \\ d\overline{w}(t) \\ du(t) \end{bmatrix}, \end{aligned} \tag{8.22}$$

where

$$\begin{aligned} \tilde{D}_{12} &= \left[0^T_{(m+2N_{uvw}+2n)\times n_u} \;\; \sqrt{\tfrac{4}{\varepsilon_5}}rI \;\; \sqrt{1+4\varepsilon_8}D_{12}^T \right]^T; \\ \tilde{D}_{21} &= \left[0_{n_y\times(m+2N_{uvw}+2n)} \;\; 2\sqrt{\varepsilon_6}I \;\; 0_{n_y\times n} \;\; g^{-1}\sqrt{(1+\varepsilon_7)}D_{21} \right]; \\ J_1 &= [2\sqrt{\varepsilon_1}\Theta E^T \;\; 2\sqrt{\varepsilon_2}\Theta\Re(\Lambda)^T \;\; 2\sqrt{\varepsilon_2}\Theta\Im(\Lambda)^T \\ &\quad 2\sqrt{(\varepsilon_3+\varepsilon_4)rI} \;\; \sqrt{\varepsilon_5}I \;\; 0_{n\times n_y}]; \\ J_2 &= \left[\sqrt{(\tfrac{4}{\varepsilon_7}+4)rI} \;\; \sqrt{(1+\varepsilon_7)}B_1 \right]; \\ J_3 &= \begin{bmatrix} \sqrt{\frac{1}{\varepsilon_1}}E \\ \sqrt{(\frac{1}{\varepsilon_2}+\frac{1}{\varepsilon_4})rI} \\ \sqrt{\frac{1}{\varepsilon_3}}\Re(\Lambda) \\ \sqrt{\frac{1}{\varepsilon_3}}\Im(\Lambda) \\ \sqrt{(4+\frac{1}{\varepsilon_6}+\frac{1}{\varepsilon_8})rI} \\ 0_{n_u\times n} \\ \sqrt{1+4\varepsilon_8}C_1 \end{bmatrix}. \end{aligned} \tag{8.23}$$

Here, $A, B_0, B_1, B_2, C_1, C_2, D_{20}, D_{21}$ and D_{22} are the same as in (8.10), $\varepsilon_i > 0$ ($i = 1, \ldots, 8$) are scaling parameters to be chosen and $g > 0$ is the disturbance attenuation that is required to achieve.

The closed-loop system connected by the system (8.22) and the controller (8.17) can be written as

$$\begin{aligned} d\eta(t) &= \tilde{A}\eta(t)dt + \left[\tilde{J}_1 \; g^{-1}\tilde{J}_2\right] d\overline{w}(t) + \tilde{G}d\zeta(t); \\ d\overline{z}(t) &= \tilde{J}_3\eta(t)dt + \tilde{J}_4 d\zeta(t), \end{aligned} \tag{8.24}$$

where

$$\begin{aligned} \tilde{J}_1 &= \begin{bmatrix} 2\sqrt{\varepsilon_1}E\Theta^T & 0_{m\times n_K} \\ 2\sqrt{\varepsilon_2}\Re(\Lambda)\Theta^T & 0_{N_{uvw}\times n_K} \\ 2\sqrt{\varepsilon_2}\Im(\Lambda)\Theta^T & 0_{N_{uvw}\times n_K} \\ 2\sqrt{(\varepsilon_3+\varepsilon_4)r}I_{n\times n} & 0_{n\times n_K} \\ \sqrt{\varepsilon_5}I_n & 0_{n\times n_K} \\ 0_{n_y\times n} & 2\sqrt{\varepsilon_6}B_K^T \end{bmatrix}^T; \\ \tilde{J}_2 &= \begin{bmatrix} \sqrt{(\frac{4}{\varepsilon_7}+4)r}I_n & 0_{n\times n_K} \\ \sqrt{(1+\varepsilon_7)}B_1^T & \sqrt{(1+\varepsilon_7)}(B_K D_{21})^T \end{bmatrix}^T; \\ \tilde{J}_3 &= \begin{bmatrix} \sqrt{\frac{1}{\varepsilon_1}}E & 0_{m\times n_K} \\ \sqrt{(\frac{1}{\varepsilon_2}+\frac{1}{\varepsilon_4})r}I & 0_{n\times n_K} \\ \sqrt{\frac{1}{\varepsilon_3}}\Re(\Lambda) & 0_{N_{uvw}\times n_K} \\ \sqrt{\frac{1}{\varepsilon_3}}\Im(\Lambda) & 0_{N_{uvw}\times n_K} \\ \sqrt{(4+\frac{1}{\varepsilon_6}+\frac{1}{\varepsilon_8})r}I & 0_{n\times n_K} \\ 0_{n_u\times n} & \sqrt{\frac{4}{\varepsilon_5}}rC_K \\ \sqrt{1+4\varepsilon_8}C_1 & \sqrt{1+4\varepsilon_8}D_{12}C_K \end{bmatrix}; \\ \tilde{J}_4 &= \begin{bmatrix} 0_{(m+2N_{uvw}+2n)\times n_v} & 0_{(m+2N_{uvw}+2n)\times n_{v_K}} \\ 0_{n_u\times n_v} & \sqrt{\frac{4}{\varepsilon_5}}rB_{K0} \\ 0_{n_z\times n_v} & \sqrt{1+4\varepsilon_8}D_{12}B_{K0} \end{bmatrix}. \end{aligned} \tag{8.25}$$

Here, we develop a connection between the robust H^∞ control problem for the uncertain quantum system (8.10) and an H^∞ control problem for the scaled system (8.22) in the following theorem:

Theorem 8.2 *Let $g > 0$ be a prescribed level of disturbance attenuation and consider a linear dynamic controller of the form (8.17). Then the system (8.10) is robustly strict bounded real with disturbance attenuation $g > 0$ via the coherent controller (8.17) if there exist $\varepsilon_i > 0$ ($i = 1, \ldots, 8$) such that the system (8.22) is strictly bounded real with unitary disturbance attenuation via the same coherent controller (8.17).*

The proof can be found in [13]. In order to solve the H^∞ controller design problem for the uncertain quantum system (8.10), we can then solve the H^∞ design problem for the scaled system (8.22) via existing H^∞ control techniques.

8.3.2 A Solution to the H^∞ Controller Design Problem

Based on the connection between the uncertain quantum system (8.10) and the scaled H^∞ system (8.22), we may construct a controller for the scaled system (8.22) to achieve H^∞ performance with unitary attenuation instead of designing a controller for uncertain system (8.10) to achieve the H^∞ objective for a given disturbance attenuation g. To present the results on quantum H^∞ control for the auxiliary system (8.22), we use a similar method to the non-singular H^∞ control approach in classical linear systems, e.g., [8]. We know that the following terms are positive definite:

$$\tilde{D}_{12}^T\tilde{D}_{12} = \frac{4}{\varepsilon_5}rI + (1+4\varepsilon_8)D_{12}^TD_{12} = \tilde{E}_1 > 0;$$
$$\tilde{D}_{21}\tilde{D}_{21}^T = 4\varepsilon_6 I + g^{-2}(1+\varepsilon_7)D_{21}D_{21}^T = \tilde{E}_2 > 0.$$

Then, the system (8.22) is required to satisfy the following assumption:

Assumption 8.2 (1) The matrix $\begin{bmatrix} A - iwI & B_2 \\ J_3 & \tilde{D}_{12} \end{bmatrix}$ is full column rank for all $w \geq 0$.

(2) The matrix $\begin{bmatrix} A - iwI & J_{12} \\ C_2 & \tilde{D}_{21} \end{bmatrix}$ is full row rank for all $w \geq 0$, where $J_{12} = [J_1 \quad g^{-1}J_2]$.

The solution to the H^∞ control problem for the scaled system (8.22) is given in terms of the following pair of algebraic Riccati equations:

$$\begin{aligned}&(A - (1+4\varepsilon_8)B_2\tilde{E}_1^{-1}D_{12}^TC_1)^TX + X(A - (1+4\varepsilon_8)B_2\tilde{E}_1^{-1}D_{12}^TC_1)\\ &+ X(J_{12}J_{12}^T - B_2\tilde{E}_1^{-1}B_2^T)X + J_3^TJ_3 - (1+4\varepsilon_8)^2C_1^TD_{12}\tilde{E}_1^{-1}D_{12}^TC_1 = 0;\end{aligned} \tag{8.26}$$

$$\begin{aligned}&(A - g^{-2}(1+\varepsilon_7)B_1D_{21}^T\tilde{E}_2^{-1}C_2)Y + Y(A - g^{-2}(1+\varepsilon_7)B_1D_{21}^T\tilde{E}_2^{-1}C_2)^T\\ &+ Y(J_3^TJ_3 - C_2^T\tilde{E}_2^{-1}C_2)Y + J_{12}J_{12}^T - g^{-4}(1+\varepsilon_7)^2B_1D_{21}^T\tilde{E}_2^{-1}D_{21}B_1^T = 0,\end{aligned} \tag{8.27}$$

where X and Y are positive-definite symmetric matrices and

$$\begin{aligned}J_{12}J_{12}^T =\ & 4\varepsilon_1\Theta E^TE\Theta^T + 4\varepsilon_2\Theta(\Re(\Lambda)^T\Re(\Lambda) + \Im(\Lambda)^T\Im(\Lambda))\Theta^T\\ &+ (4\varepsilon_3 r + 4\varepsilon_4 r + \varepsilon_5 + g^{-2}(\frac{4}{\varepsilon_7} + 4)r)I + (1+\varepsilon_7)B_1B_1^T.\end{aligned} \tag{8.28}$$

The solutions to these Riccati equations need to satisfy the following assumption:

Assumption 8.3 (1) $A - (1+4\varepsilon_8)B_2\tilde{E}_1^{-1}D_{12}^T C_1 + (J_{12}J_{12}^T - B_2\tilde{E}_1^{-1}B_2^T)X$ is a stability matrix.
(2) $A - g^{-2}(1+\varepsilon_7)B_1 D_{21}^T \tilde{E}_2^{-1}C_2 + Y(J_3^T J_3 - C_2^T\tilde{E}_2^{-1}C_2)$ is a stability matrix.
(3) The matrix XY has a spectral radius strictly less than one.

Now, we present the main results on robust H^∞ coherent controller synthesis.

Theorem 8.3 Necessity. *Consider the system (8.22) and suppose that Assumption 8.2 is satisfied. If there exists a controller of the form (8.17) such that the resulting closed-loop system (8.24) is strictly bounded real with unitary disturbance attenuation, then the Riccati equations (8.26) and (8.27) will have stabilizing solutions $X \geq 0$ and $Y \geq 0$ satisfying Assumption 8.3.*

Sufficiency. Suppose the Riccati equations (8.26) and (8.27) have stabilizing solutions $X \geq 0$ and $Y \geq 0$ satisfying Assumption 8.3. If the controller (8.17) is such that the matrices A_K, B_K, C_K are designed from the Riccati solutions in the following form:

$$\begin{aligned} A_K =& A + B_2C_K - B_KC_2 + J_{12}J_{12}^T X - g^{-2}(1+\varepsilon_7)B_K D_{21}B_1^T X; \\ B_K =& (I - YX)^{-1}(YC_2^T + g^{-2}(1+\varepsilon_7)B_1D_{21}^T)\tilde{E}_2^{-1}; \\ C_K =& -\tilde{E}_1^{-1}(B_2^T X + (1+4\varepsilon_8)D_{12}^T C_1), \end{aligned} \tag{8.29}$$

then the resulting closed-loop system (8.24) is strictly bounded real with unitary disturbance attenuation.

When the quantum system is only subject to some of the uncertainties, a modified scaled H^∞ system can be easily obtained. We can apply the same method as Theorem 8.3 to the corresponding scaled H^∞ system and obtain the required controller matrices; see, e.g., [13].

Example 8.1 We illustrate the robust H^∞ coherent controller design method using a quantum optical plant which comprises of optical cavities coupled to optical fields [1]. In particular, we consider an optical cavity as a ring cavity consisting of three partially transmitting mirrors in which a light beam can be trapped and circulated inside the cavity to form a standing wave; see, e.g., [3, 13]. This optical mode is described by a harmonic oscillator with annihilation operator a. The partially transmitting mirrors enable this optical mode to interact with an external free field.

As can be seen from Fig. 8.1, there are three optical channels $v,\ w,\ u$ coupling to the optical cavity. Our control objective is to attenuate the effect of the disturbance w on the output z. The dynamics of the system are represented by the following state equations:

$$\begin{aligned} \mathrm{d}a(t) =& -\frac{\gamma}{2}a(t)\mathrm{d}t - \sqrt{\kappa_1}\mathrm{d}A_1(t) - \sqrt{\kappa_2}\mathrm{d}A_2(t) - \sqrt{\kappa_3}\mathrm{d}A_3(t); \\ \mathrm{d}a^*(t) =& -\frac{\gamma}{2}a^*(t)\mathrm{d}t - \sqrt{\kappa_1}\mathrm{d}A_1^*(t) - \sqrt{\kappa_2}\mathrm{d}A_2^*(t) - \sqrt{\kappa_3}\mathrm{d}A_3^*(t); \\ \mathrm{d}B_2(t) =& \sqrt{\kappa_2}a(t)\mathrm{d}t + \mathrm{d}A_2(t); \\ \mathrm{d}B_3(t) =& \sqrt{\kappa_3}a(t)\mathrm{d}t + \mathrm{d}A_3(t). \end{aligned} \tag{8.30}$$

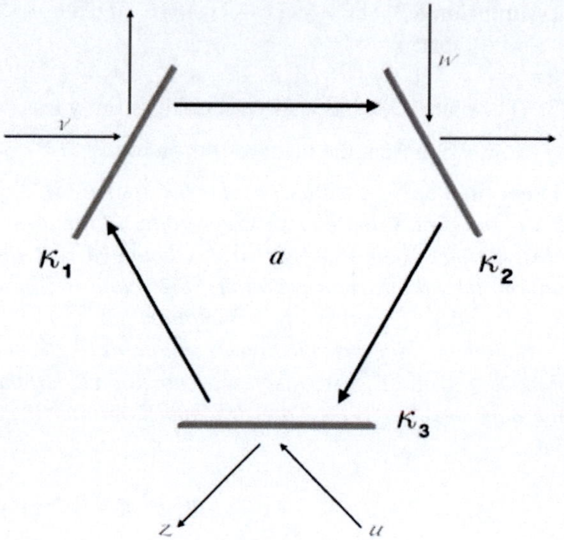

Fig. 8.1 Optical cavity system [3]

Here $\gamma = \kappa_1 + \kappa_2 + \kappa_3$. Also, $A_1(t),\ A_2(t),\ A_3(t)$ describe the input fields in channels $v,\ w,\ u$, respectively, and $B_2(t),\ B_3(t)$ represent the output fields in channels w, u, respectively. In this model, we have the commutation relation between annihilation and creation operators

$$\left[\begin{bmatrix} a \\ a^\# \end{bmatrix}, \begin{bmatrix} a \\ a^\# \end{bmatrix}^\dagger\right] = \begin{bmatrix} 1 & 0 \\ 0 & -1 \end{bmatrix}. \tag{8.31}$$

In this annihilation/creation operator representation, there often exist complex-valued coefficients. For convenience, we rewrite (8.30) in the real-valued quadrature form as in (8.10) which is represented by position/momentum operators. In the quadrature form, $x_1(t) = q(t) = a(t) + a^*(t)$ is the position operator of the cavity mode (also called the amplitude quadrature) and $x_2(t) = p(t) = (a(t) - a^*(t))/\mathrm{i}$ is the momentum operator of the cavity mode (also called the phase quadrature). Consequently, the corresponding coefficients for the system of the form (8.10) are as follows:

$$\begin{aligned} A = -\frac{\gamma}{2}I;\ B_0 = -\sqrt{\kappa_1}I;\ B_1 = -\sqrt{\kappa_2}I;\ B_2 = -\sqrt{\kappa_3}I; \\ C_1 = \sqrt{\kappa_3}I;\ D_{12} = I;\ C_2 = \sqrt{\kappa_2}I;\ D_{21} = I. \end{aligned} \tag{8.32}$$

The quadrature mode commutation relation for this plant can be represented by $[x(t), x(t)^T] = 2\mathrm{i}J$, and the quantum noises $v, \tilde{w}$ have Hermitian Itô matrices $F_v = F_{\tilde{w}} = I + \mathrm{i}J$.

We choose the total cavity decay rate $\gamma = 5.6$ and the coupling coefficients $\kappa_1 = 3.5, \kappa_2 = 2, \kappa_3 = 0.1$. Note that

$$L = \begin{bmatrix} \sqrt{\kappa_1} I \\ \sqrt{\kappa_2} I \\ \sqrt{\kappa_3} I \end{bmatrix} a$$

is the coupling operator for the given system, where κ_i $(i = 1, 2, 3)$ are the so-called mirror coupling coefficients.

In practical applications, the quantum optical system may be subject to uncertain perturbations in the system Hamiltonian. We first consider an example of a detuned cavity with the following dynamics:

$$\begin{aligned}
\mathrm{d}a(t) &= (-\frac{\gamma}{2} - 2\mathrm{i}\hat{\Delta})a(t)\mathrm{d}t - \sqrt{\kappa_1}\mathrm{d}A_1(t) - \sqrt{\kappa_2}\mathrm{d}A_2(t) - \sqrt{\kappa_3}\mathrm{d}A_3(t);\\
\mathrm{d}a^*(t) &= (-\frac{\gamma}{2} + 2\mathrm{i}\hat{\Delta})a^*(t)\mathrm{d}t - \sqrt{\kappa_1}\mathrm{d}A_1^*(t) - \sqrt{\kappa_2}\mathrm{d}A_2^*(t) - \sqrt{\kappa_3}\mathrm{d}A_3^*(t);\\
\mathrm{d}B_2(t) &= \sqrt{\kappa_2}a(t)\mathrm{d}t + \mathrm{d}A_2(t);\\
\mathrm{d}B_3(t) &= \sqrt{\kappa_3}a(t)\mathrm{d}t + \mathrm{d}A_3(t).
\end{aligned} \tag{8.33}$$

Here, the uncertainty $\hat{\Delta}$ represents the "detuning" and describes the difference between the nominal external field frequency and the cavity mode frequency. The corresponding quadrature form of the system matrices is: $A = -\frac{\gamma}{2}I + 2J\hat{\Delta}$, and $B_0, B_1, B_2, C_1, C_2, D_{12}, D_{21}$ are the same as in (8.32). Hence, we have $H_2 = \frac{1}{2}x^T E^T \hat{\Delta} E x$, where $E = I$.

We assume that the uncertainty $\hat{\Delta}$ is in the range of $\hat{\Delta} \in [-1, 1]$ and the required disturbance attenuation constant is $g = 0.25$. Now, we apply the robust H^∞ coherent controller design method to this uncertain quantum system. As indicated in Theorem 8.3, by solving corresponding Riccati equations, we obtain the solution $X = 0.0046I$, $Y = 96.1250I$, which satisfies Assumption 8.3. The corresponding controller matrices are

$$A_K = -18.315I;\ B_K = 12.741I;\ C_K = -0.3148I.$$

Since the controller is designed to be a coherent controller, the physical realization conditions need to be satisfied. It follows from Theorem 2.1 that the following system coefficients are obtained:

$$B_{K1} = \begin{bmatrix} 0.3148 & 0 & 8.1 & -8.1 \\ 0 & 0.3148 & -8.1 & -7.4311 \end{bmatrix};\ B_{K0} = [I \quad 0].$$

To make a performance comparison between the method in this paper and the method proposed in [3], we apply the approach in [3] and obtain the following results:

$$\begin{aligned}
&X = Y = 0_{2\times 2};\\
&A_K = -0.7I;\ B_K = -1.4142I;\ C_K = -0.3162I;\\
&B_{K1} = \begin{bmatrix} 0.3162 & 0 & 1 & 1.7 \\ 0 & 0.3162 & 1 & 1 \end{bmatrix};\ B_{K0} = [I \quad 0].
\end{aligned} \tag{8.34}$$

Then we can compare the performance between the proposed method and the approach in [3] for the same uncertain quantum system. Roughly speaking, the controller designed using the approach in [3] performs better for small uncertainty while the controller designed using the proposed method has better robustness performance as the uncertainty increases. A more detailed comparison can be found in [13].

8.4 Fault-Tolerant Coherent H^∞ Control for Linear Quantum Systems

8.4.1 Fault-Tolerant Coherent Control Design

As discussed in Sects. 8.2 and 8.3, the dynamics of linear quantum systems can often be described by time-invariant linear quantum stochastic differential equations. However, when the system suffers from a fault process, the equations may no longer be time-invariant. In this section, we consider the following time-varying linear quantum system:

$$\begin{aligned} \mathrm{d}x(t) &= A(t)x(t)\mathrm{d}t + B\mathrm{d}\omega(t);\, x(0) = x_0, \\ \mathrm{d}y(t) &= Cx(t)\mathrm{d}t + D\mathrm{d}\omega(t), \end{aligned} \tag{8.35}$$

where $A(t) \in \mathbb{R}^{n\times n}$, $B \in \mathbb{R}^{n\times n_\omega}$, $C \in \mathbb{R}^{n_y\times n}$, $D \in \mathbb{R}^{n_y\times n_\omega}$, and $x(t)$ is a vector of self-adjoint possibly non-commutative system variables. Here, $A(t)$ may be taken as a generalization to time-varying linear quantum systems from time-invariant linear quantum systems with constant matrix A. The fault process that we consider for linear quantum systems here is introduced in the time-varying system matrix $A(t)$. We can write $A(t) = A(f(t))$, where $f(t)$ represents the fault signal. We first consider the bounded real properties for the time-varying quantum system (8.35). The strictly bounded real lemma states a relation between a storage function and supply functions in terms of system energy, which will be used in the fault-tolerant controller synthesis. We consider a quantum system described as follows:

$$\begin{aligned} \mathrm{d}x(t) &= A(t)x(t)\mathrm{d}t + \begin{bmatrix} B\ G \end{bmatrix}\begin{bmatrix} \mathrm{d}\omega(t) \\ \mathrm{d}\nu(t) \end{bmatrix}, \\ \mathrm{d}z(t) &= Cx(t)\mathrm{d}t + \begin{bmatrix} D\ H \end{bmatrix}\begin{bmatrix} \mathrm{d}\omega(t) \\ \mathrm{d}\nu(t) \end{bmatrix}. \end{aligned} \tag{8.36}$$

Here, $\mathrm{d}\omega(t) = \beta_\omega(t)\mathrm{d}t + \mathrm{d}\tilde{\omega}$ represents the disturbance input with the quantum noise $\mathrm{d}\tilde{\omega}$, and $\mathrm{d}\nu$ represents other inputs including unexpected quantum noises in quantum systems.

We first define a storage function $V(x(t)) = x(t)^T P(t)x(t)$, where $P(t)$ is a time-varying positive definite symmetric matrix, and then define the following operator valued quadratic function:

$$\gamma(x, \beta_\omega) = \begin{bmatrix} x^T & \beta_\omega^T \end{bmatrix}^T R \begin{bmatrix} x \\ \beta_\omega \end{bmatrix},$$

as the supply function, where R is a constant real symmetric matrix.

Definition 8.1 [3] The quantum system (8.36) is said to be bounded real with disturbance attenuation g if there exists a positive time-varying storage function $V(x(t)) = x(t)^T P(t)x(t)$, a constant $\lambda > 0$, and a supply rate $\gamma(x, \beta_\omega)$ such that the following inequality holds:

$$\langle V(x(t))\rangle + \int_0^t \langle \gamma(x(s), \beta_\omega(s))\rangle ds \leq \langle V(x(0))\rangle + \lambda t, \forall t > 0. \tag{8.37}$$

Here, $\langle V(x(t))\rangle$ represents the expectation of the operator $V(x(t))$, and $\beta_z(t) = Cx(t) + D\beta_\omega(t)$. The supply rate function has the form

$$\gamma(x, \beta_\omega) = \beta_z^T \beta_z - g^2 \beta_\omega^T \beta_\omega = \begin{bmatrix} x^T & \beta_\omega^T \end{bmatrix} \begin{bmatrix} C^T C & C^T D \\ D^T C & D^T D - g^2 \mathrm{I} \end{bmatrix} \begin{bmatrix} x \\ \beta_\omega \end{bmatrix}. \tag{8.38}$$

Also, we say that the system (8.36) is strictly bounded real with disturbance attenuation g if there exists a constant $\varepsilon > 0$ such that inequality (8.37) holds for the supply function with $R + \varepsilon \mathrm{I}$.

With this definition, the following theorem states the relationship between bounded real properties and Riccati differential equations, as well as the H^∞ control problem, which will be used to design a coherent controller:

Theorem 8.4 *For the system* (8.36), *the following four statements are equivalent:*

(1) The system (8.36) *is strictly bounded real with disturbance attenuation g.*
(2) There exists a positive definite matrix $\tilde{P}(t)$ such that

$$\begin{aligned} &\dot{\tilde{P}}(t) + A(t)^T \tilde{P}(t) + \tilde{P}(t)A(t) + C^T C \\ &+ (C^T D + \tilde{P}(t)B)(g^2 I - D^T D)^{-1}(D^T C + B^T \tilde{P}(t)) < 0, \quad \forall t \geq 0. \end{aligned}$$

(3) The Riccati differential equation

$$\begin{aligned} &\dot{P}(t) + A(t)^T P(t) + P(t)A(t) + C^T C \\ &+ (C^T D + P(t)B)(g^2 I - D^T D)^{-1}(D^T C + B^T P(t)) = 0 \end{aligned}$$

has a stabilizing solution $P(t) \geq 0$.
(4) The homogeneous system $\dot{x}(t) = A(t)x(t)$ is exponentially stable, and the operator mapping ω to z satisfies $\| T_{z\omega} \|_\infty < g$.

The proof of Theorem 8.4 can be found in [4].

Now, we consider a linear quantum system suffering from abrupt variations in its parameters, structure or system dynamics such that the system dynamics may randomly transit between a finite number of different modes, named faulty modes. It is then appropriate to model the fault process on a probability space $(\Omega, \mathscr{F}, \mathscr{P})$ by a continuous-time Markov chain $\{f(t)\}_{t\geq 0}$ [2], which results in a Markovian jump linear quantum system. To be specific, $f(t)$ takes values within a finite set $\mathbb{S} = \{e_1, e_2, \ldots, e_N\}$ for an integer N. The transition rate matrix is known a prior as $\Pi = (\pi_{jk}) \in \mathbb{R}^{N\times N}$, with $\pi_{jj} = -\sum_{j\neq k} \pi_{jk}$, and $\pi_{jk} \geq 0, j \neq k$. The transition rate matrix describes the instantaneous rate at which a continuous-time Markov chain transitions between states. Here, we develop a coherent H^∞ control design for a class of linear quantum systems whose Hamiltonian is dependent on the fault process $f(t)$.

The system with a disturbance input and a control input is described as

$$\begin{aligned} \mathrm{d}x(t) &= A(f(t))x(t)\mathrm{d}t + B_1\mathrm{d}\omega(t) + B_2\mathrm{d}u(t), \\ \mathrm{d}z(t) &= C_1x(t)\mathrm{d}t + D_1\mathrm{d}u(t), \\ \mathrm{d}y(t) &= C_2x(t)\mathrm{d}t + D_2\mathrm{d}\omega(t), \end{aligned} \tag{8.39}$$

where $A(F(t))$ takes finite values in $(A_1, A_2, \ldots, A_N), A_i = A(e_i)$ since the fault process $f(t)$ has been assumed to be a Markov chain, which has values within a finite set $\mathbb{S} = \{e_1, e_2, \ldots, e_N\}$.

Assume that the controller is described by the following dynamical equations:

$$\begin{aligned} \mathrm{d}\xi(t) &= \mathbf{A}(t)\xi(t)\mathrm{d}t + \mathbf{B}(t)\mathrm{d}y(t) + \mathbf{E}(t)\mathrm{d}\nu_K(t), \\ \mathrm{d}u(t) &= \mathbf{C}(t)\xi(t)\mathrm{d}t + \mathbf{D}(t)\mathrm{d}\nu_K(t), \end{aligned} \tag{8.40}$$

where $\xi(t) = \left[\xi_1(t)\ \xi_2(t)\ \ldots\ \xi_{n_k}\right]^T$ is a vector of self-adjoint controller variables. The input ν_K is introduced to ensure the physically realizable conditions. ν_K is assumed to be a vector of non-commutative Wiener processes satisfying the Itô table with canonical Hermitian Itô matrix F_{ν_K}. To coincide with the Markovian jump linear plant, the controller also is assumed to jump between different modes with $(\mathbf{A}_1, \mathbf{B}_1, \mathbf{C}_1), \ldots, (\mathbf{A}_N, \mathbf{B}_N, \mathbf{C}_N)$.

We obtain the closed-loop systems by identifying $\beta_u(t) = \mathbf{C}(t)\xi(t)$ and denoting $\eta = \left[x(t)\ \xi(t)\right]^T$ as

$$\begin{aligned} \mathrm{d}\eta(t) &= \begin{bmatrix} A_i & B_2\mathbf{C}_i \\ \mathbf{B}_iC_2 & \mathbf{A}_i \end{bmatrix}\eta(t)\mathrm{d}t + \begin{bmatrix} B_1 \\ \mathbf{B}_iD_2 \end{bmatrix}\mathrm{d}\omega(t) + \begin{bmatrix} B_2\mathbf{D}_i \\ \mathbf{E}_i \end{bmatrix}\mathrm{d}\nu_K(t), \\ \mathrm{d}z(t) &= \begin{bmatrix} C_1 & D_1\mathbf{C}_i \end{bmatrix}\eta(t)\mathrm{d}t + D_1\mathbf{D}_i\mathrm{d}\nu_K(t). \end{aligned} \tag{8.41}$$

The control objective here is to design a controller (8.40) such that the closed-loop system (8.41) is strictly bounded real with a given disturbance attenuation g; that is, there exists a positive definite matrix $P(t)$ such that

$$
\begin{aligned}
&\left\langle \eta^T(t)P(t)\eta(t)\right\rangle \\
&+\int_0^t \left\langle \beta_z^T(s)\beta_z(s) - g^2\beta_\omega^T(s)\beta_\omega(s) + \varepsilon\eta^T(s)\eta(s) + \varepsilon\beta_\omega^T(s)\beta_\omega(s)\right\rangle \mathrm{d}s \\
&\leq \langle \eta^T(0)P_0\eta(0)\rangle + \lambda t, \forall t > 0.
\end{aligned} \tag{8.42}
$$

Substituting $\mathrm{d}\omega(t) = \beta_\omega(t)\mathrm{d}t + \mathrm{d}\tilde{\omega}(t)$ and $\mathrm{d}u(t) = \beta_u(t)\mathrm{d}t + \mathrm{d}\tilde{u}(t)$ into (8.41), we have

$$
\begin{aligned}
\mathrm{d}\eta(t) &= \tilde{A}_i\eta(t)\mathrm{d}t + \tilde{B}_{1i}\beta_\omega(t)\mathrm{d}t + \tilde{B}_{1i}\mathrm{d}\tilde{\omega}(t) + \tilde{B}_{2i}\mathrm{d}v_K(t), \\
\mathrm{d}z(t) &= \tilde{C}_i\eta(t)\mathrm{d}t + \tilde{D}_i\mathrm{d}v_K(t).
\end{aligned} \tag{8.43}
$$

Here,

$$
\tilde{A}_i = \begin{bmatrix} A_i & B_2\mathbf{C}_i \\ \mathbf{B}_iC_2 & \mathbf{A}_i \end{bmatrix}, \quad \tilde{B}_{1i} = \begin{bmatrix} B_1 \\ \mathbf{B}_iD_2 \end{bmatrix}, \tilde{B}_{2i} = \begin{bmatrix} B_2\mathbf{D}_i \\ \mathbf{E}_i \end{bmatrix},
$$

$$
\tilde{C}_i = \begin{bmatrix} C_1 & D_1\mathbf{C}_i \end{bmatrix}, \quad \tilde{D}_i = D_1\mathbf{D}_i.
$$

For this quantum system, we define $\langle V(\eta(t))\rangle = \langle \eta^T(t)P(t)\eta(t)\rangle$ with a positive-definite matrix $P(t)$ and

$$
\begin{aligned}
\gamma(\eta(t), \beta_\omega(t)) &= \beta_z(t)^T\beta_z(t) - g^2\beta_\omega^T(t)\beta_\omega(t) + \varepsilon\eta^T(t)\eta(t) + \varepsilon\beta_\omega^T(t)\beta_\omega(t) \\
&= \eta^T(t)[\tilde{C}_i^T\tilde{C}_i + \varepsilon\mathrm{I}]\eta(t) - (g^2 - \varepsilon)\mathrm{I}\beta_\omega^T(t)\beta_\omega(t).
\end{aligned} \tag{8.44}
$$

Substituting (8.44) into (8.42), the control objective can be written as:

$$
\begin{aligned}
&\langle \eta^T(t)P(t)\eta(t)\rangle + \int_0^t \left\langle \eta^T(s)\left[\tilde{C}_i^T\tilde{C}_i + \varepsilon\mathrm{I}\right]\eta(s)\right\rangle \mathrm{d}s \\
&- \int_0^t \left\langle (g^2 - \varepsilon)\beta_\omega^T(s)\beta_\omega(s)\right\rangle \mathrm{d}s \leq \langle \eta_0^T P_0\eta_0\rangle + \lambda t.
\end{aligned} \tag{8.45}
$$

Now, we have the following result for the quantum systems under consideration whose proof can be found in [4]:

Theorem 8.5 *If there exists a controller of the form* (8.40) *such that the closed-loop system* (8.41) *is strictly bounded real with disturbance attenuation g, then the linear matrix inequalities (LMIs)* (8.46)–(8.47) *have feasible solutions* X_i, Y_i *and* L_i, F_i,

$$
\begin{bmatrix} A_i^TX_i + X_iA_i + L_iC_2 + C_2^TL_i^T + C_1^TC_1 + \sum_{j=1}^N \pi_{ij}X_j & X_iB_1 + L_iD_2 \\ B_1X_i + D_2^TL_i^T & -g^2I \end{bmatrix} < 0. \tag{8.46}
$$

$$\begin{bmatrix} Y_i & I \\ I & X_i \end{bmatrix} > 0,$$

$$\begin{bmatrix} A_iY_i + Y_iA_i^T + B_2F_i + F_i^TB_2^T + \pi_{ii}Y_i + g^{-2}B_1B_1^T & (C_1Y_i + D_1F_i)^T & R_i(Y) \\ C_1Y_i + D_1F_i & -I & 0 \\ R_i^T(Y) & 0 & S_i(Y) \end{bmatrix} < 0. \quad (8.47)$$

Here for $i = 1, \ldots, N$, *we define*

$$S_i(Y) = -\text{diag}\left(Y_1, \ldots, Y_{i-1}, Y_{i+1}, \ldots, Y_N\right),$$

and

$$R_i(Y) = \left[\sqrt{\pi_{1i}}Y_i \ \ldots \ \sqrt{\pi_{(i-1)i}}Y_i \ \sqrt{\pi_{(i+1)i}}Y_i \ \ldots \ \sqrt{\pi_{Ni}}Y_i\right].$$

In this case, the corresponding controller is given by

$$\mathbf{C}_i = F_iY_i^{-1}, \quad (8.48)$$

$$\mathbf{B}_i = (Y_i^{-1} - X_i)^{-1}L_i, \quad (8.49)$$

$$\mathbf{A}_i = (Y_i^{-1} - X_i)^{-1}M_iY_i^{-1}, \quad (8.50)$$

where

$$\begin{aligned} M_i = & -A_i^T - X_iA_iY_i - X_iB_2F_i - L_iC_2Y_i - C_1^T(C_1Y_i + D_{12}F_i) \\ & -\frac{1}{g^2}(X_iB_1 + L_iD_{21})B_1^T - \sum_{j=1}^{N} \pi_{ij}Y_j^{-1}Y_i. \end{aligned}$$

Similarly, if the LMIs (8.46) and (8.47) have feasible solutions and the controller is defined as in (8.48)–(8.50), then the closed-loop system (8.41) is strictly bounded real with the disturbance attenuation g.

8.4.2 Fault-Tolerant Control of Quantum Optical Systems

We consider possible applications of fault-tolerant coherent H^∞ control of linear quantum systems in quantum optics. The system is a dynamical squeezer composed of three mirrors, and its simplified diagram is shown as in Fig. 8.2. When this system suffers from a fault signal, an H^∞ coherent feedback controller can be designed to deal with the fault process as well as the disturbance input. The aim of this coherent controller is to make the quantum system fault-tolerant and robust against external disturbance inputs.

We obtain the differential equations of motion for this dynamical squeezer as

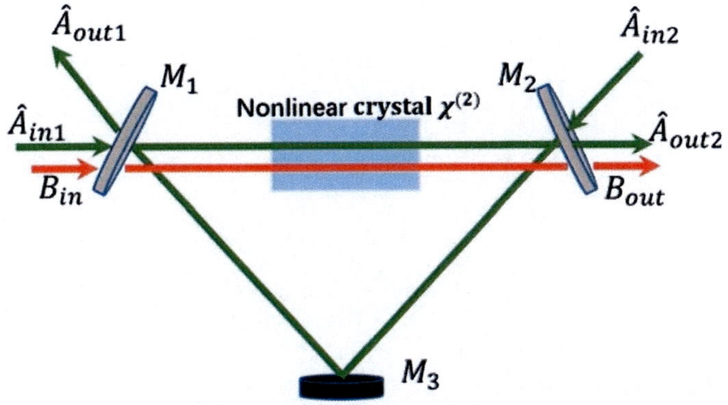

Fig. 8.2 Schematic of an OPO composed of a nonlinear crystal and three mirrors. M_1 and M_2 have partial transmissivity for the fundamental field and high transimissivity for the pumping field; M_3 is a fully reflective mirror for the fundamental field. $\hat{A}_{\text{in1}}$ and $\hat{A}_{\text{in2}}$ are the disturbance input and the control input fields, and $\hat{A}_{\text{out1}}$ and $\hat{A}_{\text{out2}}$ are the corresponding outputs. B_{in} and B_{out} are the input and output pumping fields, respectively [4]

$$\begin{aligned} d\hat{a}(t) &= \left[-\frac{\kappa}{2}\hat{a}(t) - \chi\hat{a}^{\dagger}(t)\right]dt + \sqrt{\kappa_1}d\hat{A}_{\text{in1}}(t) + \sqrt{\kappa_2}d\hat{A}_{\text{in2}}(t), \\ d\hat{a}^{\dagger}(t) &= \left[-\frac{\kappa}{2}\hat{a}^{\dagger}(t) - \chi\hat{a}\right]dt + \sqrt{\kappa_1}d\hat{A}_{\text{in1}}^{\dagger}(t) + \sqrt{\kappa_2}d\hat{A}_{\text{in2}}^{\dagger}(t), \end{aligned} \tag{8.51}$$

$$\begin{aligned} d\hat{A}_{\text{out1}}(t) &= \sqrt{\kappa_1}\hat{a}(t)dt - d\hat{A}_{\text{in1}}(t), \\ d\hat{A}_{\text{out2}}(t) &= \sqrt{\kappa_2}\hat{a}(t)dt - d\hat{A}_{\text{in2}}(t), \end{aligned} \tag{8.52}$$

where κ_1 and κ_2 are decay rates for the mirror M_1 and mirror M_2, and $\kappa = \kappa_1 + \kappa_2$. The system has two input fields $\hat{A}_{\text{in1}}, \hat{A}_{\text{in2}}$, and two output fields $\hat{A}_{\text{out1}}$ and $\hat{A}_{\text{out2}}$. Here, $\hat{A}_{\text{in1}}$ represents the disturbance input. B_{in} and B_{out} are the input and output of the pumping field.

To ensure that the operators are self-adjoint and work with real-valued coefficients, we write the amplitude and phase quadrature as

$$x = \begin{bmatrix} \hat{a}(t) + \hat{a}^{\dagger}(t) \\ -i(\hat{a}(t) - \hat{a}^{\dagger}(t)) \end{bmatrix},$$

and denote

$$\omega(t) = \begin{bmatrix} \hat{A}_{\text{in1}}(t) + \hat{A}_{\text{in1}}^{\dagger}(t) \\ -\mathrm{i}(\hat{A}_{\text{in1}}(t) - \hat{A}_{\text{in1}}^{\dagger}(t)) \end{bmatrix},$$

$$u(t) = \begin{bmatrix} \hat{A}_{\text{in2}}(t) + \hat{A}_{\text{in2}}^{\dagger}(t) \\ -\mathrm{i}(\hat{A}_{\text{in2}}(t) - \hat{A}_{\text{in2}}^{\dagger}(t)) \end{bmatrix},$$

$$z(t) = \begin{bmatrix} \hat{A}_{\text{out2}}(t) + \hat{A}_{\text{out2}}^{\dagger}(t) \\ -\mathrm{i}(\hat{A}_{\text{out2}}(t) - \hat{A}_{\text{out2}}^{\dagger}(t)) \end{bmatrix},$$

$$y(t) = \begin{bmatrix} \hat{A}_{\text{out1}}(t) + \hat{A}_{\text{out1}}^{\dagger}(t) \\ -\mathrm{i}(\hat{A}_{\text{out1}}(t) - \hat{A}_{\text{out1}}^{\dagger}(t)) \end{bmatrix}.$$

The system may be described by

$$\begin{aligned} \mathrm{d}x(t) &= A(t)x(t)\mathrm{d}t + B_1\mathrm{d}\omega(t) + B_2\mathrm{d}u(t), \\ \mathrm{d}z(t) &= C_1x(t)\mathrm{d}t + D_1\mathrm{d}u(t), \\ \mathrm{d}y(t) &= C_2x(t)\mathrm{d}t + D_2\mathrm{d}\omega(t), \end{aligned} \tag{8.53}$$

where

$$\begin{aligned} A(t) &= \begin{bmatrix} -\frac{\kappa}{2} - \chi(t) & 0 \\ 0 & \chi(t) - \frac{\kappa}{2} \end{bmatrix}, \\ B_1 &= \sqrt{\kappa_1}\mathrm{I},\ B_2 = \sqrt{\kappa_2}I, \\ C_1 &= \sqrt{\kappa_2}\mathrm{I},\ D_1 = -I, \\ C_2 &= \sqrt{\kappa_1}\mathrm{I},\ D_2 = -I. \end{aligned}$$

We write $\chi(t) = \chi = \chi^{(2)}\beta(t)$ as a time-varying parameter, where $\beta(t)$ represents the fault signal due to the unstable voltage of the laser generator.

Since the pump laser is treated in a classical way, if the macroscopic laser device is subject to an undesired fault signal, a time-varying Hamiltonian is introduced. In some practical applications, we may assume that the amplitude of the laser is not changing with time continuously and only jumps among several values. This makes it reasonable to model the fault process as a Markovian chain. Therefore, the whole system is a Markovian jump linear quantum system. To deal with this fault process, as well as the disturbance input that the dynamical squeezer itself suffers from, a coherent feedback controller is designed and connected to the plant directly without any measurement. After applying a controller to the plant, the whole closed-loop system is shown in Fig. 8.3.

We choose $\kappa_1 = 0.8264$, $\kappa_2 = 0.0011$, and $\kappa = 0.8275$, and only consider the case that the system is acting as an amplifier ($\chi \leq \kappa$) [4]. In the numerical example, we take three different values $\chi \in \{0.1\kappa, 0.2\kappa, 0.3\kappa\}$, which results in three modes for the system with

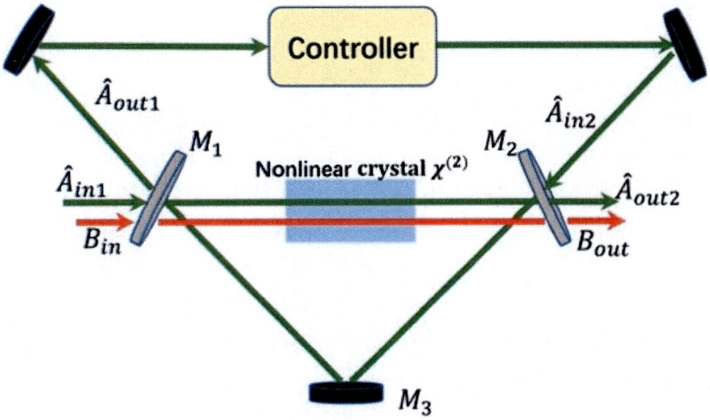

Fig. 8.3 OPO with a coherent controller

$$A_1 = \begin{bmatrix} -0.4551 & 0 \\ 0 & -0.3724 \end{bmatrix},$$
$$A_2 = \begin{bmatrix} -0.4965 & 0 \\ 0 & -0.3310 \end{bmatrix}, \tag{8.54}$$
$$A_3 = \begin{bmatrix} -0.5379 & 0 \\ 0 & -0.2896 \end{bmatrix}.$$

We first consider the case where the transition rate matrix is known as

$$\begin{bmatrix} -0.02 & 0.01 & 0.01 \\ 0.01 & -0.01 & 0 \\ 0.01 & 0 & -0.01 \end{bmatrix}.$$

By solving the LMIs in (8.46) and (8.47), we obtain the controller as

$$\mathbf{A}_1 = \begin{bmatrix} -1.7535 & 0 \\ 0 & -2.1226 \end{bmatrix}, \mathbf{B}_1 = \begin{bmatrix} 1.2524 & 0 \\ 0 & 1.8944 \end{bmatrix}, \mathbf{C}_1 = \begin{bmatrix} -0.0331 & 0 \\ 0 & -0.0331 \end{bmatrix}; \tag{8.55}$$

$$\mathbf{A}_2 = \begin{bmatrix} -1.5796 & 0 \\ 0 & -2.2738 \end{bmatrix}, \mathbf{B}_2 = \begin{bmatrix} 0.9713 & 0 \\ 0 & 2.2099 \end{bmatrix}, \mathbf{C}_2 = \begin{bmatrix} -0.0331 & 0 \\ 0 & -0.0331 \end{bmatrix}; \tag{8.56}$$

$$\mathbf{A}_3 = \begin{bmatrix} -1.3992 & 0 \\ 0 & -2.4340 \end{bmatrix}, \mathbf{B}_3 = \begin{bmatrix} 0.7024 & 0 \\ 0 & 2.5600 \end{bmatrix}, \mathbf{C}_3 = \begin{bmatrix} -0.0331 & 0 \\ 0 & -0.0331 \end{bmatrix}. \tag{8.57}$$

Here, $\{\mathbf{A}_i, \mathbf{B}_i, \mathbf{C}_i\}$ are the parameters of the i-th mode of the controller. It can be checked that the controller for each $i = 1, 2, 3$ is not physically realizable without additional quantum inputs. In [4], additional quantum inputs have been used to construct the matrices $\mathbf{E}_{i1}$ and $\mathbf{E}_{i2}$ such that the controller $\{\mathbf{A}_i, \tilde{\mathbf{B}}_i, \mathbf{C}_i\}$ satisfies the physical realizability conditions, where $\tilde{\mathbf{B}}_i = \left[\mathbf{B}_i \ \mathbf{E}_{i1} \ \mathbf{E}_{i2}\right]$. More details can be found in [4]. Moreover, an implementation of the quantum controller using a dynamic squeezer, an OPO and a phase shifter has been proposed in [4].

8.5 Fault-Tolerant Control of Measurement-Based Feedback Quantum Systems

In Sects. 8.2–8.4, we consider quantum coherent feedback control. In this section, we present a systematic fault-tolerant control design method for a class of measurement-based quantum feedback control systems subject to faults. In particular, we present an estimator-based approach for fault-tolerant control design of quantum stochastic systems where we assume that the classical fault signal is independent of quantum noise. Although our work has close connection with the result of robust observer design for quantum systems in [14], different from uncertainties in the system Hamiltonian and the system operator considered there, we consider the classical fault signals in quantum systems and design a controller to compensate the effect of fault signals. We aim to develop a fault-tolerant control design approach with a reduced-order dynamic estimator for a class of measurement-based quantum feedback systems subject to faults.

8.5.1 *Fault-Tolerant Control Problem Formulation*

Consider a quantum plant with fault signals described by a non-commutative stochastic model of the following form:

$$\begin{aligned} \mathrm{d}x(t) &= Ax(t)\mathrm{d}t + B_w \mathrm{d}w(t) + B_u y_u(t)\mathrm{d}t + B_f f(t)\mathrm{d}t, \\ \mathrm{d}y(t) &= Cx(t)\mathrm{d}t + D\mathrm{d}w(t), \end{aligned} \tag{8.58}$$

where $A \in \mathbb{R}^{n\times n}$, $B_w \in \mathbb{R}^{n\times n_w}$, $B_u \in \mathbb{R}^{n\times n_u}$, $B_f \in \mathbb{R}^{n\times n_f}$, $C \in \mathbb{R}^{n_y\times n}$, $D \in \mathbb{R}^{n_y\times n_w}$ (n, n_w, n_u and n_y are even). A, $B = [B_w \ B_u]$, C and D should satisfy physical realizability conditions. x represents a vector of plant variables and w is the vector of vacuum quantum fields. B_f is known and the real column vector $f(t)$ represents the unknown fault signal to be estimated. The relevant measurement-based feedback control system is illustrated in Fig. 8.4, where a quantum optical plant subject to fault signals f is measured using a homodyne detector (HD). The output from the HD y_m is used to establish a classical estimator and a classical controller for compensating the effect of f on the quantum plant.

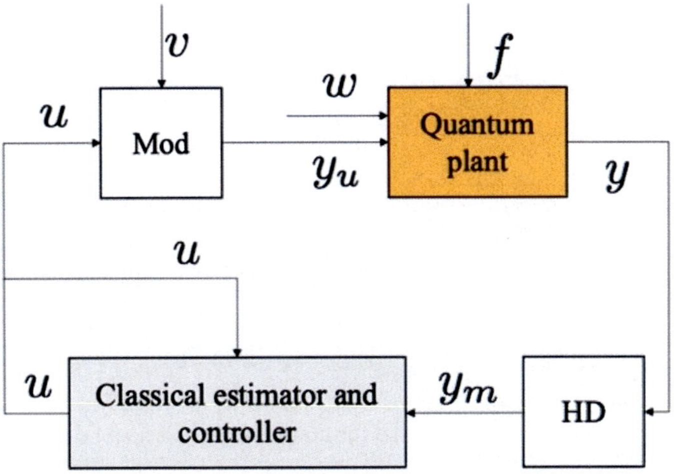

Fig. 8.4 A measurement-based quantum feedback control system with an estimator-based fault-tolerant controller. HD represents a homodyne detector for measurements, and Mod represents a modulator [10]

For example, $f(t)$ may originate from the voltage fluctuation in controlling the laser generator, malfunction of beam splitters, or phase shifters. The signal $y_u(t)$ is a control input of the form

$$y_u(t) = u(0) + \int_0^t u(s)\mathrm{d}s + v(t), \tag{8.59}$$

where $v \in \mathbb{R}^{n_u}$ and w are independent and n_u is even; u and v are the signal and quantum noise parts of y_u, respectively. When the quantum output signals $y(t)$ are measured by homodyne detectors, classical signals $y_m(t) = Gy(t) \in \mathbb{R}^{n_{y_m}}$ are produced. The matrix G corresponding to measurement processes satisfies the condition below [6]:

$$G\Theta_y G^T = 0 \tag{8.60}$$

with $\mathrm{rank}(G) \leq \frac{n_y}{2}$, where G represents a static linear transformation (measurement processes) that converts boson fields into classical signals.

Now, we give the following assumption:

Assumption 8.4 The fault signal $f(t)$ satisfies $\| f(t) \| \leq \alpha$ and $\| \dot{f}(t) \| \leq \beta$, where $\alpha > 0, \beta > 0$.

Given a quantum optical plant with a fault signal of the form (8.58), there always exists a permutation matrix T such that the transformed system is given as [10]

$$\begin{aligned} \mathrm{d}\tilde{x}(t) &= \tilde{A}\tilde{x}(t)\mathrm{d}t + \tilde{B}_w \mathrm{d}w(t) + \tilde{B}_u y_u(t)\mathrm{d}t + \tilde{B}_f f(t)\mathrm{d}t, \\ \mathrm{d}y(t) &= \tilde{C}\tilde{x}(t)\mathrm{d}t + \tilde{D}\mathrm{d}w(t) \end{aligned} \tag{8.61}$$

with system matrices

$$\tilde{A} = TAT^{-1} = \begin{bmatrix} \tilde{A}_{11} & \tilde{A}_{12} \\ \tilde{A}_{21} & \tilde{A}_{22} \end{bmatrix},$$

$\tilde{B}_w = TB_w = \left[\tilde{B}_{w_1}^T \ \tilde{B}_{w_2}^T\right]^T$, $\quad \tilde{B}_u = TB_u = \left[\tilde{B}_{u_1}^T \ \tilde{B}_{u_2}^T\right]^T$, $\quad \tilde{B}_f = TB_f = \left[\tilde{B}_{f_1}^T \ \tilde{B}_{f_2}^T\right]^T$, $\tilde{C} = CT^{-1} = [\tilde{C}_1 \quad \tilde{C}_2]$, $\tilde{D} = D$ and new defined system variables

$$\tilde{x}(t) = Tx(t) = \begin{bmatrix} \tilde{x}_{uo}(t) \\ \tilde{x}_o(t) \end{bmatrix},$$

where $\tilde{x}_o \in \mathbb{R}^{n_o}$ represents n_o components of $\tilde{x}(t)$ to be estimated while $\tilde{x}_{uo} \in \mathbb{R}^{n-n_o}$ represents unestimated components. Here $n_o \leq \frac{n}{2}$ and $\tilde{x}_o$ should satisfy $[\tilde{x}_o(t), \tilde{x}_o(t)^T] = 0$ for all $t \geq 0$, so that the components of $\tilde{x}_o$ can be simultaneously observed.

Let $\eta = \left[\tilde{x}_{uo}^T \ \tilde{x}_o^T \ f^T\right]^T$ and $h(t) = \frac{\mathrm{d}f(t)}{\mathrm{d}t}$. We first design an augmented system for (8.61) given by

$$\begin{aligned} \mathrm{d}\eta(t) &= \mathbf{A}\eta(t)\mathrm{d}t + \mathbf{B}_w \mathrm{d}w(t) + \mathbf{B}_u \mathrm{d}y_u(t) + \mathbf{B}_h h(t)\mathrm{d}t, \\ \mathrm{d}y_m(t) &= G(\mathbf{C}\eta(t)\mathrm{d}t + \tilde{D}\mathrm{d}w), \end{aligned} \tag{8.62}$$

with the augmented matrices

$$\mathbf{A} = \begin{bmatrix} \tilde{A}_{11} & \tilde{A}_{12} & \tilde{B}_{f_1} \\ \tilde{A}_{21} & \tilde{A}_{22} & \tilde{B}_{f_2} \\ 0 & 0 & 0 \end{bmatrix},$$

$$\mathbf{B}_w = \begin{bmatrix} \tilde{B}_{w_1} \\ \tilde{B}_{w_2} \\ 0 \end{bmatrix}, \quad \mathbf{B}_u = \begin{bmatrix} \tilde{B}_{u_1} \\ \tilde{B}_{u_2} \\ 0 \end{bmatrix}, \quad \mathbf{B}_h = \begin{bmatrix} 0 \\ 0 \\ I \end{bmatrix},$$

$\mathbf{C} = [\tilde{C}_1 \ \tilde{C}_2 \ 0]$.

In order to estimate the plant observables $\tilde{x}_o$ and the fault signal f together, we have from (8.62)

$$\mathrm{d}\xi(t) = \hat{A}\xi(t)\mathrm{d}t + \hat{A}_{uo}\tilde{x}_{uo}(t)\mathrm{d}t + \hat{B}_w \mathrm{d}w(t) + \hat{B}_u \mathrm{d}y_u(t) + \hat{B}_h h(t)\mathrm{d}t, \tag{8.63}$$

where $\xi(t) = \left[\tilde{x}_o^T \ f^T\right]^T \in \mathbb{R}^{\hat{n}}$ with $\hat{n} = n_o + n_f$,

$$\hat{A} = \begin{bmatrix} \tilde{A}_{22} & \tilde{B}_{f_2} \\ 0 & 0 \end{bmatrix},$$

$$\bar{A}_{uo} = \begin{bmatrix} \tilde{A}_{21} \\ 0 \end{bmatrix} \in \mathbb{R}^{\hat{n}\times(n-n_o)},$$

$\hat{B}_w = \begin{bmatrix} \tilde{B}_{w_2}^T & 0 \end{bmatrix}^T \in \mathbb{R}^{\hat{n}\times n_w}$, $\hat{B}_u = \begin{bmatrix} \tilde{B}_{u_2}^T & 0 \end{bmatrix}^T \in \mathbb{R}^{\hat{n}\times n_u}$, $\hat{B}_h = [0\ I]^T \in \mathbb{R}^{\hat{n}\times n_f}$.

We aim to build a classical linear estimator-based fault-tolerant controller for (8.63) given by

$$\begin{aligned} d\hat{\xi}(t) &= \hat{A}\hat{\xi}(t)dt + \hat{B}_u u(t)dt + L\left(dy_m(t) - G\hat{C}\hat{\xi}(t)dt\right), \\ u(t) &= K\hat{\xi}(t), \end{aligned} \tag{8.64}$$

where the estimate $\hat{\xi}(t) = [\hat{\tilde{x}}_o(t)^T \quad \hat{f}(t)^T]^T \in \mathbb{R}^{\hat{n}}$, $\hat{C} = [\tilde{C}_2 \quad 0]$ and $K = [K_x \quad K_f]$. The matrices L and K are gain parameters to be designed.

Define $e = \xi(t) - \hat{\xi}(t) = [e_o(t)^T \quad e_f(t)^T]^T$ with $e_o(t) = \tilde{x}_o(t) - \hat{\tilde{x}}_o(t)$ and $e_f(t) = f(t) - \hat{f}(t)$. Now, we have the error system:

$$de(t) = A_e e(t)dt + B_e dw_e(t) + E\tilde{x}_{uo}(t)dt + \hat{B}_h h(t)dt, \tag{8.65}$$

where $A_e = \hat{A} - LG\hat{C}$, $B_e = [\hat{B}_w - LG\tilde{D} \quad \hat{B}_u]$, $E = \hat{A}_{uo} - LG\tilde{C}_1$ and $w_e(t) = \begin{bmatrix} w(t) \\ v(t) \end{bmatrix}$ satisfying $dw_e(t)dw_e(t)^T = F_{w_e}dt$.

Interconnecting (8.61) and (8.65), we obtain the following closed-loop system:

$$dz(t) = \bar{A}z(t)dt + \bar{B}_w dw_e(t) + \bar{B}_f f(t)dt + \bar{B}_h h(t)dt, \tag{8.66}$$

where $z(t) = \begin{bmatrix} \tilde{x}(t) \\ e(t) \end{bmatrix}$, which commutes with the fault signal $f(t)$,

$$\bar{A} = \begin{bmatrix} \tilde{A}_{11} & \tilde{A}_{12} + \tilde{B}_{u1}K_x & -\tilde{B}_{u1}K \\ \tilde{A}_{21} & \tilde{A}_{22} + \tilde{B}_{u2}K_x & -\tilde{B}_{u2}K \\ E & 0 & A_e \end{bmatrix} = \begin{bmatrix} \tilde{A} + \begin{bmatrix} 0 & \tilde{B}_u K [I \quad 0]^T \end{bmatrix} & -\tilde{B}_u K \\ [E \ 0] & A_e \end{bmatrix},$$

$$\bar{B}_w = \begin{bmatrix} \tilde{B}_{w1} & \tilde{B}_{u1} \\ \tilde{B}_{w2} & \tilde{B}_{u2} \\ \hat{B}_w - LG\tilde{D} & \tilde{B}_u \end{bmatrix} = \begin{bmatrix} \tilde{B}_w & \tilde{B}_u \\ \hat{B}_w - LG\tilde{D} & \tilde{B}_u \end{bmatrix},$$

$$\bar{B}_f = \begin{bmatrix} \tilde{B}_{f1} + \tilde{B}_{u1}K_f \\ \tilde{B}_{f2} + \tilde{B}_{u2}K_f \\ 0 \end{bmatrix} = \begin{bmatrix} \tilde{B}_f + \tilde{B}_u K \begin{bmatrix} 0 \\ I \end{bmatrix} \\ 0 \end{bmatrix},$$

and $\bar{B}_h = [0 \quad 0 \quad \hat{B}_h^T]^T$.

8.5.2 Stability Results and Controller Synthesis

We consider the following stability definition:

Definition 8.2 The system (8.66) is said to be *mean square bounded stable* if there exists a real function $g(t) = \langle V(t) \rangle$ satisfying inequality

$$g(t) \leq e^{-ct} g(0) + \frac{\tau}{c} \quad \forall t \geq 0, \tag{8.67}$$

where c and τ are positive real numbers, $V(t)$ represents an abstract internal energy for the system (8.66) at time t.

Now we may relate the stability of system (8.66) to certain linear matrix inequalities.

Theorem 8.6 *The system* (8.66) *is mean square bounded stable in the sense of Definition* 8.2 *with*

$$g(t) = \langle V_z(t) \rangle = \langle z(t)^T S z(t) \rangle \tag{8.68}$$

if there exists a real positive definite matrix $S > 0$ satisfying the following relation:

$$\bar{A}^T S + S\bar{A} + S\bar{B}_h \bar{B}_h^T S + S\bar{B}_f \bar{B}_f^T S \leq 0. \tag{8.69}$$

Theorem 8.7 *Under Assumption* 8.4, *if there exists a constant matrix P, such that*

$$\begin{bmatrix} \bar{A}P + P\bar{A}^T + 4P & \bar{B} \\ \bar{B}^T & -I \end{bmatrix} \leq 0, \tag{8.70}$$

with $\bar{B} = [\sqrt{2}\alpha \bar{B}_f \ \sqrt{2}\beta \bar{B}_h \ \bar{B}_w]$, then (8.64) *generates an estimate of ξ satisfying*

$$\lim_{t \to \infty} \left\langle (\xi(t) - \hat{\xi}(t))(\xi(t) - \hat{\xi}(t))^T \right\rangle \leq \mathrm{Tr}(Y_2), \tag{8.71}$$

where $P = \begin{bmatrix} Y_1 & N \\ N^T & Y_2 \end{bmatrix}$, Y_1 is a $n \times n$ symmetric matrix and Y_2 is a $\hat{n} \times \hat{n}$ symmetric matrix.

The proof of Theorems 8.6 and 8.7 can be found in [10]. From the conclusions of Theorems 8.6 and 8.7, we have the following corollary [10]:

Corollary 8.1 *If the following relation holds:*

$$4P + (2\alpha^2 - 1)\bar{B}_f \bar{B}_f^T + (2\beta^2 - 1)\bar{B}_h \bar{B}_h^T + \bar{B}_w \bar{B}_w^T \geq 0, \tag{8.72}$$

then condition (8.70) *implies* (8.69) *for $S = P^{-1}$.*

Now the fault-tolerant control problem may be summarized as Problem 8.1:

Problem 8.1 Given a quantum optical plant with faults of the form (8.58) that can be transformed into (8.61) and for an estimation error upper bound γ (expected to be close to 0, i.e., small error bound), find a classical linear estimator-based fault-tolerant controller of the form (8.64) with parameters L and K such that the following conditions hold for fixed G satisfying (8.60):

1. There exists a symmetric matrix $P > 0$ satisfying (8.70) and (8.72).
2. $0 < \mathrm{Tr}(Y_2) \leq \gamma$.

In [10], a numerical procedure has been presented to solve this problem and the following example can be used to illustrate the method:

Example 8.2 Consider a quantum optical plant with faults as follows:

$$\begin{aligned} dx(t) &= \begin{bmatrix} -1 & 0 \\ 3 & -1 \end{bmatrix} x(t)dt + \begin{bmatrix} 0 & 0 \\ 2 & -1 \end{bmatrix} dw(t) + \begin{bmatrix} 2 & 1 \\ 4 & 3 \end{bmatrix} dy_u(t) + \begin{bmatrix} 0 \\ 1 \end{bmatrix} f(t)dt, \\ dy(t) &= \begin{bmatrix} -3 & 1 \\ 4 & -2 \end{bmatrix} x(t)dt + \begin{bmatrix} 1 & 0 \\ 0 & 1 \end{bmatrix} dw(t), \end{aligned} \tag{8.73}$$

where $f(t)$ is represented as

$$f(t) = \begin{cases} 0.25\cos t, & 0 \leq t \leq 10, \\ 0.5 + 0.4\sin(t - 10), & 10 < t \leq 20. \end{cases}$$

Applying the proposed numerical procedure in [10] to the quantum plant (8.73) for a given estimation error upper bound $\gamma = 0.001$, we obtain

$$L = \begin{bmatrix} -4.07 & 1.03 \\ 9.22 & -30.21 \end{bmatrix}, K = \begin{bmatrix} -0.1500 & -2.6643 \\ 0.3000 & 2.6857 \end{bmatrix}.$$

Now we check that if the resulting solutions satisfy the constraints listed in Problem 8.1. It can be verified that P is a positive symmetric matrix and conditions (8.70) and (8.72) are satisfied. Furthermore, $\mathrm{Tr}(Y_2) = 0.001$. That means that we design an acceptable fault-tolerant controller for the system.

8.6 Summary and Further Reading

This chapter presented several results on H^∞ control and fault-tolerant control for several classes of quantum systems. H^∞ control and robust H^∞ control for linear quantum systems were investigated. Fault-tolerant coherent H^∞ control was developed for a class of linear quantum systems subject to Markovian jump faults and applied to quantum optical systems for coherent controller design. Fault-tolerant control was also developed for measurement-based quantum feedback systems subject to classical faults using an estimator-based approach.

Further reading may include [3] for H^∞ control of linear quantum systems, [7] for control analysis and design of linear quantum systems, [4, 5] for optical implementation and physical realizability of fault-tolerant coherent controllers, [10] for numerical procedure to design an estimator-based fault-tolerant controller, [2] for fault-tolerant filtering of quantum systems, and [11, 12] for robust control of quantum systems with interaction uncertainties.

References

1. Bachor HA, Ralph TC (2019) A guide to experiments in quantum optics, 3rd edn. Wiley-VCH
2. Gao Q, Dong D, Petersen IR (2016) Fault tolerant filtering and fault detection for quantum systems. Automatica 71:125–134
3. James MR, Nurdin HI, Petersen IR (2008) H^∞ control of linear quantum stochastic systems. IEEE Trans Autom Control 53:1787–1803
4. Liu Y, Dong D, Petersen IR, Gao Q, Ding SX, Yokoyama S, Yonezawa H (2022) Fault-tolerant coherent H_∞ control for linear quantum systems. IEEE Trans Autom Control 67:5087–5101
5. Liu Y, Dong D, Petersen IR, Yonezawa H (2022) Fault-tolerant H^∞ control for optical parametric oscillators with pumping fluctuations. Automatica 140:110236
6. Nurdin HI (2011) Network synthesis of mixed quantum-classical linear stochastic systems. In: Proceedings of the Australian control conference, pp 68–75, Melbourne, Australia, 10–11 Nov 2011
7. Nurdin HI, Yamamoto N (2017) Linear dynamical quantum systems. Springer International Publishing
8. Petersen IR, Anderson BDO, Jonckheere EA (1991) A first principles solution to the non-singular H^∞ control problem. Int J Robust Nonlin Control 1:171–185
9. Petersen IR, Ugrinovskii VA, Savkin AV (2000) Robust control design using H^∞ methods. Springer, London
10. Wang S, Dong D (2017) Fault-tolerant control of linear quantum stochastic systems. IEEE Trans Autom Control 62:2929–2935
11. Xiang C (2020) Robust stability control for a class of uncertain quantum systems through direct and indirect couplings. IEEE Access 8:157253–157260
12. Xiang C, Ma S, Kuang S, Dong D (2021) Coherent H^∞ control for linear quantum systems with uncertainties in the interaction Hamiltonian. IEEE/CAA J Autom Sinica 8:432–440
13. Xiang C, Petersen IR, Dong D (2017) Coherent robust H^∞ control of linear quantum systems with uncertainties in the Hamiltonian and coupling operators. Automatica 81:8–21
14. Yamamoto N (2006) Robust observer for uncertain linear quantum systems. Phys Rev A 74:032107
15. Zhou K, Doyle J, Glover K (1996) Robust and Optimal Control. Prentice-Hall, Upper Saddle River

Chapter 9
Concluding Remarks

9.1 Conclusions

In this monograph, we presented a number of results on learning control and robust control of quantum systems. Quantum learning control is mainly covered in Chaps. 3, 4 and 5, where quantum optimal and robust control are two control objectives. In particular, control, discrimination and classification of inhomogeneous quantum ensembles were investigated using a sampling-based learning control approach in Chap. 3. In Chap. 4, a sampling-based learning control method has been applied to quantum state transfer and quantum gate generation for quantum superconducting systems, spin systems and molecular systems with uncertainties. Chapter 5 discussed the potential of machine learning (including evolutionary computation and reinforcement learning) for achieving various quantum control tasks including quantum autoencoders, femtosecond laser control of molecules, quantum ensemble control and quantum network control. Chapters 6–8 further discussed robust control of quantum systems. In Chap. 6, sliding mode control was developed for two-level and multi-level quantum systems as well as open quantum systems. Robust stability and performance analysis have been explored for several classes of stochastic quantum systems using the small gain theorem and the Popov approach in Chap. 7. H^{∞} control and fault-tolerant control of quantum systems were presented under the framework of quantum feedback control in Chap. 8.

9.2 Outlook

Active control of quantum systems has become a critical ingredient to facilitate the emerging area of quantum technology, which presents many new challenging problems different from classical control systems and also provides many exciting opportunities for researchers in the area of systems and control [4]. Although some

D. Dong and I. R. Petersen, *Learning and Robust Control in Quantum Technology*, Communications and Control Engineering,
https://doi.org/10.1007/978-3-031-20245-2_9

results on quantum learning control and quantum robust control have been presented, many topics are worth further exploring. For example, closed-loop learning control [5, 9] provides a useful data-driven control method for guiding quantum physicists and quantum chemists to design efficient experiments, and there are many potential opportunities for systems and control scientists and engineers to collaborate with them by developing efficient and robust quantum learning control algorithms. Another potential research direction is to develop quantum machine learning solutions [3, 6, 7] to challenging high-dimensional quantum control problems. For quantum robust control, besides these results mentioned in this book, there is plenty of room to further develop systematic robust control theory for various quantum control applications. On the one hand, the most robust control theory and approaches discussed in this book are still at an early stage and far from mature, and it is worth further developing these quantum robust control methods, especially for those based on quantum feedback control [11]. On the other hand, many new quantum control methods inspired by classic control theory (e.g., adaptive control, model predictive control, nonlinear regulation) are worth developing by fully considering the unique quantum characteristics for improving the robustness of quantum systems. Another future research direction is to develop learning and robust control methods for quantum networks, attosecond dynamical systems [10, 13] and other emerging applications such as quantum computers [14], quantum error correction [8] and quantum sensing [1, 2, 12]. It needs the collaborative efforts of researchers from various areas of systems control, quantum physics, quantum engineering, physical chemistry and quantum information to systematically solve these challenging problems and accelerate the development of quantum technology [4].

References

1. Bao L, Qi B, Dong D (2021) Fundamental limits for reciprocal and nonreciprocal non-Hermitian quantum sensing. Phys Rev A 103:042418
2. Bao L, Qi B, Dong D (2022) Exponentially-enhanced quantum non-Hermitian sensing via optimized coherent drive. Phys Rev Appl 17:014034
3. Biamonte J, Wittek P, Pancotti N, Rebentrost P, Wiebe N, Lloyd S (2017) Quantum machine learning. Nature 549:195–202
4. Dong D, Petersen IR (2022) Quantum estimation, control and learning: opportunities and challenges. Ann Rev Control 54:243–251
5. Dong D, Xing X, Ma H, Chen C, Liu Z, Rabitz H (2020) Learning-based quantum robust control: algorithm, applications, and experiments. IEEE Trans Cybern 50:3581–3593
6. Dunjko V, Briegel HJ (2018) Machine learning & artificial intelligence in the quantum domain: a review of recent progress. Rep Progr Phys 81(7):074001
7. Li J-A, Dong D, Wei Z, Liu Y, Pan Y, Nori F, Zhang X (2020) Quantum reinforcement learning during human decision-making. Nat Hum Behav 4:294
8. Nielsen MA, Chuang IL (2010) Quantum computation and quantum information. Cambridge University Press, Cambridge
9. Rabitz H, De Vivie-Riedle R, Motzkus M, Kompa K (2000) Whither the future of controlling quantum phenomena? Science 288(5467):824–828

10. Shu CC, Guo Y, Yuan KJ, Dong D, Bandrauk AD (2020) Attosecond all-optical control and visualization of quantum interference between degenerate magnetic states by circularly polarized pulses. Opt Lett 45(4):960–963
11. Wiseman HM, Milburn GJ (2010) Quantum Measurement and Control. Cambridge University Press, Cambridge
12. Yu Q, Wang Y, Dong D, Petersen IR (2021) On the capability of a class of quantum sensors. Automatica 129:109612
13. Yuan KJ, Shu CC, Dong D, Bandrauk AD (2017) Attosecond dynamics of molecular electronic ring currents. J Phys Chem Lett 8:2229–2235
14. Zhong H-S, Wang H, Deng Y-H, Chen M-C, Peng L-C, Luo Y-H et al (2020) Quantum computational advantage using photons. Science 370(6523):1460–1463

Index

D. Dong and I. R. Petersen, *Learning and Robust Control in Quantum Technology*, Communications and Control Engineering,
https://doi.org/10.1007/978-3-031-20245-2